The Art of Finding Hidden Risks

Sidney Resnick

The Art of Finding Hidden Risks

Hidden Regular Variation in the 21st Century

 Springer

Sidney Resnick
ORIE
Cornell University
Ithaca, NY, USA

ISBN 978-3-031-57598-3   ISBN 978-3-031-57599-0   (eBook)
https://doi.org/10.1007/978-3-031-57599-0

Mathematics Subject Classification: 60F17, 60F05, 60G51, 60G57, 60G70, 62G32, 62H12, 62H22, 62H20, 62E20, 62F10, 62G30, 62P30, 62P05, 90B18, 90B15, 05C82, 05C20, 91D30, 91G70

© The Editor(s) (if applicable) and The Author(s), under exclusive license to Springer Nature Switzerland AG 2024

This work is subject to copyright. All rights are solely and exclusively licensed by the Publisher, whether the whole or part of the material is concerned, specifically the rights of translation, reprinting, reuse of illustrations, recitation, broadcasting, reproduction on microfilms or in any other physical way, and transmission or information storage and retrieval, electronic adaptation, computer software, or by similar or dissimilar methodology now known or hereafter developed.

The use of general descriptive names, registered names, trademarks, service marks, etc. in this publication does not imply, even in the absence of a specific statement, that such names are exempt from the relevant protective laws and regulations and therefore free for general use.

The publisher, the authors and the editors are safe to assume that the advice and information in this book are believed to be true and accurate at the date of publication. Neither the publisher nor the authors or the editors give a warranty, expressed or implied, with respect to the material contained herein or for any errors or omissions that may have been made. The publisher remains neutral with regard to jurisdictional claims in published maps and institutional affiliations.

This Springer imprint is published by the registered company Springer Nature Switzerland AG
The registered company address is: Gewerbestrasse 11, 6330 Cham, Switzerland

If disposing of this product, please recycle the paper.

Preface

Why is this book different from other books on heavy tails?[1] Well, years pass, ideas develop and mature, methods improve, explanations simplify and the subject becomes clearer and better focussed.

My previous book on heavy tails [156] centered on a convergence concept and notion of *extreme* using variants of a one point uncompactification. In retrospect, this was a bit of a horror and can be improved. It was awkward dealing with different cases and examples and occasionally led to limits lacking uniqueness. The other focus of the 2007 book was the centrality of the Poisson random measure as a tool of theoretical analysis. The point process technique was considered a transform method. Some people complained about the probability and topological overhead. So while I consider the Poisson transform interesting, useful and natural, for many purposes it is unnecessary and does not get used here.

The present volume concentrates mostly on multivariate problems. The one-dimensional Karamata theory of regularly varying functions is briefly summarized in Appendix A but is not the focus. (This appendix is shamelessly cribbed from [156] which was cribbed from [150] and heavily dependent on the seminal [49].) One-dimensional function theory is not an essential part of this story and if dealing with analysis and function theory gives you hives, feel free to assume a one-dimensional heavy tail means Pareto tail or power law. Ignoring, for instance, slowly varying functions or pretending they are constants will not cause you to be left back a grade at school.

While you can read this book without worrying too much about niceties of slow variation, you are presumed to know something about one-dimensional statistical techniques for heavy-tailed data such as QQ-plots, Hill plots, altHill plots, standardization of multivariate data using the rank transformation, angular density plots. This is covered in [156, Chapters 4, 9, 11], and other excellent references are [7, 27, 51, 110].

---

[1] As we (sort of) say on Passover: Ma nishtanah hasefer hazeh mikol hasfareem?

Hindsight reveals that analysts and function theorists often attempted to lift one-dimensional theory to higher dimensions by succumbing to a tendency to generalize intervals into squarish regions. This led to difficulties since such an approach may or may not be natural, elegant or particularly useful. One awkward issue was that on the positive real line $\mathbb{R}_+$ it was clear what *extreme* meant, namely "near ∞". In $\mathbb{R}_+^p$, for $p > 1$, there is no unique definition of *extreme* and it is preferable to have flexibility depending on the application.

A way forward is to abandon the centrality of functions and focus on measures. This allows consideration of different regions beyond squarish ones but still requires

(a) Specifying what you mean by *extreme*;
(b) Specifying a class of regions where you know with certainty that measures of a specified class are finite.

The first requirement is handled by a user specifying a theatrically named set called the *forbidden zone* and then declaring that *extreme* means "far from the forbidden zone." This allows a user to flexibly taylor the definition of extreme to the application. The second requirement can be coped with by insisting that the considered class of measures be finite on regions bounded away from the forbidden zone. This setup allows a unifying setting of metric spaces with closed sets removed. It also allows for several (or in the case of *steroidal regular variation* infinitely many) regular variation properties with the same state space; having avoided a forbidden zone and specified a scaling function, you are always free to change the scaling and avoid a bigger forbidden zone. For example, in $\mathbb{R}_+^2$ does extreme mean "at least one component is large" or maybe "both components are large simultaneously?" In the first case, the forbidden zone is the origin, and in the second case, it is the two axes.

This formulation allows definition of regular variation of measures that is dependent on

1. Choice of *forbidden zone* (hence choice of definition of *extreme*) and
2. Scaling function.

Identifying forbidden zone and scaling as essential ingredients leads to recognition that multiple heavy tails can co-exist in multivariate problems with each depending on its own different pair of forbidden zone and scaling. This potential multiplicity of heavy tail regimes suggests caution about which heavy tail is chosen to base a risk calculation on; a poor choice will result in a risk estimate of 0 but an alternate choice may give a positive risk estimate.

Earlier treatments struggled with requirements (a), (b) above and tried to use compactness and vague convergence. This leads to tying yourself in knots. For instance, in $\mathbb{R}_+ = [0, \infty)$, it is natural to consider sets of the form (x, ∞) where $x > 0$ because probabilities of such sets are probabilities some quantity is large. However, (x, ∞) is not relatively compact. To fix this, one solution is to compactly at ∞ and make the state space $[0, \infty]$. Then $(x, \infty]$ is relatively compact. But then $\nu_\alpha(x, \infty] = x^{-\alpha}$, the quintessential heavy tailed measure, is infinite on

[0, ∞] and hence non-Radon. To fix this we have to uncompactify by removing {0}. This works but is clunky. (Feel free to blame me!) In higher dimensions, you have similar (ahem) awkwardness. The vague topology approach becomes embarrassingly clumsy if you consider several power law regimes on the same state space as in *hidden* or *steroidal* regular variation scenarios.

Here are the Netflix-style coming attractions. Chapter 1 discusses how to handle the basic metric space setup and specifies needed ideas about defining convergence of measures and how to prove convergence. The hero of this story is the power of continuity applied to mappings to extract new convergence results from existing ones. Chapter 2 builds on Chap. 1 material on convergence concepts and pursues the topic of regular variation of measures as a way to discuss multidimensional heavy tails. The importance of choosing a *forbidden zone* and scaling is emphasized and methods of constructing regularly varying measures are given relying on *generalized polar coordinates*. A background theme is the ability to change coordinate systems, and there is also an examination of methods of proving regular variation of measures as well as many examples. Even in $\mathbb{R}_+^2$, the positive quadrant, we have flexibility to consider a variety of forbidden zones:

- The origin
- Both axes $\{(x, 0) : x \geq 0\} \cup \{(0, y) : y \geq 0\}$
- One axis $\{(x, 0) : x > 0\}$
- A ray $\{(x, y) \in \mathbb{R}_+^2 : y/x = m\}$, for some $m > 0$
- A pizza-shaped wedge $\{(x, y) \in \mathbb{R}_+^2 : x/(x + y) \in [a, b]\}$, for some $[a, b] \subsetneq [0, 1]$

In addition to $\mathbb{R}_+^2$, we consider examples in $\mathbb{R}_+^p$ for $p > 1$; $\mathbb{R}_+^\infty$; $\mathbb{D}(0, \infty)$, the space of right continuous functions with finite left limits on $(0, \infty)$.

On to Chap. 3 with definitions and examples of both hidden and steroidal regularly varying measures. Such measures have the property that there are multiple choices of scaling and forbidden zone pairs resulting in multiple regular variation properties. Hidden regularly varying measures have at least two different heavy-tail properties while steroidally regularly varying measures have infinitely many distinct choices. Chapter 3 on steroidal theory is exemplified by two examples: The first is the case of the distribution of iid non-negative random variables whose common distribution has a regularly varying tail. The second is the distribution of an $\mathbb{R}_+^\infty$ random element consisting of a sequence of Poisson points whose mean measure on $\mathbb{R}_+ \setminus \{0\}$ has a regularly varying tail.

The Poisson example is basic in Chap. 4 to treating the distribution of a Lévy process $\{X(t), t \geq 0\}$ with a Lévy measure that is regularly varying at ∞. The treatment in Chap. 4 shows that the distribution of such processes is another example of steroidal regular variation of measures, this time on the function space of right continuous functions on $[0, \infty)$ with finite left limits on $(0, \infty)$.

Chapter 5 shifts tone to focus on statistical techniques. It begins by urging caution and humility since having a heavy tail is an asymptotic property but inferring something from a finite data set is based on a view of the distribution that is a long way from asymtopia. Then there is discussion of the fraught activity of choosing a threshold beyond which data is presumed to come from the asymptotic model. We give pros and cons of one threshold choice guideline called the minimum distance selection procedure that is popular in computer and network science.

Next comes description of the two graphical techniques:

(i) Diamond plots, which assess component dependencies
(ii) Hillish plots, which provide visual evidence consistent with the heavy tail assumption.

These techniques are illustrated in finite dimensions with several examples. In two dimensions $\mathbb{R}_+^2$, we highlight four cases of importance:

(i) Asymptotic independence, where the limit measure of regular variation fails to charge sets with two or more components simultaneously large
(ii) Full dependence, where the limit measure of regular variation concentrates on a ray from the origin
(iii) Strong dependence, where the limit measure concentrates on a wedge of the form

$$[\text{wedge}] := \{(x, y) \in \mathbb{R}_+^2 : x/(x+y) \in [a, b] \subsetneq [0, 1]\}$$

for some a, b such that $[a, b] \subsetneq [0, 1]$.
(iv) Weak dependence, where the limit measure concentrates on all of $\mathbb{R}_+^2$.

We move on to a dependence measure that can be informative called the *extremal dependence measure* and illustrate its use with examples and real data. Inference based on asymptotic normality is prefaced by questioning limitations.

The chapter closes with brief exposition of network models of social network growth containing peferential attachment. Adding the property of reciprocity has a dramatic effect on the support of the limit measure, and this class of models illustrates concepts introduced earlier.

Appendices contain a summary of Karamata theory of regularly varying functions and the all important notation summary, an essential aid to any aging brain. Each chapter contains exercises, some quite clever thanks to former students. Use your own judgement about how much effort you give exercises depending on your needs and goals.

What do you need to know to profitably read this book? The book should be accessible to someone with a US first year PhD student background. A basic object of study is the regularly varying measure, so if you anticipate getting a heart attack at mention of the word *measure*, check first with your cardiologist. You need a little metric space theory for the setup in Chap. 1 and an understanding of the concept of continuity of mappings. Some familiarity with weak convergence at the level of [13] would increase your reading joy. If you do not know or care about Lévy processes, skip Chap. 4; otherwise treat it as a stochastic process payoff for the work in earlier chapters, especially the discussion of steroidal regular variation. The last Chap. 5 on some heavy tail statistical methods requires you to be sympathetic to the difficulty of extracting information from recalcitrant data and to know one-dimensional statistical techniques.

Acknowledgements

Cornell University's School of Operations Research and Information Engineering has been my professional home since 1987. The move to Cornell in Ithaca, NY, after three prior jobs was a turning point both professionally and personally.

Ithaca is a great place to live and have your kids prosper, and people are continually surprised when I claim it is arguably one of the nicest small towns in the United States. True, I would not object if climate were a bit warmer or if the location were a little closer to New York City. However, the outdoor environment is gorgeous, the social atmosphere is cosmopolitan and the cultural life is good. It is in the middle of wine country, has good restaurants and has a reasonably functional and convenient airport. For faculty, time lost to commuting is negligible and going around town by bike or the good bus system very doable. I still schlep into Cornell by bike, weather permitting. Ithaca does not get as much snow as people think though Syracuse and Buffalo do.

Cornell is a place of excellence where faculty and staff are familiar with the phrase "work/life balance" but also have plenty of healthy professional ambition. Students are excellent, salaries relatively good, seminars and advanced courses plentiful and opportunities for professional growth abound. My brush with Cornell administrative life as the ORIE School's Director for 5 years as well as several stints on the Faculty Senate left me with the conviction that most faculty, staff and administrators earnestly, rationally and unselfishly try to promote positive academic goals and university activities. If you are wondering, nobody ever accused me of being Pollyanna.

Aside from general appreciation for the advantages of working at a excellent university in a first rate department in a nice geographical place, I am particularly grateful for the sensitive and tactful manner Cornell handled my transition to emeritus status in 2019. (Coming just before Covid hit, this was the best timed transition in the history of Cornell.) I remain connected and active in my current state of perpetual sabbatical. Former ORIE Director David Shmoys could not have been more considerate and sensitive managing the process and current Director Mark Lewis makes me feel welcome and well housed.

Over the years, I have been blessed with many excellent colleagues, students and academic friends whose positive influence is apparent throughout these pages and hopefully is sufficiently acknowledged. Many of these people became lifelong friends.

Family history as described in literature:

- (1987) I ... thank Minna, Nathan and Rachel Resnick for a cheery, happy family life. Minna and Rachel bought me the mechanical pencil that made this project possible, and Rachel generously shared her erasers with me as well as providing a back-up mechanical pencil from her stockpile when the original died after 400 manuscript pages. I appreciate the fact that Nathan was only moderately aggressive about attacking my Springer-Verlag correspondence with a hole puncher. [150]

- (1992) Minna, Rachel and Nathan provided a warm, loving family life and generously shared the home computer with me. They were also very consoling as I coped with two hard disk crashes and a monitor melt-down. [152]
- (1998) Rachel, who grew into a terrific adult, no longer needs to share her mechanical pencils with me. Nathan has stopped attacking my manuscripts with a hole puncher and gives ample evidence of the fine adult he will soon be. Minna is the ideal companion on the random path of life. [153]
- (2006) Nathan graduated Cornell, moved to New York City and found a job, and Rachel (now a Director!) will soon marry Randy. Nathan and Randy gang up on poor, defenseless me which Minna claims I deserve. Minna and I, the artist and the ninja math geek, continue to explore a wonderful life together. [156]
- (2023) After meeting as Freshmen, Nathan eventually married Jackie, moved to Brooklyn (oh the bagels and brisket!) and subsequently miraculously managed to buy into a difficult housing market in New Jersey and become a Managing Director at work. They begat two kids; one is the answer to the energy crisis and the other will give any form of AI a run for its money. Rachel (now Executive VP!) and Randy begat Gigantor #1, #2. Gigantor #1 is the ace guitarist and occasional drummer for his rock band and Gigantor #2 throws no-hitters for his baseball team and asks for suggestions on basketball dribbling technique. (Eat your heart out Steph!) If they get much bigger, evacuation of the east coast is advised. Minna is a bit grayer but cute and still happily producing art and having shows; I'm a bit balder and less cute. We are both a bit creaky but trucking happily onward together.

See you next time.

Ithaca, NY, USA
June 2024

Sidney Resnick

Contents

1 **Foundation** 1
   1.1 Leaving the Comfort of the Real Line 1
   1.2 TABOF Spaces 2
        1.2.1 Examples of TABOF Spaces 3
        1.2.2 Generating TABOF Spaces 4
   1.3 The Forbidden Zone and the $\mathbb{BA}$ Class 5
   1.4 Measures on TABOF Spaces: The Space $\mathbb{M}(\mathbb{S}_F) = \mathbb{M}(\mathbb{S} \setminus F)$ 7
        1.4.1 A Topology on $\mathbb{M}(\mathbb{S}_F)$; $\mathbb{M}$-convergence 7
        1.4.2 Equivalent Forms of Convergence; the Portmanteau Theorem 9
        1.4.3 Taking the Definitions for a Test Drive 13
   1.5 New Results from Existing Ones: Mapping Theorems 15
        1.5.1 The Problem and Context 15
        1.5.2 The First Mapping Theorem 15
        1.5.3 The Second Mapping Theorem and Uniform Continuity 17
        1.5.4 Examples for Uniform Continuity 18
   1.6 Maneuvers Around Roadblocks 21
        1.6.1 More Subtle Use of Compactness 21
        1.6.2 Restriction of the State Space 22
   1.7 Problems 24

2 **Regular Variation of Measures** 31
   2.1 Regular Variation of Measures in $\mathbb{M}(\mathbb{S} \setminus F)$ 31
        2.1.1 Measures on $\mathbb{R}_+$ 31
        2.1.2 More on Scaling 32
        2.1.3 Definition of Regular Variation of Measures 33
   2.2 Building Regularly Varying Distributions 37
        2.2.1 Constructing Regularly Varying Distributions in $\mathbb{R}_+^p$ and Moving Between Cartesian and Polar Coordinates . 37
        2.2.2 Construction of Regular Variation on $\mathbb{S} \setminus F$ Using the Generalized Polar Coordinate Transform 39

	2.3	How to Prove Regular Variation of Measures?	46
		2.3.1 Proof Method 1: Great Oaks From Little Acorns Grow	46
		2.3.2 Proof Method 2: Map Your Way to Happiness by Using the Great(er) Oaks and the Mapping Theorems; Some Examples	53
		2.3.3 Proof Method 3: Reduce Problems in $\mathbb{R}^\infty$ to Finite Dimensions	65
		2.3.4 Proof Method 4: Prove Convergence on a Sufficiently Rich Class of Sets	74
	2.4	Problems	82
3	**Hidden Regular Variation**		93
	3.1	Hidden Regular Variation	94
		3.1.1 Simple Diagnostics for HRV in $\mathbb{R}_+^p$, $p > 1$	95
	3.2	Steroidal Regular Variation: Roid Rage Strikes	99
		3.2.1 Steroidal Regular Variation in $\mathbb{R}_+^\infty$ for the iid Model	100
		3.2.2 Poisson Points as Random Elements of $\mathbb{R}_+^\infty$	103
		3.2.3 Different Tail Rates in Different Directions	111
	3.3	Problems	113
4	**Lévy Processes with Regularly Varying Distributions: Where Do the Jumps Go?**		121
	4.1	Background	122
		4.1.1 Skorohod Metric	122
		4.1.2 The Lévy Process X	122
	4.2	Steroidal Regular Variation of $P[X \in \cdot]$	124
		4.2.1 The Plan	126
		4.2.2 The Mapping	127
		4.2.3 Filling the Gap: Sweating the Small Stuff to Get Regular Variation of $P[X \in \cdot]$	135
	4.3	Heavy Lifting Done? Time for a Cruise	138
		4.3.1 First Cruise Stop: The Supremum Functional	138
		4.3.2 Second Cruise Layover: The Largest Jump Functional	141
		4.3.3 Next Cruise Stop: Smoothing	145
	4.4	Problems	146
5	**Exploring Data Cautiously**		149
	5.1	Reasons for Caution: Zombies and Fairy Tales	150
	5.2	Threshold Selection	152
		5.2.1 The Minimum Distance Threshold Method	153
	5.3	Two-Dimensional Visualization of Multivariate Dependence	158
		5.3.1 The Diamond Plot	161
	5.4	Evidence Consistent with Multivariate Regular Variation from Hillish Analysis	171
		5.4.1 Hillish Analysis for Heavy Tails on $\mathbb{R}_+^2 \setminus \{0\}$	171
		5.4.2 Hillish Examples	175

	5.5	Beyond $\mathbb{R}^2 \setminus \{\mathbf{0}\}$: Other Cones as Forbidden Zones	178
		5.5.1 Parameterizing a Cone	179
		5.5.2 A Coordinate System for Regular Variation on $\mathbb{R}_+^2 \setminus$ [wedge]	179
		5.5.3 Hillish Analysis for Heavy Tails in $\mathbb{R}_+^2 \setminus$ [wedge]	182
		5.5.4 Further Analysis of Facebook Data	182
	5.6	More on Asymptotic Independence	187
		5.6.1 Definitions and Basics	188
		5.6.2 Preservation of Asymptotic Independence Under Mappings: What Could Possibly Go Wrong?	191
	5.7	Measuring Extremal Dependence with the EDM	193
		5.7.1 Exploring Pairwise Extremal Dependence with the Dependence Graph	195
		5.7.2 Hang on! Is the Estimated EDM Asymptotically Normal?	197
		5.7.3 Unclunking the Clunk Function $g(\cdot)$	204
		5.7.4 Is Relying on Asymptotic Normality Wise?	209
		5.7.5 Preferential Attachment with Reciprocity Gives Asymptotic Full Dependence	214
	5.8	Problems	222
A	A Crash Course on Regularly Varying Functions		225
	A.1	Preliminaries from Analysis	225
		A.1.1 Uniform Convergence	225
		A.1.2 Inverses of Monotone Functions	226
		A.1.3 Convergence of Monotone Functions	227
		A.1.4 Cauchy's Functional Equation	228
	A.2	Regular Variation: Definition and First Properties	228
	A.3	A Maximal Domain of Attraction	231
	A.4	Regular Variation: Deeper Results, Karamata's Theorem	232
		A.4.1 Uniform Convergence	232
		A.4.2 Integration and Karamata's Theorem	233
		A.4.3 Karamata's Representation	237
		A.4.4 Differentiation	239
	A.5	Regular Variation: Further Properties	241
B	Notation Summary		245
References			251
Index			259

Chapter 1
Foundation

The subject of multivariate heavy tail analysis rests on the univariate theory of regularly varying functions. These are functions that near infinity look roughly like power functions of the form x^ρ, $x > 0$ for some $\rho \in \mathbb{R}$. If you are not familiar with the subject of regularly varying functions, consider browsing Appendix A, page 225 or one of the excellent references noted there until you are operational. Alternatively, depending on your goals, think of regularly varying functions as functions that "near infinity look roughly like power functions".

A general treatment of heavy tails is basically a story of convergence of a family of measures that are distributions of a scaled random element in a metric space such as $\mathbb{R}_+$, $\mathbb{R}_+^p$ for $p > 1$, or $\mathbb{R}_+^\infty$. The convergence concept is affected by which sets we designate as *remote* (synonym *extreme*). We begin with the somewhat abstract set-up which becomes increasingly friendly and pertinent.

1.1 Leaving the Comfort of the Real Line

Comfy?

Traditionally, regularly varying distributions in higher dimensions have been defined using a clumsy device called the one-point uncompactification.[1] This approach rested on vague convergence and relative compactness and required one to compactify the state space and then remove a point such as the origin, thereby uncompactifying the state space. This became awkward if you either needed the polar coordinate transformation or needed to remove more than a point but it provided a way to designate regions away from the

---

[1] Mea culpa. I helped purvey this horror.

origin as tail regions. A more direct and flexible approach is to decide that *tail regions* are regions bounded away from a designated region (such as the origin or maybe the axes or maybe something else). The idea of a tail region can thus be application dependent.

This leads to $\mathbb{M}$-convergence whose antecedents are mysterious to me but the definitions appear in [34]. The idea was picked up in [95–97]. Thanks are due to A. Mitra and B. Das who convinced me of the appropriateness of the approach leading to [38] and F. Lindskog helped make the approach flexible and coherent in [125].

1.2 TABOF Spaces

Consider a complete separable metric space (CSMS) $\mathbb{S}$ with metric $d(x, y)$ for $x, y \in \mathbb{S}$. Measures are defined on nice subsets of $\mathbb{S}$ and random elements will live in $\mathbb{S}$. A TABOF[2] space is a CSMS $\mathbb{S}$ with a closed subset $F \in \mathcal{F}(\mathbb{S})$ removed where $\mathcal{F}(\mathbb{S})$ are the closed subsets of $\mathbb{S}$. Denote this TABOF space by

$$\mathbb{S}_F = \mathbb{S}(F) = \mathbb{S} \setminus F.$$

(Applications to regular variation will require that F also be a closed cone; more later.) The deleted set F is called the *forbidden zone*. The TABOF space $\mathbb{S}_F$ is given the subset topology so open subsets of $\mathbb{S}_F$ are $\mathcal{G}(\mathbb{S}) \cap \mathbb{S}_F = \mathcal{G}(\mathbb{S}) \cap F^c$, that is, the open subsets $\mathcal{G}(\mathbb{S})$ of $\mathbb{S}$ intersected with F^c. The space $\mathbb{S}_F$ is also a complete, separable metric space with metric $d(x, y)$ for $x, y \in \mathbb{S}_F$ and the metric on $\mathbb{S}_F$ is just the metric on $\mathbb{S}$ restricted to $\mathbb{S} \setminus F$.

For $x \in \mathbb{S}_F$, the distance from x to F is

$$d(x, F) := \inf_{y \in F} d(x, y).$$

Note $x \in \mathbb{S}_F$, meaning $x \notin F$, implies $d(x, F) > 0$ since otherwise if $d(x, F) = 0$, we would have $x \in F$ because F is closed. If A, B are two subsets of $\mathbb{S}$, write

$$d(A, B) = \inf_{\substack{a \in A, \\ b \in B}} d(a, b).$$

We will allow $F = \emptyset$, the empty set. Then $\mathbb{S}_\emptyset = \mathbb{S}$ and we follow the convention that for $x \in \mathbb{S}$, $d(x, \emptyset) = \infty$.

---

[2] Take **a bite out of**.

1.2 TABOF Spaces

1.2.1 Examples of TABOF Spaces

Here are some common examples of TABOF spaces.

1. $\mathbb{S} = \mathbb{R}_+, F = \{0\}$. Then $\mathbb{S}_F = \mathbb{R}_+ \setminus \{0\}$. Write $\mathbb{S}_F = (0, \infty)$ at your peril.
2. $\mathbb{S} = \mathbb{R}_+^p, p \geq 1, F = \{\mathbf{0}_p\}, \mathbb{S}_F = \mathbb{R}_+^p \setminus \{\mathbf{0}_p\}$.
3. $\mathbb{S} = \mathbb{R}_+^2, F = \{(x, 0) : 0 \leq x < \infty\} = \mathbb{R}_+ \times \{0\}$. Then

$$\mathbb{S}_F = \{(x, y) : x \geq 0, y > 0\} = \mathbb{R}_+^2 \setminus [\text{x-axis}] =: \mathbb{D}_{\sqcap}, \quad (1.1)$$

the first quadrant with the x-axis removed. This is the space used for discussions of the conditional extreme value (CEV) model [39, 40, 62, 88, 89] and also products of independent random variables, one of which is heavy tailed [17, 131].

4. $\mathbb{S} = \mathbb{R}_+^2, F = [\text{axes}] := \{(x, 0) : x \geq 0\} \cup \{(0, x) : x \geq 0\}$ and $\mathbb{S}_F = (0, \infty)^2 = \mathbb{R}_+^2 \setminus [\text{axes}]$ which is the first quadrant with both the x-axis and y-axis removed. As elsewhere, we prefer to write $\mathbb{S}_F$ as $\mathbb{R}_+^2 \setminus [\text{axes}]$ over $(0, \infty)^2$ since it makes the forbidden zone clear and explicit. This space arises in the traditional approach to *hidden regular variation*. See [38, 130, 136, 137, 154, 158].

5. $\mathbb{S} = \mathbb{R}_+^2, F = [\text{diag}] = \{(x, x) : 0 \leq x < \infty\}$, and $\mathbb{S}_F$ is the first quadrant with the diagonal removed.

6. $\mathbb{S} = \mathbb{R}_+^\infty$. The metric is the usual one,

$$d_\infty(\mathbf{x}, \mathbf{y}) = \sum_{i=1}^\infty \frac{|x_i - y_i| \wedge 1}{2^i}, \quad (1.2)$$

where $\mathbf{x} = (x_1, x_2, \ldots)$ is a sequence and similarly for $\mathbf{y}$. Set

$$\mathbf{0}_\infty = (0, 0, \ldots), \quad \mathbf{1}_\infty = (1, 1, 1, \ldots), \quad \mathbf{e}_i = (0, 0, \ldots, 0, 1, 0, \ldots),$$

and $\mathbf{e}_i$ is sequence all of whose components are 0 except for the ith component. Let $\epsilon_x(\cdot)$ be the Dirac probability measure putting all mass at x. (See (B.1), page 246.) For $j \geq 0$ define

$$F_j = \{\mathbf{x} : \sum_{l=1}^\infty \epsilon_{x_l}(0, \infty) \leq j\},$$

which is the collection of sequences which have at most j non-zero components. Note $F_0 = \{\mathbf{0}_\infty\}$ and

$$F_1 = \bigcup_{i=1}^\infty \{x\mathbf{e}_i : x \geq 0\} \cup \{\mathbf{0}_\infty\},$$

which are the axes and the origin. See [125]. This setting is appropriate for studying regular variation in $\mathbb{R}^\infty$.

7. $\mathbb{S} = \mathbb{D}[0, 1]$ = the space of all càdlàg functions on $[0, 1]$, that is right continuous functions which have finite left limits everywhere on $(0, 1]$. The metric is the Skorohod metric [14, 157, 175] and set

$$F_j = \{x(\cdot) \in D[0, 1] :$$
$$x(\cdot) \text{ is a non-decreasing step function with at most } j \text{ jumps.}\}.$$

This setting is used to study the jumps of Lévy processes. See [95, 97, 125].

1.2.2 Generating TABOF Spaces

One way to generate higher dimensional TABOF spaces is by taking products of lower dimensional TABOF spaces.[3] For instance, given two complete, separable metric spaces $\mathbb{S}_i$, $i = 1, 2$ and closed subsets $F_i \in \mathcal{F}(\mathbb{S}_i)$, $i = 1, 2$ form the TABOF spaces $\mathbb{S}_i(F_i) = \mathbb{S}_i \setminus F_i$, $i = 1, 2$. Then the product

$$\mathbb{S}_1(F_1) \times \mathbb{S}_2(F_2) = (\mathbb{S}_1 \setminus F_1) \times (\mathbb{S}_2 \setminus F_2) \tag{1.3}$$

is also a TABOF space (Fig. 1.1).

To see this, write the product in (1.3) as $F_1^c \times F_2^c$ where each complement is taken in the appropriate factor space. Then

$$(F_1^c \times F_2^c)^c = \{(s_1, s_2) : s_1 \notin F_1^c \text{ or } s_2 \notin F_2^c\}$$
$$= \{(s_1, s_2) : s_1 \in F_1 \text{ or } s_2 \in F_2\}$$
$$= F_1 \times \mathbb{S}_2 \cup \mathbb{S}_1 \times F_2,$$

which is closed in the product space $\mathbb{S}_1 \times \mathbb{S}_2$. Therefore,

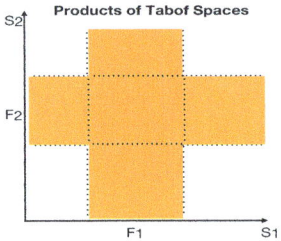

Fig. 1.1 Shaded area: $(F_1^c \times F_2^c)^c$

$$\mathbb{S}_1(F_1) \times \mathbb{S}_2(F_2) = F_1^c \times F_2^c$$
$$= \mathbb{S}_1 \times \mathbb{S}_2 \setminus (F_1^c \times F_2^c)^c = \mathbb{S}_1 \times \mathbb{S}_2 \setminus (F_1 \times \mathbb{S}_2 \cup \mathbb{S}_1 \times F_2), \tag{1.4}$$

which is the required form for a TABOF space and the forbidden zone of the product space is the closed set $F_1 \times \mathbb{S}_2 \cup \mathbb{S}_1 \times F_2 \in \mathcal{F}(\mathbb{S}_1 \times \mathbb{S}_2)$.

---

[3] Catchy mnemonic: TABOF × TABOF=TABOF.

1.2.2.1 Examples for Products of TABOF Spaces

1. Take $\mathbb{S}_i \setminus F_i = \mathbb{R}_+ \setminus \{0\}$ for $i = 1, 2$. Then

$$(\mathbb{S}_1 \setminus F_1) \times (\mathbb{S}_2 \setminus F_2) = \mathbb{S}_1 \times \mathbb{S}_2 \setminus \big((\mathbb{S}_1 \times F_2) \cup (F_1 \times \mathbb{S}_2)\big)$$
$$= \mathbb{R}_+^2 \setminus \big((R_+ \times \{0\}) \cup (\{0\} \times \mathbb{R}_+)\big) = \mathbb{R}_+^2 \setminus [\text{axes}].$$

This is the space used in the original definition of *hidden regular variation* [38, 130, 136, 137, 154, 158].

2. Take $\mathbb{S}_1 = \mathbb{S}_2 = \mathbb{R}_+$ and $F_1 = \emptyset$, and $F_2 = \{0\}$. Then

$$\mathbb{S}_1(F_1) = \mathbb{R}_+, \quad \mathbb{S}_2(F_2) = \mathbb{R}_+ \setminus \{0\}$$

and

$$\mathbb{S}_1(F_1) \times \mathbb{S}_2(F_2) = \mathbb{S}_1 \times \mathbb{S}_2 \setminus \big((\mathbb{S}_1 \times F_2) \cup (F_1 \times \mathbb{S}_2)\big)$$
$$= \mathbb{R}_+^2 \setminus \big(R_+ \times \{0\} \cup \emptyset \times \mathbb{R}_+\big).$$

Since the second expression in the union is empty, this is

$$\mathbb{R}_+^2 \setminus (R_+ \times \{0\}) = [0, \infty) \times (0, \infty) =: \mathbb{D}_\sqcap.$$

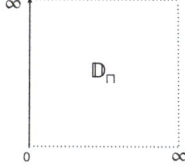

Fig. 1.2 The space $\mathbb{D}_\sqcap$

This space $\mathbb{D}_\sqcap$, the first quadrant with an axis removed, is used for studying products of independent random variables when one variable has a heavy tail and also for the *conditional extreme value model* [39, 40, 87, 89] (Fig. 1.2).

1.3 The Forbidden Zone and the $\mathbb{BA}$ Class

For the metric space $\mathbb{S}$ and $F \in \mathcal{F}(\mathbb{S})$, form the TABOF space $\mathbb{S}_F$ and remember that the set F deleted from $\mathbb{S}$ is called the *forbidden zone*. Sets far away from the forbidden zone are what we consider as the tail or extreme or remote regions.

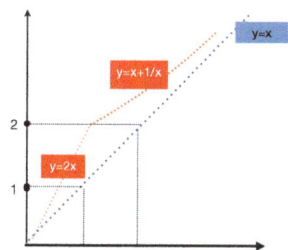

Fig. 1.3 Example

Definition 1.1 A set $A \subset \mathbb{S}_F$ is bounded away (BA) from the forbidden zone F if $d(A, F) > 0$. Denote the collection of all measurable sets bounded away from F by $\mathbb{BA}(\mathbb{S}_F)$.

The sets $\mathbb{BA}(\mathbb{S}_F)$ form what [5] calls a *boundedness* but the sets of $\mathbb{BA}(\mathbb{S}_F)$ need not be metrically bounded.

Recall that $F^\delta = \{x \in \mathbb{S} : d(x, F) < \delta\}$, and the following are equivalent:

1. $d(A, F) > 0$. (1.5)
2. There exists $\delta > 0$ such that $A \subset \left(F^\delta\right)^c$. (1.6)

To verify the equivalences:

1. (1.5) $\Rightarrow$ (1.6). If $d(A, F) > 0$, $\exists \eta > 0$ such that $d(a, z) \geq \eta$, for all $z \in F$ and $a \in A$. So for any $a \in A$, $d(a, F) \geq \eta$ and $a \in (F^\eta)^c$ so $A \subset (F^\eta)^c$.
2. (1.6) $\Rightarrow$ (1.5). If $A \subset (F^\delta)^c$ for $\delta > 0$, then for all $a \in A$, $d(a, F) \geq \delta$ so $d(A, F) = \inf_{a \in A} d(a, F) \geq \delta$.

Warning: If $\bar{A} \cap F = \emptyset$, this does NOT imply $d(\bar{A}, F) > 0$.

Example See Fig. 1.3: Take $\mathbb{S} = \mathbb{R}_+^2$, $F = [\text{diag}] = \{(x, x) : 0 \leq x < \infty\}$ and let

$$A = \{(x, x + \frac{1}{x}), x \geq 1\} \cup \{(x, 2x) : 0 \leq x \leq 1\}.$$

Then A is closed so $A = \bar{A}$, and $A \cap F = \emptyset$ but $d(A, F) = 0$.

What makes this example work is that A is not compact. If we add the requirement of compactness, then with a little metric space canoodling we get the following.

Lemma 1.1 *If $A \subset \mathbb{S}_F$ is compact, then the following are equivalent:*

1. $A \cap F = \emptyset$.
2. $d(A, F) > 0$.

If $d(A, F) > 0$ then $A \cap F = \emptyset$. Conversely, if $A \cap F = \emptyset$ but $d(A, F) = 0$ we get a contradiction as follows: If $d(A, F) = 0$ there exist $a_n \in A$ such that $d(a_n, F) \to 0$. Compactness of A means there is a convergent subsequence $\{a_{n'}\}$ converging to some $a_\infty \in A$ and $d(a_\infty, F) = 0$ so also $a_\infty \in F$ contradicting disjointness.

1.4 Measures on TABOF Spaces: The Space $\mathbb{M}(\mathbb{S}_F) = \mathbb{M}(\mathbb{S} \setminus F)$

Let $\mathbb{M}(\mathbb{S}_F)$ be all measures $\mu(\cdot)$ on $\mathbb{S}_F$ satisfying $\mu(D) < \infty$ for all $D \in \mathbb{BA}(\mathbb{S}_F)$. If $F = \emptyset$, then by convention all sets are bounded away including $\mathbb{S}$ so measures in $\mathbb{BA}(\mathbb{S} \setminus \emptyset)$ are finite. The BA sets play the role of compact sets when discussing Radon measures and vague convergence.

Warning: The class of measures $\mathbb{M}(\mathbb{S}_F)$ is not the same as Radon measures on $\mathbb{S}_F$; that is the measures finite on relatively compact subsets of $\mathbb{S}$. Let $\mathbb{S} = \mathbb{R}_+$, $F = \{0\}$ so $\mathbb{S}_F = \mathbb{R}_+ \setminus \{0\}$. Then $\mu \in \mathbb{M}(\mathbb{S}_F)$ means $\mu(x, \infty) < \infty$ if $x > 0$ since $(x, \infty) \in \mathbb{BA}(\mathbb{R}_+ \setminus \{0\})$. However, μ Radon with the usual meaning is that $\mu[a, b] < \infty$ for $0 < a < b < \infty$. So Lebesgue measure Leb is Radon but not in $\mathbb{M}(\mathbb{S}_F)$. This is explored further in Problem 1.6, page 25.

1.4.1 A Topology on $\mathbb{M}(\mathbb{S}_F)$; M-convergence

A topology is needed for $\mathbb{M}(\mathbb{S}_F)$ and once this is specified, a notion of convergence will exist. In fact $\mathbb{M}(\mathbb{S}_F)$ is a metric space. If the word *topology* freezes your heart, the convergence notion (1.9) below is what you need and is readily understood. This notion of convergence is the basis for the definition of regular variation of measures. The mathematically inclined will find more perspective and applications in [35], [96], [45], [125], [74], [107, Chapter 4], [5], [114].

Think of a measure $\mu \in \mathbb{M}(\mathbb{S}_F)$ as identified by its values $\{\mu(A), A \in \sigma(\mathbb{S}_F)$, where $\sigma(\mathbb{S}_F)$ are the Borel subsets of $\mathbb{S}_F$. We do not need to know $\mu(A)$ for all A; an appropriate subclass of sets should uniquely specify μ. Instead of identifying the measure with its values on a distinguished class of sets, we could uniquely specify $\mu(\cdot)$ by integrating against a class of $\mathbb{R}_+$-valued test functions $f : \mathbb{S}_F \mapsto \mathbb{R}_+$ that approximate indicator functions of sets. So it is natural to consider

$$\mu \mapsto \mu(f) := \int_\mathbb{S} f(s) \mu(ds).$$

For instance on $\mathbb{R}$, think of f as a smooth approximation to an interval and then the leap from sets to functions becomes more natural. We seek a convenient but rich enough class of test functions $\mathbb{C}(\mathbb{S}_F)$ so that knowing $\mu(f), f \in \mathbb{C}(\mathbb{S}_F)$ both determines μ and leads to a useful definition of convergence of measures. We then think of the object μ as having components $\{\mu(f), f \in \mathbb{C}(\mathbb{S}_F)\}$ in the same way as elements of $\mathbb{R}^\infty$ have components.

1.4.1.1 Test Functions

Define the class of test functions $\mathbb{C}(\mathbb{S}_F)$ as the set of all bounded, non-negative, continuous functions $f : \mathbb{S}_F \mapsto \mathbb{R}_+$ such that there exists $\delta > 0$ and if $d(x, F) < \delta$, then $f(x) = 0$. So $f(x) = 0$ if x is too close to the forbidden zone and the support of f is in $\mathbb{BA}(\mathbb{S}_F)$. Because $\mu \in \mathbb{M}(\mathbb{S}_F)$ is finite on sets bounded away from the forbidden zone and since test functions $f \in \mathbb{C}(\mathbb{S}_F)$ are bounded with BA supports, we have

$$\mu(f) := \int_{\mathbb{S}_F} f\,d\mu < \infty.$$

If $F = \emptyset$, $\mathbb{C}(\mathbb{S} \setminus \emptyset)$ consists of bounded continuous functions on $\mathbb{S}$.

1.4.1.2 Topology and Convergence

The topology and therefore the notion of convergence in $\mathbb{M}(\mathbb{S}_F)$ can be specified in a variety of ways. Each relies on the definition of the test functions $\mathbb{C}(\mathbb{S}_F)$. The briefest but not most appealing description is to announce that the topology on $\mathbb{M}(\mathbb{S}_F)$ is the smallest topology making the map from $\mathbb{M}(\mathbb{S}_F) \mapsto \mathbb{R}_+$,

$$\mu \mapsto \mu(f), \quad f \in \mathbb{C}(\mathbb{S}_F),$$

continuous. A more fulsome description is to say that a basic neighborhood of $\mu \in \mathbb{M}(\mathbb{S}_F)$ is a subset of $\mathbb{M}(\mathbb{S}_F)$ of the form

$$\{\nu(\cdot) \in \mathbb{M}(\mathbb{S}_F) : |\nu(f_i) - \mu(f_i)| < \epsilon,\ i = 1, \ldots, k\}, \qquad (1.7)$$

where $\epsilon > 0$, and $f_i \in \mathbb{C}(\mathbb{S}_F)$ for $i = 1, \ldots, k$. The topology is generated by the basic neighborhoods. This means sub-basis sets are of the form

$$\{\nu \in \mathbb{M}(\mathbb{S}_F) : \nu(f) \in G\}, \quad G \in \mathcal{G}(\mathbb{R}_+),\ f \in \mathbb{C}(\mathbb{S}_F). \qquad (1.8)$$

If we identify the measure $\mu(\cdot)$ with its components $\{\mu(f), f \in \mathbb{C}(\mathbb{S}_F)\}$, then a sub-base set is a set of measures whose f-th component is restricted to values in the open set G.

Convergence The notion of convergence given by this topology is that if $\mu_n \in \mathbb{M}(\mathbb{S}_F)$ for $n \geq 0$, we have $\mu_n \to \mu_0$ if for all $f \in \mathbb{C}(\mathbb{S}_F)$ we have

$$\mu_n(f) \to \mu_0(f) \quad (n \to \infty). \qquad (1.9)$$

Think of this as component-wise convergence. When $F = \emptyset$, we have *weak convergence*.

1.4.1.3 Metrizability

We will not use metrizability of $\mathbb{M}(\mathbb{S}_F)$ but a quick way to metrize $\mathbb{M}(\mathbb{S}_F)$ as a complete, separable metric space is to use the *Prohorov distance*, $p_r(\cdot, \cdot)$ ([35, p. 398], [13, p. 72]) on bounded measures on $\mathbb{S} \setminus F^{1/r}$ where remember $F^\delta = \{x \in \mathbb{S} : d(x, F) < \delta\}$. For $\mu, \nu \in \mathbb{M}(\mathbb{S}_F)$, define $\mu^{(r)}, \nu^{(r)}$ as the finite restrictions of μ, ν to $\mathbb{S} \setminus F^{1/r}$. (More in Sect. 1.6.2.1.) Then the metric $d_{\mathbb{M}(\mathbb{S}_F)}(\mu, \nu)$ is

$$d_{\mathbb{M}(\mathbb{S}_F)}(\mu, \nu) = \int_0^\infty e^{-r} \frac{p_r(\mu^{(r)}, \nu^{(r)})}{1 + p_r(\mu^{(r)}, \nu^{(r)})} dr. \qquad (1.10)$$

There are other equivalent ways to metrize $\mathbb{M}(\mathbb{S}_F)$ ([114, p. 498], [66, Theorem 11.3.3]) but the takeaway message is convergence in $\mathbb{M}(\mathbb{S}_F)$ is metric.

1.4.2 Equivalent Forms of Convergence; the Portmanteau Theorem

Every good theory about convergence needs a collection of equivalences that allow flexibility in how you approach proofs of convergence. The phrase *portmanteau*[4] seems to originate with P. Billingsley [11] and appears in his well thought out and successful books.

An important conclusion from the following result is that the class of functions used to test for convergence can be reduced from $\mathbb{C}(\mathbb{S}_F)$ to bounded and *uniformly continuous* functions.

Theorem 1.1 *Suppose $\mu_n \in \mathbb{M}(\mathbb{S}_F)$ for $n \geq 0$. The following are equivalent.*

1. $\mu_n \to \mu_0$.
2. *For all $f \in \mathbb{C}(\mathbb{S}_F)$ which are uniformly continuous,*

$$\mu_n(f) = \int f d\mu_n \to \mu_0(f) = \int f d\mu_0. \qquad (1.11)$$

3. *For all $A \in \mathcal{F}(\mathbb{S}_F) \cap \mathbb{BA}(\mathbb{S}_F)$,*

$$\limsup_{n \to \infty} \mu_n(A) \leq \mu_0(A); \qquad (1.12)$$

and for all $G \in \mathcal{G}(\mathbb{S}_F) \cap \mathbb{BA}(\mathbb{S}_F)$,

$$\liminf_{n \to \infty} \mu_n(G) \geq \mu_0(G). \qquad (1.13)$$

---

[4] Often, non-literary types assume there is a Mr. or Ms. Portmanteau who created the results. No.

4. Let ∂A be the boundary of A. (See Appendix B.) For all $A \in \mathbb{BA}(\mathbb{S}_F)$ with $\mu_0(\partial A) = 0$, we have

$$\mu_n(A) \to \mu_0(A).$$

Remark 1.1 An additional equivalence that relates $\mathbb{M}$-convergence to weak convergence could have been included in this list. For $\delta > 0$ and a measure $\mu \in \mathbb{M}(\mathbb{S} \setminus F)$, let $\mu^\delta(\cdot)$ be the restriction of μ to $\mathbb{S} \setminus F^\delta$ so that μ is a finite measure on $\mathbb{S} \setminus F^\delta$. Then $\mu_n \to \mu_0$ in $\mathbb{M}(\mathbb{S} \setminus F)$ iff for any $\delta > 0$ such that $\mu_0\big(\partial(\mathbb{S} \setminus F^\delta)\big) = 0$ we have $\mu_n^\delta \to \mu_0^\delta$ weakly, meaning for any bounded, continuous function $f(s)$ on $\mathbb{S} \setminus F^\delta$, $\mu_n^\delta(f) \to \mu_0^\delta(f)$. See Sect. 1.6.2.

Proof We show (1) $\Rightarrow$ (2) $\Rightarrow$ (3) $\Rightarrow$ (4) $\Rightarrow$ (1).

(1) $\Rightarrow$ (2): The uniformly continuous functions in $\mathbb{C}(\mathbb{S}_F)$ are a subset of $\mathbb{C}(\mathbb{S}_F)$ so this is clear.

(2) $\Rightarrow$ (3): This discussion is facilitated by a lemma.

Lemma 1.2 *Suppose $F, A \in \mathcal{F}(\mathbb{S})$ are two closed sets and $d(A, F) = 2\eta$ for $\eta > 0$. Then for all $\delta < \eta$, there exists a bounded, uniformly continuous function $f_\delta : \mathbb{S} \mapsto [0, 1]$ with $f_\delta \in \mathbb{C}(\mathbb{S}_F)$ and*

$$1_A \leq f_\delta \leq 1_{A^\delta}. \tag{1.14}$$

The lemma rests on the fact that $d(x, A)$ is uniformly continuous in x which can be seen by writing for $x, y \in \mathbb{S}$,

$$d(x, A) \leq d(x, y) + d(y, A).$$

Reverse the role of x and y and subtract to get

$$|d(x, A) - d(y, A)| \leq d(x, y), \tag{1.15}$$

which makes $d(\cdot, A)$ uniformly continuous.

For $\delta < \eta$, define

$$f_\delta(x) = \big(1 - \delta^{-1} d(x, A)\big)^+$$

so that $0 \leq f_\delta \leq 1$ and because $d(\cdot, A)$ is uniformly continuous, so is f_δ.

For $x \in A$, $d(x, A) = 0$ and $f_\delta(x) = 1$. Therefore $1_A \leq f_\delta$. For $x \in \mathbb{S} \setminus A^\delta$ we have $d(x, A) \geq \delta$ which means $f_\delta(x) = 0$. Therefore

$$\mathbb{S} \setminus A^\delta \subset [f_\delta = 0]. \tag{1.16}$$

In particular, since $F^\delta \subset \mathbb{S} \setminus A^\delta$, $F^\delta \subset [f_\delta = 0]$ which means $\mathrm{supp}(f_\delta)$, the support of f_δ, satisfies $\mathrm{supp}(f_\delta) \subset \mathbb{S} \setminus F^\delta$. Therefore, $f_\delta \in \mathbb{C}(\mathbb{S}_F)$. Since (1.16) is equivalent

1.4 Measures on TABOF Spaces: The Space $\mathbb{M}(\mathbb{S}_F) = \mathbb{M}(\mathbb{S} \setminus F)$

to
$$[f_\delta > 0] \subset A^\delta,$$

which means $f_\delta \leq 1_{A^\delta}$ and shows (1.14). This completes the proof of the lemma.

Now assume (2) and let $A \in \mathcal{F}(\mathbb{S}_F) \cap \mathrm{BA}(\mathbb{S}_F)$. Then

$$\limsup_{n\to\infty} \mu_n(A) \leq \limsup_{n\to\infty} \mu_n(f_\delta) \qquad \text{(from (1.14))}$$
$$= \mu_0(f_\delta) \qquad \text{(from (2))}$$
$$\leq \mu_0(A^\delta) \qquad \text{(from (1.14))}$$
$$\downarrow \mu_0(A) \qquad \text{(as } \delta \downarrow 0\text{)},$$

since $A^\delta \downarrow A$ as $\delta \downarrow 0$. This verifies (1.12).

For (1.13), let $G \in \mathcal{G}(\mathbb{S})$ with $d(G, F) = 2\eta$. For $\epsilon < \eta$, define the closed set,

$$A_\epsilon = \left((G^c)^\epsilon\right)^c \in \mathcal{F}(\mathbb{S}).$$

Note the supescript c means *complement*; it does not refer to the ϵ-swelling of the set but the superscript ϵ does. We have the following properties for A_ϵ.

1. $A_\epsilon \subset G$ since $A_\epsilon^c = (G^c)^\epsilon \supset G^c$.
2. $d(A_\epsilon, G^c) \geq \epsilon$ since for $a \in A_\epsilon$, $a \notin (G^c)^\epsilon$ which means $d(a, G^c) \geq \epsilon$. Hence

$$d(A_\epsilon, G^c) = \inf_{a \in A_\epsilon} d(a, G^c) \geq \epsilon.$$

3. $d(A_\epsilon, F) \geq 2\eta$ since $d(A_\epsilon, F) \geq d(G, F) = 2\eta$.
4. $d(A_\epsilon^{\epsilon/2}, G^c) \geq \epsilon/2$. To verify this, consider two cases. For $a \in A_\epsilon^{\epsilon/2}$, either (i) $a \in A_\epsilon$ or (ii) $a \in A_\epsilon^{\epsilon/2} \setminus A_\epsilon$. In the first case, $d(a, G^c) \geq \inf_{x \in A_\epsilon} d(x, G^c) = d(A_\epsilon, G^c) \geq \epsilon$ from item 2. In case (ii), there exists $a^* \in A_\epsilon$ with $d(a, a^*) < \epsilon/2$. From (1.15),

$$|d(a, G^c) - d(a^*, G^c)| \leq d(a, a^*) \leq \epsilon/2$$

and thus

$$d(a^*, G^c) \geq d(a, G^c) - \epsilon/2 \geq \epsilon - \epsilon/2 = \epsilon/2.$$

5. Owing to item 4, $\overline{A_\epsilon^{\epsilon/2}} \cap G^c = \emptyset$ which implies

$$A_\epsilon^{\epsilon/2} \subset \overline{A_\epsilon^{\epsilon/2}} \subset G.$$

6. From Lemma 1.2, there exists $f_{\epsilon/2} \in \mathbb{C}(\mathbb{S}_F)$ such that

$$1_{A_\epsilon} \leq f_{\epsilon/2} \leq 1_{A_\epsilon^{\epsilon/2}}. \tag{1.17}$$

Therefore,

$$\liminf_{n\to\infty} \mu_n(G) \geq \liminf_{n\to\infty} \mu_n(A_\epsilon^{\epsilon/2}) \qquad \text{(from item 5)}$$

$$\geq \liminf_{n\to\infty} \mu_n(f_{\epsilon/2}) = \mu_0(f_{\epsilon/2}) \qquad \text{(from (1.11))}$$

$$\geq \mu_0(A_\epsilon).$$

Now as $\epsilon \downarrow 0$, $(G^c)^\epsilon \downarrow G^c$ and thus $\big((G^c)^\epsilon\big)^c \uparrow (G^c)^c = G$ so $A_\epsilon \uparrow G$ and $\mu_0(A_\epsilon) \uparrow \mu_0(G)$ which proves (1.13).

(3)$\Rightarrow$(4): Assume $A \in \mathbb{BA}(\mathbb{S}_F)$ with $\mu_0(\partial A) = 0$. Then

$$\limsup \mu_n(A) \leq \limsup \mu_n(\bar{A}) \leq \mu_0(\bar{A})$$

$$= \mu_0(A^0) \leq \liminf_{n\to\infty} \mu_n(A^0) \leq \liminf_{n\to\infty} \mu_n(A).$$

This implies $\mu_n(A) \to \mu_0(A)$.

(4)$\Rightarrow$(1): Given (4), let $f \in \mathbb{C}(\mathbb{S}_F)$ and we must show $\mu_n(f) \to \mu_0(f)$. Since f is bounded, there is a positive number $\|f\| > 0$ such that $f(x) \leq \|f\|$ and furthermore, since $f \in \mathbb{C}(\mathbb{S}_F)$, there is a $\delta > 0$ such that

$$[f > 0] \subset (F^\delta)^c, \tag{1.18}$$

and we are free to choose the δ so that it also satisfies the requirement that $(F^\delta)^c$ is a μ_0-continuity set; that is, $\mu_0\big(\partial((F^\delta)^c)\big) = 0$.

Next, write

$$\mu_n(f) = \int_\mathbb{S} f d\mu_n = \int_{[f>0]} f d\mu_n = \iint_{\{(s,t)\in\mathbb{S}\times(0,\|f\|]: f(s)>0,\, 0<t\leq f(s)\}} dt\, \mu_n(ds)$$

and by Fubini's theorem this is

$$= \int_0^{\|f\|} \mu_n\{s \in \mathbb{S} : f(s) \in (t, \|f\|]\} dt. \tag{1.19}$$

Because f is continuous, for $t > 0$,

$$\partial\{s \in \mathbb{S} : f(s) \in (t, \|f\|]\} \subset \{s : f(s) = t\}$$

and the measure $\mu_0 \circ f^{-1}$ has at most countably many atoms so avoiding these we have for almost all t,

1.4 Measures on TABOF Spaces: The Space $\mathbb{M}(\mathbb{S}_F) = \mathbb{M}(\mathbb{S} \setminus F)$

$$\mu_0\big(\partial\{s \in \mathbb{S} : f(s) \in (t, \|f\|]\}\big) = 0.$$

Thus, by assuming (4), we have for for almost all t,

$$\mu_n\{s \in \mathbb{S} : f(s) \in (t, \|f\|]\} \to \mu_0\{s \in \mathbb{S} : f(s) \in (t, \|f\|]\}.$$

We just have to finish with a dominated convergence argument. Using (1.18),

$$\mu_n\{s \in \mathbb{S} : f(s) \in (t, \|f\|]\} \leq \mu_n([f > 0]) \leq \mu_n\big((F^\delta)^c\big)$$

and since $\delta > 0$ was chosen to make $(F^\delta)^c$ a μ_0-continuity set, there is an $\eta > 0$ such that for all large n

$$\mu_n\big((F^\delta)^c\big) \leq \mu_0\big((F^\delta)^c\big) + \eta,$$

and this allows application of dominated convergence to (1.19). □

1.4.3 Taking the Definitions for a Test Drive

We have now laid out basic definitions and rules of convergence and so we pause to get a sense of the power and flexibility of these basics. We use some standard weak convergence theory plus the definitions to produce a result elaborated in Chap. 2, Example 2.1, page 40.

Proposition 1.1 *For $i = 1, 2$, we are given two CSMSs $\mathbb{S}_i$ with metrics $d_i(\cdot, \cdot)$. For closed sets $F_i \in \mathcal{F}(\mathbb{S}_i)$ form two TABOF spaces $\mathbb{S}_i \setminus F_i$. Suppose for $n \geq 0$, $\mu_n^{(i)} \in \mathbb{M}(\mathbb{S}_i \setminus F_i)$ and $\mu_n^{(i)} \to \mu_0^{(i)}$ in $\mathbb{M}(\mathbb{S}_i \setminus F_i)$. Then*

$$\mu_n^{(1)} \times \mu_n^{(2)} \to \mu_0^{(1)} \times \mu_0^{(2)} \tag{1.20}$$

in $\mathbb{M}\big((\mathbb{S}_1 \setminus F_1) \times (\mathbb{S}_2 \setminus F_2)\big) = \mathbb{M}\big(\mathbb{S}_1 \times \mathbb{S}_2 \setminus (F_1 \times \mathbb{S}_2 \cup \mathbb{S}_1 \times F_2)\big).$

For a reminder about products of TABOF spaces, see (1.3), page 4. If for $i = 1, 2$, $\mathbb{S}_i = \mathbb{R}_+$ and $F_i = \{0\}$, then the convergence claimed in (1.20) is in $\mathbb{M}(\mathbb{R}_+^2 \setminus [\text{axes}])$, measures on the first quadrant with the axes removed. This is different than claiming convergence in $\mathbb{M}(\mathbb{R}_+^2 \setminus \{\mathbf{0}\})$.

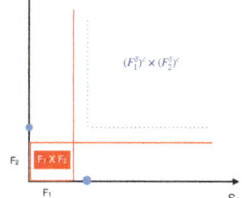

Proof Suppose $f \in \mathbb{C}\big((\mathbb{S}_1 \setminus F_1) \times (\mathbb{S}_2 \setminus F_2)\big)$ and f is uniformly continuous with modulus of continuity $\omega_f(\gamma)$. Choose $\delta > 0$ such that for $i = 1, 2$,

(i) $\mu_0^{(i)}(\partial(\mathbb{S}_i \setminus F_i^\delta)) = 0$.
(ii) $\mathrm{supp}(f) \subset (F_1^\delta)^c \times (F_2^\delta)^c$.
(iii) $\mu_n^{(i)} \to \mu_0^{(i)}$, weakly in $\mathbb{M}(\mathbb{S}_i \setminus F_i^\delta)$.

For (iii), see Remark 1.1, page 10.

We must show $\mu_n^{(1)} \times \mu_n^{(2)}(f) \to \mu_0^{(1)} \times \mu_0^{(2)}(f)$ or

$$LHS = \int_{\mathbb{S}_2 \setminus F_2^\delta} \mu_n^{(2)}(dy) \int_{\mathbb{S}_1 \setminus F_1^\delta} f(x,y) \mu_n^{(1)}(dx)$$

$$\to \int_{\mathbb{S}_1 \times \mathbb{S}_2} f(x,y) \mu_0^{(1)}(dx) \mu_0^{(2)}(dy) =: RHS.$$

Now

$$LHS = \int_{\mathbb{S}_2 \setminus F_2^\delta} \mu_n^{(2)}(dy) \left[\int_{\mathbb{S}_1 \setminus F_1^\delta} f(x,y) \mu_n^{(1)}(dx) - \int_{\mathbb{S}_1 \setminus F_1^\delta} f(x,y) \mu_0^{(1)}(dx) \right]$$

$$+ \int_{\mathbb{S}_2 \setminus F_2^\delta} \mu_n^{(2)}(dy) \int_{\mathbb{S}_1 \setminus F_1^\delta} f(x,y) \mu_0^{(1)}(dx)$$

$$= A + B.$$

Write

$$h_0(y) = \int_{\mathbb{S}_1 \setminus F_1^\delta} f(x,y) \mu_0^{(1)}(dx)$$

and observe $h_0 \in \mathbb{C}(\mathbb{S}_2 \setminus F_2)$ so

$$B \to \int_{\mathbb{S}_2 \setminus F_2^\delta} \mu_0^{(2)}(dy) \int_{\mathbb{S}_1 \setminus F_1^\delta} f(x,y) \mu_0^{(1)}(dx) = RHS$$

and it remains to show $A \to 0$. To check this quickly, we resort to the Skorohod representation theorem [13, page 70]. Assume temporarily that all $\mu_n^{(1)}$ are probability measures on $\mathbb{S}_1 \setminus F_1^\delta$. There exists a probability space $(\Omega, \mathcal{F}, P)$ on which are defined random elements X_n, $n \geq 0$ such that X_n has distribution $\mu_n^{(i)}$ and $X_n \to X_0$ almost surely-P. Then a bound on A is

$$|A| \leq \int_{\mathbb{S}_2 \setminus F_2^\delta} \mu_n^{(2)}(dy) E \left| f(X_n, y) - f(X_0, y) \right|$$

$$\leq \int_{\mathbb{S}_2 \setminus F_2^\delta} \mu_n^{(2)}(dy) E \left(\omega_f(d_1(X_n, X_0)) \right).$$

For $\gamma > 0$, break the expectation according to whether $d_1(X_n, X_0) > \gamma$ or $< \gamma$ and the above is bounded by

$$\leq \omega_f(\gamma)\mu_n^{(2)}(\mathbb{S}_2 \setminus F_2^\delta) + (const)\mu_n^{(2)}(\mathbb{S}_2 \setminus F_2^\delta)P[d_1(X_0, X_n) > \gamma] \to 0$$

as $n \to \infty$ and then $\gamma \to 0$.

If the $\mu_n^{(1)}$ measures are bounded but not probability measures, write $P_n = \mu_n^{(1)}/\mu_n^{(1)}(\mathbb{S}_1 \setminus F_1^\delta)$ and with some extra algebraic finagling, proceed as before. □

1.5 New Results from Existing Ones: Mapping Theorems

Mapping theorems capitalize on the power of continuity for proving limit theorems and approximations by taking existing or initial, known convergence results, applying a mapping and generating corollary convergences. Mapping theorems often save considerable effort.

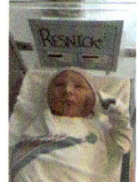

New result

1.5.1 The Problem and Context

Suppose $\mathbb{S}_i$, $i = 1, 2$ are two complete separable metric spaces with metrics $d_1(\cdot, \cdot)$ and $d_2(\cdot, \cdot)$ respectively. Let $F_i \in \mathcal{F}(\mathbb{S}_i)$, $i = 1, 2$ be two closed sets and form the TABOF spaces $\mathbb{S}_i(F_i) = \mathbb{S}_i \setminus F_i$. Suppose $h : \mathbb{S}_1(F_1) \mapsto \mathbb{S}_2(F_2)$ and h can be used to induce a measure on $\mathbb{S}_2(F_2)$ by $\hat{h}(\mu) := \mu \circ h^{-1}$. If $\mu \in \mathbb{M}(\mathbb{S}_1(F_1))$, we *hope* $\hat{h}(\mu) \in \mathbb{M}(\mathbb{S}_2(F_2))$. If h is continuous, when is $\hat{h}$ a continuous map?

The first result benefits from the approach of Hult and Lindskog [95] and [124].

1.5.2 The First Mapping Theorem

Continuity of h carries over to continuity of $\hat{h}$ provided h *respects the forbidden zone*.[5]

---

[5] Pop culture enthusiasts will want to call this the *Aretha* property.

Definition 1.2 The map $h : \mathbb{S}_1(F_1) \mapsto \mathbb{S}_2(F_2)$ respects the forbidden zone $F_1 \subset \mathbb{S}_1$ if $A_2 \in \mathbb{BA}(\mathbb{S}_2(F_2)) \cap h(\mathbb{S}_1(F_1))$ implies $h^{-1}(A_2) \in \mathbb{BA}(\mathbb{S}_1(F_1))$.

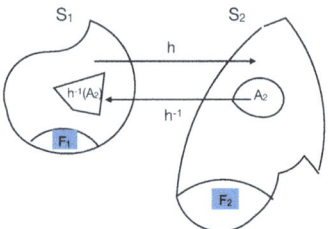

So if A_2 in the range of h is bounded away from F_2, its pre-image must be bounded away from F_1.

Theorem 1.2 (First Mapping Theorem) *If $h : \mathbb{S}_1(F_1) \mapsto \mathbb{S}_2(F_2)$ is continuous and respects the forbidden zone, then $\hat{h} : \mathbb{M}(\mathbb{S}_1(F_1)) \mapsto \mathbb{M}(\mathbb{S}_2(F_2))$ is also continuous where $\hat{h}(\mu) = \mu \circ h^{-1}$.*

Proof To prove $\hat{h}$ is continuous, we will use the characterization of continuity that the inverse image of an open set should be open. It suffices to consider the inverse image of a sub-base set so pick $f_2 \in \mathbb{C}(\mathbb{S}_2(F_2))$ and referring to (1.8) a sub-base set has the form,

$$\{\mu_2 \in \mathbb{M}(\mathbb{S}_2(F_2)) : \mu_2(f_2) \in G\}, \qquad G \in \mathcal{G}(\mathbb{R}_+).$$

Then

$$\hat{h}^{-1}(\{\mu_2 \in \mathbb{M}(\mathbb{S}_2(F_2)) : \mu_2(f_2) \in G\}) = \{\mu_1 \in \mathbb{M}(\mathbb{S}(F_1)) : \hat{h}(\mu_1)(f_2) \in G\}$$
$$= \{\mu_1 \in \mathbb{M}(\mathbb{S}(F_1)) : \mu_1 \circ h^{-1}(f_2) \in G\}$$
$$= \{\mu_1 \in \mathbb{M}(\mathbb{S}(F_1)) : \mu_1(f_2 \circ h) \in G\}.$$

This is sub-base set of $\mathbb{M}(\mathbb{S}_1(F_1))$ *provided $f_2 \circ h \in \mathbb{C}(\mathbb{S}_1(F_1))$* and this requires we check

(a) $f_2 \circ h$ is bounded and continuous; this is clear since f_2 is bounded and continuous and h is continuous.
(b) The support of $f_2 \circ h$ is bounded away from F_1.

Requirement (b) is why h must respect the forbidden zone as described in Definition 1.2. The support of $f_2 \circ h$, supp$(f_2 \circ h)$, is the closure of $\{x_1 \in \mathbb{S}_1(F_1) : f_2(h(x_1)) > 0\}$. If $f_2(h(x_1)) > 0$ then $h(x_1) \in \text{supp}(f_2)$ (which is closed in $\mathbb{S}_2(F_2)$) and thus $x_1 \in h^{-1}(\text{supp}(f_2))$. Since h is continuous, $h^{-1}(\text{supp}(f_2))$ is closed in $\mathbb{S}_1(F_1)$. Therefore,

$$\{x_1 \in \mathbb{S}_1(F_1) : f_2(h(x_1)) > 0\} \subset h^{-1}(\text{supp}(f_2))$$

and taking closure on the left

$$\text{supp}(f_2 \circ h) = \overline{\{x_1 \in \mathbb{S}_1(F_1) : f_2(h(x)) > 0\}} \subset h^{-1}(\text{supp}(f_2)).$$

It follows that

1.5 New Results from Existing Ones: Mapping Theorems

$$d_1(\text{supp}(f_2 \circ h), F_1) \geq d_1(h^{-1}(\text{supp}(f_2)), F_1) > 0,$$

since h respects the forbidden zone. This shows $f_2 \circ h \in \mathbb{C}(\mathbb{S}_1(F_1))$ as required. □

Corollary 1.1 (Flexibility) *Fix $\mu \in \mathbb{M}(\mathbb{S}_1(F_1))$ and suppose $h : \mathbb{S}_1(F_1) \mapsto \mathbb{S}_2(F_2)$ and h respects the forbidden zone. If*

$$\mu(D_h) := \mu\{x_1 \in \mathbb{S}_1(F_1) : h \text{ is discontinuous at } x_1\} = 0 \tag{1.21}$$

then $\hat{h}$ is continuous at μ.

So h does not need to be continuous everywhere which provides some flexibility in concrete circumstances. The standard proof uses part 4 of Theorem 1.1, page 9.

Proof First of all, for the nervous, D_h is measurable; see [13, p. 243]. Assume $\mu_n \in \mathbb{M}(\mathbb{S}_1(F_1))$ and $\mu_n \to \mu$ and to show $\hat{h}\mu_n \to \hat{h}\mu$ in $\mathbb{M}(\mathbb{S}_2(F_2))$, we use part 4 of Theorem 1.1 and assume A_2 is a measurable subset of $\mathbb{S}_2(F_2)$ that is bounded away from F_2 satisfying $\hat{h}\mu(\partial A_2) = \mu \circ h^{-1}(\partial A_2) = 0$. It suffices to show $\hat{h}\mu_n(A_2) \to \hat{h}\mu(A_2)$. It is readily verified that

$$\partial h^{-1}(A_2) \subset h^{-1}(\partial A_2) \cup D_h$$

so

$$\mu\big(\partial h^{-1}(A_2)\big) \leq \mu\big(h^{-1}(\partial A_2)\big) + \mu(D_h) = 0, \tag{1.22}$$

from (1.21) and (1.22). Since h preserves forbidden zones, we have, $h^{-1}(A_2)$ bounded away from F_1 with 0-mass on its boundary and hence, $\hat{h}\mu_n(A_2) = \mu_n\big(h^{-1}(A_2)\big) \to \mu\big(h^{-1}(A_2)\big) = \hat{h}\mu(A_2)$ as required. □

1.5.3 The Second Mapping Theorem and Uniform Continuity

... The beat goes on[6].

The second mapping theorem, when applicable, is simpler to apply compared with Theorem 1.2 since one does not have to check how respectful the map h might be. The cost for ease of application is that the map h must be *uniformly* continuous rather than just continuous.

---

[6] A 1967 song by Sonny and Cher.

Theorem 1.3 (Second Mapping Theorem) *Suppose $h : \mathbb{S}_1 \mapsto \mathbb{S}_2$ is uniformly continuous. Fix $F_1 \in \mathcal{F}(\mathbb{S}_1)$ and assume $h(F_1) \in \mathcal{F}(\mathbb{S}_2)$. Then*

$$\hat{h} : \mathbb{M}(\mathbb{S}_1 \setminus F_1) \mapsto \mathbb{M}(\mathbb{S}_2 \setminus h(F_1))$$

defined by $\hat{h}(\mu) = \mu \circ h^{-1}$ is continuous.

Proof We show h respects forbidden zones. Suppose $A_2 \in \mathbb{BA}(\mathbb{S}_2 \setminus h(F_1))$ so that $d_2(A_2, h(F_1)) > 0$. We claim that in $\mathbb{S}_1$ we have $d_1(h^{-1}(A_2), F_1) > 0$ and h respects forbidden zones. If not and $d_1(h^{-1}(A_2), F_1) = 0$, there exist

$$x_n \in h^{-1}(A_2), \quad y_n \in F_1 \tag{1.23}$$

with $d_1(x_n, y_n) \to 0$. Uniform continuity of h gives $d_2(h(x_n), h(y_n)) \to 0$. From (1.23) we get

$$h(x_n) \in A_2, \quad h(y_n) \in h(F_1), \tag{1.24}$$

which means $d_2(A_2, h(F_1)) = 0$. This contradicts the BA assumption of A_2. □

Compact is wonderful

Compactness of $\mathbb{S}_1$ solves many problems (assuming one can prove the compactness).

Corollary 1.2 *Suppose $h : \mathbb{S}_1 \mapsto \mathbb{S}_2$ is continuous, $\mathbb{S}_1$ is compact, $F_1 \in \mathcal{F}(\mathbb{S}_1)$. Then*

$$\hat{h} : \mathbb{M}(\mathbb{S}_1 \setminus F_1) \mapsto \mathbb{M}(\mathbb{S}_2 \setminus h(F_1))$$

defined by $\hat{h}(\mu) = \mu \circ h^{-1}$ is continuous.

Proof If $\mathbb{S}_1$ is compact and h is continuous, h is automatically uniformly continuous. Also, $F_1 \in \mathcal{F}(\mathbb{S}_1)$ is compact and $h(F_1) \in \mathcal{K}(\mathbb{S}_2)$ since the continuous image of a compact set is compact. So $h(F_1)$ is automatically closed. □

1.5.4 Examples for Uniform Continuity

Here are two useful instances where a map is uniformly continuous, namely CUMSUM and projection. Each will be used later.

1.5.4.1 The CUMSUM Map in $\mathbb{R}_+^\infty$

Suppose we have two sequences $\boldsymbol{x}, \boldsymbol{y} \in \mathbb{R}_+^\infty$. A standard metric is

1.5 New Results from Existing Ones: Mapping Theorems

$$d_\infty(x, y) = \sum_{i=1}^{\infty} \frac{|x_i - y_i| \wedge 1}{2^i}.$$

Define the map CUMSUM : $\mathbb{R}_+^\infty \mapsto \mathbb{R}_+^\infty$ by

$$\text{CUMSUM}(x) = (x_1, x_1 + x_2, x_1 + x_2 + x_3, \ldots).$$

Proposition 1.2 *The map CUMSUM* : $\mathbb{R}_+^\infty \mapsto \mathbb{R}_+^\infty$ *is uniformly continuous (and Lipschitz) in the d_∞ metric.*

Some background: Given a metric space $\mathbb{S}$, two metrics d_1, d_2 on $\mathbb{S}$ are *equivalent* if there exist $c_1 > 0, c_2 > 0$ such that

$$\forall x, y \in \mathbb{S} : c_1 d_2(x, y) \leq d_1(x, y) \leq c_2 d_2(x, y).$$

Then d_1, d_2 induce the same topology and notion of convergence on $\mathbb{S}$.

All metrics are equivalent on $\mathbb{R}_+^p$. On $\mathbb{R}_+^\infty$, we also need the metric on $\mathbb{R}_+^\infty$,

$$d'_\infty(x, y) = \sum_{p=1}^{\infty} \frac{\left(\sum_{l=1}^{p} |x_l - y_l|\right) \wedge 1}{2^p} = \sum_{p=1}^{\infty} \frac{\|x_{|p} - y_{|p}\|_1 \wedge 1}{2^p},$$

where $\|\cdot\|_1$ is the usual L_1 norm on $\mathbb{R}_+^p$ and for $x = (x_1, x_2, \ldots,) \in \mathbb{R}_+^\infty$, $x_{|p} = (x_1, \ldots, x_p) \in \mathbb{R}_+^p$.

Lemma 1.3 *The metrics d_∞ and d'_∞ are equivalent on $\mathbb{R}_+^\infty$ and*

$$d_\infty(x, y) \leq d'_\infty(x, y) \leq 2 d_\infty(x, y).$$

Proof First of all,

$$d'_\infty(x, y) = \sum_{i=1}^{\infty} \frac{\left(\sum_{l=1}^{i} |x_l - y_l|\right) \wedge 1}{2^i} \geq \sum_{i=1}^{\infty} \frac{|x_i - y_i| \wedge 1}{2^i} = d_\infty(x, y).$$

Furthermore, observe that

$$d'_\infty(x, y) = \sum_{i=1}^{\infty} \frac{\left(\sum_{l=1}^{i} |x_l - y_l|\right) \wedge 1}{2^i} \leq \sum_{i=1}^{\infty} \frac{\sum_{l=1}^{i} \left(|x_l - y_l| \wedge 1\right)}{2^i}$$

$$= \sum_{l=1}^{\infty} \sum_{i=l}^{\infty} 2^{-i} \left(|x_l - y_l| \wedge 1\right) = \sum_{l=1}^{\infty} 2 \cdot 2^{-l} \left(|x_l - y_l| \wedge 1\right) = 2 d_\infty(x, y),$$

which proves the other inequality.

Note here we used the little number theoretic result for $\xi_i \geq 0$, $i \geq 1$,

$$(\sum_{i=1}^{p} \xi_i) \wedge 1 \leq \sum_{i=1}^{p} (\xi_i \wedge 1).$$

This can be verified by looking at cases (i) $\sum_{i=1}^{p} \xi_i \leq 1$ and (ii) $\sum_{i=1}^{p} \xi_i > 1$. Case (ii) can be further broken into (iia) $\wedge_{i=1}^{p} \xi \leq 1$ and (iib) $\wedge_{i=1}^{p} \xi_i > 1$. □

Proof Proof of Proposition 1.2. We write

$$d_\infty\big(\text{CUMSUM}(x), \text{CUMSUM}(y)\big) = \sum_{i=1}^{\infty} \frac{\left|\sum_{l=1}^{i} x_l - \sum_{l=1}^{i} y_l\right| \wedge 1}{2^i}$$

$$\leq \sum_{i=1}^{\infty} \frac{\left(\sum_{l=1}^{i} |x_l - y_l|\right) \wedge 1}{2^i} = d'_\infty(x, y) \leq 2 d_\infty(x, y).$$

Thus CUMSUM is uniformly continuous because it is Lipschitz. □

We now apply the Second Mapping Theorem 1.3.

Corollary 1.3 *Let* $\mathbb{S}_1 = \mathbb{S}_2 = \mathbb{R}_+^\infty$ *and suppose both F and CUMSUM(F) are closed in* $\mathbb{R}_+^\infty$. *If for $n \geq 0$, $\mu_n \in \mathbb{M}(\mathbb{R}_+^\infty \setminus F)$ and $\mu_n \to \mu_0$ in $\mathbb{M}(\mathbb{R}_+^\infty \setminus F)$, then $\mu_n \circ \text{CUMSUM}^{-1} \to \mu_0 \circ \text{CUMSUM}^{-1}$ in $\mathbb{M}(\mathbb{R}_+^\infty \setminus \text{CUMSUM}(F))$.*

For example, if $F = \{\mathbf{0}_\infty\}$, then $\text{CUMSUM}(F) = \{\mathbf{0}_\infty\} \in \mathcal{F}(\mathbb{R}_+^\infty)$.

1.5.4.2 The Projection Map on $\mathbb{R}_+^\infty$

For $p \geq 1$, define the projection map from from $\mathbb{R}_+^\infty \mapsto \mathbb{R}_+^p$ by

$$\text{PROJ}_p(x) := x_{|p} = (x_1, \ldots, x_p). \tag{1.25}$$

So PROJ_p takes an infinite sequence, truncates it and retains the first p components.

Proposition 1.3 $\text{PROJ}_p : \mathbb{R}_+^\infty \mapsto \mathbb{R}_+^p$ *is uniformly continuous.*

Proof Let $d_p(x_{|p}, y_{|p}) = \sum_{i=1}^{p} |x_i - y_i|$ be the usual L_1 metric. Given $0 < \epsilon < 1$, we must find $\delta > 0$ such that $d_\infty(x, y) < \delta$ implies $d_p(x_{|p}, y_{|p}) < \epsilon$. We try $\delta = 2^{-p}\epsilon$. Then

$$\delta = 2^{-p}\epsilon > d_\infty(x, y) = \sum_{i=1}^{\infty} \frac{|x_i - y_i| \wedge 1}{2^i} \geq \sum_{i=1}^{p} \frac{|x_i - y_i| \wedge 1}{2^i}$$

$$\geq 2^{-p} \sum_{i=1}^{p} |x_i - y_i| \wedge 1.$$

Therefore $1 > \epsilon \geq \sum_{i=1}^{p} |x_i - y_i| \wedge 1$, so that $\epsilon \geq \sum_{i=1}^{p} |x_i - y_i| = d_p(\boldsymbol{x}_{|p}, \boldsymbol{y}_{|p})$.
□

We now apply the Second Mapping Theorem 1.3.

Corollary 1.4 *Let* $\mathbb{S}_1 = \mathbb{R}_+^\infty$ *and* $\mathbb{S}_2 = \mathbb{R}_+^p$. *Suppose F is closed in $\mathbb{R}_+^\infty$ and $PROJ_p(F)$ is closed in $\mathbb{R}_+^p$. If for $n \geq 0$, $\mu_n \in \mathbb{M}(\mathbb{R}_+^\infty \setminus F)$ and $\mu_n \to \mu_0$ in $\mathbb{M}(\mathbb{R}_+^\infty \setminus F)$, then $\mu_n \circ PROJ_p^{-1} \to \mu_0 \circ PROJ_p^{-1}$ in $\mathbb{M}(\mathbb{R}_+^p \setminus PROJ_p(F))$.*

1.6 Maneuvers Around Roadblocks

In Corollary 1.2, page 18, we commented that if $\mathbb{S}_1$ were compact, certain assumptions of the mapping theorems automatically held. However, assuming $\mathbb{S}_1$ compact is a fairly strong assumption that is often not true. What to do? Here are two possibilities.

1.6.1 More Subtle Use of Compactness

Instead of assuming $\mathbb{S}$ is compact, we may assume F_1 is compact as in the following variant.

Proposition 1.4 (Variant) *Suppose $h : \mathbb{S}_1 \mapsto \mathbb{S}_2$ is continuous and $F_1 \in \mathcal{K}(\mathbb{S}_1)$. Then $\hat{h} : \mathbb{M}(\mathbb{S}_1(F_1)) \mapsto \mathbb{M}(\mathbb{S}_2 \setminus h(F_1))$ defined by $\hat{h}(\mu) = \mu \circ h^{-1}$ is continuous.*

Proof The continuous image of a compact set is compact [66, p. 35] so $h(F_1)$ is closed and it suffices to check that h respects forbidden zones. Suppose $A_2 \in \mathcal{F}(\mathbb{S}_2 \setminus h(F_1))$ such that $d_2(A_2, h(F_1)) > 0$. We claim $d_1(h^{-1}(A_2), F_1) > 0$. If not, then $d_1(h^{-1}(A_2), F_1) = 0$ and we look forward to a contradiction. Zero distance means

$$\exists x_n \in h^{-1}(A_2) \subset \mathbb{S}_1, \exists y_n \in F_1, \text{ and } d_1(x_n, y_n) \to 0. \tag{1.26}$$

Compactness of F_1 means there is a subsequence $\{y_{n'}\}$ converging to a limit $y_\infty \in F_1$ say or $d_1(y_{n'}, y_\infty) \to 0$. Therefore (1.26) and the triangle inequality imply $d_1(x_{n'}, y_\infty) \to 0$. Since h is continuous, $d_2(h(x_{n'}), h(y_\infty)) \to 0$. Note $h(x_{n'}) \in A_2$ and $h(y_\infty) \in h(F_1)$ so we may conclude $d_2(A_2, h(F_1)) = 0$ which is the desired contradiction since we assumed this distance was positive. □

But.... Of course, this result does not solve all difficulties and will not bring about world peace. If $\mathbb{S}_1 = \mathbb{R}_+^2$ and $F_1 = [\text{axes}]$, there is nothing compact about F_1. When considering regular variation, typically as in this example, the forbidden zone F_1 is a cone and not compact. In the case $F_1 = \{\mathbf{0}\}$ where F_1 is compact, if h is defined on all of $\mathbb{S}_1$, then all is well. However, even if $F = \{\mathbf{0}\}$, often h will only be defined on $\mathbb{R}_+^2 \setminus \{\mathbf{0}\}$ as in the example of the polar coordinate transform $h : \mathbb{R}_+^2 \setminus \{\mathbf{0}\} \mapsto (0, \infty) \times \{x \in \mathbb{R}_+^2 : \|x\| = 1\}$ defined by $x \mapsto (\|x\|, x/\|x\|)$, where $\|x\|$ is a norm of x. See page 37 below.

1.6.2 Restriction of the State Space

Sometimes compactness or uniform continuity or the bounded away property are either hard to prove or just not true. How to cross a bridge too far? Sometimes restriction of the state space helps. See [75, Proposition 3.3, page 1170] or [156, page 206]. There are several ways to formulate results.

1.6.2.1 Restriction

For a TABOF space $\mathbb{S}_1(F_1)$, suppose $B_1 \subset \mathbb{S}_1$ is Borel and then the restriction of the state space is $\mathbb{S}_1(F_1) \cap B_1 = \mathbb{S}_1 \cap B_1 \cap F_1^c = \mathbb{S}_1 B_1 \setminus F_1 = B_1 \setminus F_1 = B_1 \setminus B_1 F_1$. If we want this to be a new TABOF space, we need the restricted metric space B_1 to be complete and separable. Separability is no problem and completeness is guaranteed if we assume $B_1 \in \mathcal{F}(\mathbb{S}_1)$. This will also imply $B_1 F_1$ is closed. For $\mu \in \mathbb{M}(\mathbb{S}_1(F_1))$ the restriction to B_1 is the mapping $\hat{B}_1 : \mathbb{M}(\mathbb{S}_1(F_1)) \mapsto \mathbb{M}(\mathbb{S}_2(F_2)) := \mathbb{M}(B_1 \setminus B_1 F_1)$ given by

$$\hat{B}_1 \mu(\cdot) = \mu(\cdot \cap B_1)$$

defined on Borel subsets of B_1.

Of course it is possible for $B_1 F_1 = \emptyset$; one example (with apologies for the disgraceful forward referencing) is the sets $B_1 = \mathbb{D}_{=j}$ and $F_1 = \mathbb{D}_{\leq j-1}$ in Sect. 4.1.2.4, page 124. If $B_1 F_1 = \emptyset$ and B_1 is bounded away from F_1, $\mu(B_1) < \infty$. Compare with Problem 1.7, page 26. If $B_1 F_1 = \emptyset$, Proposition 1.5 holds without assuming B_1 is closed which is only used if we need to guarantee the forbidden zone $B_1 F_1$ is closed.

When is the mapping $\hat{B}_1$ continuous?

Proposition 1.5 *Suppose $\mu \in \mathbb{M}(\mathbb{S}_1(F_1))$, $B_1 \in \mathcal{F}(\mathbb{S}_1) \cap \mathbb{BA}(\mathbb{S}_1(F_1))$ and B_1 satisfies $\mu(\partial B_1) = 0$. Then $\hat{B}_1 : \mathbb{M}(\mathbb{S}_1(F_1)) \mapsto \mathbb{M}(B_1 \setminus \emptyset)$ is continuous at μ.*

Proof We use part 4 of Theorem 1.1, page 9. Suppose $\mu_n \to \mu$ in $\mathbb{M}(\mathbb{S}_1(F_1))$ and we must show $\hat{B}_1 \mu_n \to \hat{B}_1 \mu$ in $\mathbb{M}(B_1 \setminus \emptyset)$. Suppose $A_1 \subset B_1$ with

$$\hat{B}_1 \mu(\partial_{B_1} A_1) = 0, \tag{1.27}$$

where in (1.27), $\partial_{B_1} A_1$ refers to the boundary of A_1 in B_1. According to Theorem 1.1, it suffices to show

1.6 Maneuvers Around Roadblocks

$$\hat{B}_1\mu_n(A_1) = \mu_n(A_1) \to \hat{B}_1\mu(A_1) = \mu(A_1).$$

If (1.27) also implies

$$\mu\big(\partial_{\mathbb{S}_1}(A_1)\big) = 0, \tag{1.28}$$

then from Theorem 1.1 we would have $\mu_n(A_1) \to \mu(A_1)$ or $\hat{B}_1\mu_n(A_1) \to \hat{B}_1\mu(A_1)$ as required. The claim (1.28) follows from

$$\partial_{\mathbb{S}_1}(A_1) \subset \partial_{B_1}(A_1) \cup \partial_{\mathbb{S}_1}(B_1), \quad \forall A_1 \subset B_1. \tag{1.29}$$

If you believe (1.29), you are done since the first set on the right of (1.29) has zero measure from (1.27) and the second set has zero mass from the hypothesis in the statement of the Proposition. Those harder to convince, read the next paragraph.

Let $x \in \partial_{\mathbb{S}_1}(A_1)$. Then

(1) $\exists x_n \in A_1$ such that $x_n \to x$ and
(2) $\exists y_n \in \mathbb{S}_1 \setminus A_1$ such that $y_n \to x$.

Either

(2a) There is an infinite subsequence $y_{n_k} \in B_1$ and also $y_{n_k} \in \mathbb{S}_1 \setminus A_1$; or
(2b) There does not exist such a subsequence and for all large n, $y_n \notin B_1$ but it is still true $y_n \to x$ (Fig. 1.4).

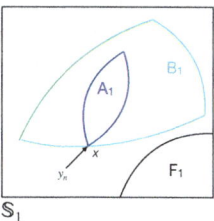

Fig. 1.4 Case (2b)

If (1) & (2a) hold, x can be approached from inside $A_1 \subset B_1$ and also x can be approached from outside A_1 but inside B_1 and therefore $x \in \partial_{B_1}(A_1)$. If (1) & (2b) hold, x can be approached from inside A_1 (and hence inside B_1) and can also be approached from outside B_1. Therefore $x \in \partial_{\mathbb{S}_1}(B_1)$. □

1.6.2.2 Convergence Reduced to Weak Convergence of Finite Measures

Given a TABOF space $\mathbb{S} \setminus F$ and $\delta > 0$, enlarge the forbidden zone from F to the δ-swelling F^δ. Define $B_\delta = \mathbb{S} \setminus F^\delta$ and $B_\delta \in \mathbb{BA}(\mathbb{S} \setminus F) \cap \mathcal{F}(\mathbb{S})$. Therefore for any measure $\mu \in \mathbb{M}(\mathbb{S} \setminus F)$, the restricted measure $\widehat{B_\delta}\mu(\cdot)$ is finite. But is $\mu(\partial(B_\delta)) = 0$ so that $\mu \mapsto \widehat{B_\delta}\mu$ is continuous according to Proposition 1.27? Suspense!

For fixed $\mu \in \mathbb{M}(\mathbb{S} \setminus F)$, consider the family

$$\Lambda := \{0 < \delta \leq 1 : \mu(\partial B_\delta) > 0\}. \tag{1.30}$$

A standard argument shows Λ is countable: The reason is that $\Lambda = \cup_{n,m} \Lambda_{n,m}$ where

$$\Lambda_{n,m} := \{1/m < \delta \le 1 : \mu(\partial B_\delta) > 1/n\}$$

must be finite since $\mu(B_{1/m}) < \infty$ on account of $B_{1/m} \in \mathbb{BA}(\mathbb{S} \setminus F)$. So we conclude there exist a sequence of $\delta_n \downarrow 0$ such that $\delta_n \in \Lambda^c$. This enables the next result [125, Theorem 2.1, p. 275].

Proposition 1.6 *The following are equivalent.*

1. $\mu_n \to \mu$ in $\mathbb{M}(\mathbb{S} \setminus F)$.
2. For all but countably many $\delta \in (0, 1]$ the restriction map is continuous at μ and for such $\delta > 0$, $\widehat{B_\delta}\mu_n \to \widehat{B_\delta}\mu$ in $\mathbb{M}(B_\delta \setminus \emptyset)$. This is weak convergence of finite measures.
3. There exists a sequence $\delta_m \downarrow 0$ such that $\delta_m \in \Lambda^c$ and for each m, as $n \to \infty$, $\widehat{B_{\delta_m}}\mu_n \to \widehat{B_{\delta_m}}\mu$ in $\mathbb{M}(B_{\delta_m} \setminus \emptyset)$.

Proof $(1 \to 2)$. Any $\delta \in \Lambda^c$ qualifies to make the restriction map continuous at μ by Proposition 1.5 and since Λ is at most countable, we are done.

$(2 \to 3)$. Given (2), pick out a countable collection $\{\delta_m\}$ with $\delta_m \downarrow 0$ such that the restriction map $\widehat{B_{\delta_m}}$ is continuous at μ because $\mu(\partial B_{\delta_m}) = 0$. Since Λ is at worst countable, there are plenty of points in Λ^c from which to choose.

$(3 \to 1)$. Pick a test function $f \in \mathbb{C}(\mathbb{S} \setminus F)$ and assume for some $\delta_0 > 0$ that $\mathrm{supp}(f) \subset \mathbb{S} \setminus F^{\delta_0}$. Then,

$$\mu_n(f) = \int_{\mathbb{S} \setminus F^{\delta_0}} f \, d\mu_n$$

and for all m such that $\delta_m < \delta_0$, this is

$$= \int_{\mathbb{S} \setminus F^{\delta_m}} f \, d\mu_n = \int_{B_{\delta_m}} f \, d\widehat{B_{\delta_m}}\mu_n$$

$$\to \int_{B_{\delta_m}} f \, d\widehat{B_{\delta_m}}\mu = \int_{\mathbb{S} \setminus F^{\delta_m}} f \, d\mu \quad \text{(from (3))}$$

$$= \int_{\mathbb{S} \setminus F} f \, d\mu.$$

Thus, as $n \to \infty$, $\mu_n(f) \to \mu(f)$ and thus (1) is true. □

1.7 Problems

1.1 Apropos of (1.4): Given two complete, separable metric spaces $\mathbb{S}_i$ with metrics $d_i(\cdot, \cdot)$, $i = 1, 2$. Define the metric on $\mathbb{S}_1 \times \mathbb{S}_2$ as

$$d\big((x, y), (z, w)\big) = d_1(x, z) \vee d_2(y, w).$$

1.7 Problems

Set
$$FZ = F_1 \times \mathbb{S}_2 \cup \mathbb{S}_1 \times F_2$$
where $F_i \in \mathcal{F}(\mathbb{S}_i)$ for $i = 1, 2$. Verify
$$(FZ)^\delta = F_1^\delta \times \mathbb{S}_2 \cup \mathbb{S}_1 \times F_2^\delta. \tag{1.31}$$

Note the notation $(FZ)^\delta$ is the δ-neighborhood in the product space and F_i^δ is the δ-neighborhood in the factor space.

1.2 Recall $\mathbb{D}_\sqcap := [0, \infty) \times (0, \infty)$ where the forbidden zone is the x-axis. Consider the map $h : \mathbb{D}_\sqcap \mapsto \mathbb{D}_\sqcap$ defined by
$$h(x, y) = (x, xy).$$
Does h respect the forbidden zone?

1.3 Suppose for $p \geq 1$ that $\mathbb{S} = \mathbb{R}_+^p$ and the forbidden zone is $F = \{\mathbf{0}\}$. For a norm $\|x\|$ on $\mathbb{R}_+^p$, define for $x \in \mathbb{R}_+^p \setminus \{\mathbf{0}\}$
$$h(x) = \left(\|x\|, \frac{x}{\|x\|}\right) = (r, a).$$
Then
$$h : \mathbb{R}_+^p \setminus \{\mathbf{0}\} \mapsto (0, \infty) \times \aleph_+$$
where
$$\aleph_+ = \{x \in \mathbb{R}_+^p : \|x\| = 1\}.$$
Is $(0, \infty) \times \aleph_+$ a TABOF space and if so, does h respect the forbidden zone?

1.4 Suppose $\mathbb{S} \setminus F = \mathbb{R}_+^p \setminus \{\mathbf{0}\}$. Define $h : \mathbb{R}_+^p \setminus \{\mathbf{0}\} \mapsto (0, \infty) = \mathbb{R}_+ \setminus \{0\}$ by $h(x) = \sum_{i=1}^p x_i$. Does h respect the forbidden zone?

1.5 Suppose as usual that $\mathbb{S}$ is a CSMS and $F \in \mathcal{F}(\mathbb{S})$. Show
$$\mathcal{K}(\mathbb{S}(F)) = \{K \in \mathcal{K}(\mathbb{S}) : K \cap F = \emptyset\}.$$
If $\mathbb{S} = \mathbb{R}_+^2$ and $F = [\text{diag}] = \{(x, x) : x \geq 0\}$, what is

1. $\mathcal{K}(\mathbb{R}_+^2 \setminus [\text{diag}])$?
2. $\text{BA}(\mathbb{R}_+^2 \setminus [\text{diag}])$?

1.6 Let $\mathbb{C}_K^+(\mathbb{S}(F))$ be the non-negative, bounded, continuous functions on $\mathbb{S}(F)$ which have compact support. Let $\mathbb{M}_+(\mathbb{S}(F))$ be the class of measures on $\mathbb{S}(F)$ which are finite on compact subsets of $\mathbb{S}(F)$.

1. Show
$$\mathbb{C}_K^+(\mathbb{S}(F)) \subset \mathbb{C}(\mathbb{S}(F)).$$

2. Show
$$\mathbb{M}(\mathbb{S}(F)) \subset \mathbb{M}_+(\mathbb{S}(F)).$$

3. Vague convergence in $\mathbb{M}_+(\mathbb{S}(F))$ means that if for $n \geq 0$, $\mu_n \in \mathbb{M}_+(\mathbb{S}(F))$, then
$$\mu_n(f) \to \mu_0(f)$$
for all $f \in \mathbb{C}_K^+(\mathbb{S}(F))$. Show that if for $n \geq 0$, $\mu_n \in \mathbb{M}(\mathbb{S}(F))$, then $\mu_n \to \mu_0$ in $\mathbb{M}(\mathbb{S}(F))$ implies μ_n converges vaguely to μ_0 in $\mathbb{M}_+(\mathbb{S}(F))$.

4. Let $\mathbb{S} = \mathbb{R}_+$ and $F = \{0\}$ so that $\mathbb{S}(F) = (0, \infty)$. Define
$$\mu_n(\cdot) = \frac{1}{n}\sum_{i=1}^{\infty} \epsilon_{i/n}(\cdot).$$
Then $\mu_n \in \mathbb{M}_+(\mathbb{S}(F))$ but $\mu_n \notin \mathbb{M}(\mathbb{S}(F))$ and $\mu_n \to$ Leb vaguely in $\mathbb{M}_+(\mathbb{S}(F))$ but $\{\mu_n\}$ does not converge in $\mathbb{M}(\mathbb{S}(F))$.

5. Let $\mathbb{S} = \mathbb{R}_+$ and $F = \{0\}$ so that $\mathbb{S}(F) = (0, \infty)$. Define
$$\mu_n(\cdot) = \frac{1}{n}\sum_{i=1}^{n^2} \epsilon_{i/n}(\cdot).$$
Then $\mu_n \in \mathbb{M}(\mathbb{S}(F)) \subset \mathbb{M}_+(\mathbb{S}(F))$ and $\mu_n \to$ Leb vaguely in $\mathbb{M}_+(\mathbb{S}(F))$ but $\{\mu_n\}$ does not converge in $\mathbb{M}(\mathbb{S}(F))$. For instance, take
$$f(x) = \begin{cases} 0, & \text{if } 0 < x \leq 1, \\ x-1, & \text{if } 1 \leq x \leq 2, \\ 1, & \text{if } x \geq 2. \end{cases}$$
Then $f \in \mathbb{C}(\mathbb{S}(F))$ but $\mu_n(f) \to \infty$.

1.7 Suppose as usual $\mathbb{S}$ is a CSMS and suppose the forbidden zone is $F = \emptyset$. Then $\mathbb{S} \setminus \emptyset = \mathbb{S}$ is a TABOF space and if $\mu \in \mathbb{M}(\mathbb{S} \setminus \emptyset)$, μ must be a finite measure on $\mathbb{S}$.

However, if $\mu \in \mathbb{M}(\mathbb{S} \setminus F)$ and for measurable B we have $B \cap F = \emptyset$ this does not necessarily imply $\hat{B}\mu = \mu(\cdot \cap B) \in \mathbb{M}(B \setminus \emptyset)$.

1.7 Problems

1.8 Suppose as $t \to \infty$ that $\mu_t \to \mu_\infty$ in $\mathbb{M}(\mathbb{S}_1 \setminus F_1)$. If $\nu(\cdot) \in \mathbb{M}(\mathbb{S}_2 \setminus \emptyset)$, show

$$\mu_t \times \nu \to \mu_\infty \times \nu$$

in $\mathbb{M}(\mathbb{S}_1 \times \mathbb{S}_2 \setminus (F_1 \times \mathbb{S}_2))$.

1.9 Adapt Proposition 1.1, page 13, to the circumstance that X_i, $i = 1, 2$ are independent, positive random variables with regularly varying distributions

$$tP\left[\frac{X_i}{b_i(t)} \in \cdot\right] \to \nu_{\alpha_i}(\cdot), \quad i = 1, 2,$$

in $\mathbb{M}(\mathbb{R}_+ \setminus \{0\})$ and recall $\nu_{\alpha_i}(x, \infty) = x^{-\alpha_i}$, $x > 0$. This shows

$$t^2 P\left[\left(\frac{X_1}{b_1(t)}, \frac{X_2}{b_2(t)}\right) \in \cdot\right] \to \nu_{\alpha_1} \times \nu_{\alpha_2}$$

in $\mathbb{M}(\mathbb{R}_+^2 \setminus [\text{axes}])$.

1.10 Write down the special case of Proposition 1.1, page 13, where for $i = 1, 2$, $\mathbb{S}_i = \mathbb{R}_+^2$ and $F_i = \{(x, x) : x \geq 0\}$.

1.11 For $i = 1, \ldots, p$, $p > 1$, suppose $\mathbb{S}_i = [0, \infty]$ is metrized by

$$d_i(x, y) = \left|\frac{x}{1+x} - \frac{y}{1+y}\right|,$$

and set $F_i = \{\infty\}$. What is the forbidden zone of $(\mathbb{S}_1 \setminus F_1) \times \cdots \times (\mathbb{S}_1 \setminus F_1)$? What is a suitable metric on the product?

1.12 Suppose $\mathbb{S}_1 = \mathbb{R}_+$ and $\mathbb{S}_2 = [0, \infty]$ and on $\mathbb{S}_1 \times \mathbb{S}_2$ define the metric

$$d\big((s_1, x_1), (s_2, x_2)\big) = |s_1 - s_2| + \left|\frac{x_1}{1+x_1} - \frac{x_2}{1+x_2}\right|.$$

Define $h(s, x) = (s, sx)$, from $\mathbb{S}_1 \times \mathbb{S}_2 \mapsto \mathbb{S}_1 \times \mathbb{S}_2$. Is h uniformly continuous?

In $\mathbb{S}_2$ define a forbidden zone $F_2 := [0, \infty] \setminus \mathbb{R}_+ = \{\infty\}$. On the TABOF space $\mathbb{S}_1 \times (\mathbb{S}_2 \setminus F_2)$, does h respect the forbidden zone?

1.13 Suppose $\mathbb{S} = \mathbb{R}^\infty$ and $F = \{0\} \times \mathbb{R}^\infty$. Verify that F is a closed cone and that $\mathbb{S} \setminus F$ is a TABOF space.

1.14 Respect the forbidden zone.

1. Let $\mathbb{N}_{++} = \{1, 2, \ldots\}$. Define

$$T : (\mathbb{R}_+^\infty \setminus \{\mathbf{0}_\infty\}) \times \mathbb{N}_{++} \mapsto \mathbb{R}_+^\infty \setminus \{\mathbf{0}_\infty\}$$

by $T(\boldsymbol{x}, n) = \sum_{i=1}^{n} x_i$. Does T respect forbidden zones as in Definition 1.2? (Hint: No!)

2. Fix a positive integer M and define

$$T_M : (\mathbb{R}_+^\infty \setminus \{\mathbf{0}_\infty\}) \times \{1, \ldots, M\} \mapsto \mathbb{R}_+^\infty \setminus \{\mathbf{0}_\infty\}$$

by $T(\boldsymbol{x}, n) = \sum_{i=1}^{n} x_i$, for $n \leq M$. Does T respect forbidden zones? (Hint: Yes.)

1.15 Apropos of Corollary 1.1, page 17, if we try to proceed with a proof as in Theorem 1.2, what goes wrong? Choosing $f_2 \in \mathbb{C}(\mathbb{S}_2(F_2))$ as on page 16, we find item (b) is fine but item (a) fails because $f_2 \circ h$ is not continuous on $\mathbb{S}_1(F_1)$.

1.16 For a TABOF space $\mathbb{S}(F)$, let $\mathbb{M}_p(\mathbb{S}(F)) \subset \mathbb{M}(\mathbb{S}(F))$ be non-negative integer-valued measures in $\mathbb{M}(\mathbb{S}(F))$. Show $\mathbb{M}_p(\mathbb{S}(F)) \in \mathcal{F}\big(\mathbb{M}(\mathbb{S}(F))\big)$ is a closed subset.

1.17 Let $\mathbb{S}_1$ be a metric space with metric $d_1(\cdot, \cdot)$. Suppose $F_1 \in \mathcal{F}(\mathbb{S}_1)$ and form the TABOF space $\mathbb{S}_1(F_1)$. Define

$$h : x \in \mathbb{S}_1 \mapsto \epsilon_x(\cdot) \in \mathbb{S}_2 := \mathbb{M}_1(\mathbb{S}_1) \subset \mathbb{M}(\mathbb{S}_1 \setminus \emptyset)$$

where $\mathbb{M}_1(\mathbb{S}_1) = \{\epsilon_x(\cdot), x \in \mathbb{S}_1\}$ and ϵ_x is the probability measure putting all probability on x. Metrize $\mathbb{M}_1(\mathbb{S}_1)$ with d_2 defined as

$$d_2(\epsilon_{x_1}, \epsilon_{x_2}) = d_1(x_1, x_2).$$

Questions and assertions to ponder:

1. Is $F_2 := h(F_1) \in \mathcal{F}(\mathbb{M}_1(S))$? (Sure!)
2. Is $\mathbb{M}_1(\mathbb{S}_1(F_1))$ a TABOF space $\mathbb{S}_2(F_2)$? Hint: Verify,

$$\mathbb{M}_1(\mathbb{S}_1(F_1)) = \mathbb{M}_1(\mathbb{S}_1) \setminus \mathbb{M}_1(F_1).$$

3. The map $h : \mathbb{S}_1 \mapsto \mathbb{M}_1(\mathbb{S}_1)$ is continuous. In fact it is uniformly continuous.
4. On $\mathbb{M}_1(\mathbb{S}_1(F_1))$, is d_2-convergence the same as $\mathbb{M}(S_1(F_1))$ convergence?
5. The map $\hat{h} : \mathbb{M}(\mathbb{S}_1(F_1)) \mapsto \mathbb{M}(\mathbb{S}_2(F_2)) = \mathbb{M}\big(\mathbb{M}_1(\mathbb{S}) \setminus h(F)\big)$ defined for $\mu \in \mathbb{M}(\mathbb{S}_1(F_1))$ by $\hat{h}(\mu) = \mu \circ h^{-1}$ is continuous.

In the next chapter, we define regular variation of measures and provided we pay attention to requirements of the first or second mapping theorems, the regular variation property is robust under transformations. For example, suppose X is a positive random variable such that there exists $b(n) \to \infty$ and

$$nP[X/b(n) \in \cdot] \to \nu_\alpha(\cdot), \qquad \nu_\alpha(x, \infty) = x^{-\alpha}, \; x > 0, \; \alpha > 0,$$

in $\mathbb{M}(\mathbb{R}_+ \setminus \{0\})$. This says the distribution of X is regularly varying and this exercise shows

$$nP[\epsilon_{X/b(n)} \in \cdot] \to \nu_\alpha \circ h^{-1}(\cdot)$$

in $\mathbb{M}(M_1(\mathbb{R}_+) \setminus \{\epsilon_0\})$. So the distribution of the random point measure ϵ_X is also regularly varying. (For more, see [58].)

1.18 If $f \in \mathbb{C}(\mathbb{R}_+^2 \setminus \{\mathbf{0}_2\})$ show both $f(0, \cdot)$ and $f(\cdot, 0)$ are in $\mathbb{C}(\mathbb{R}_+ \setminus \{0\})$.

Chapter 2
Regular Variation of Measures

2.1 Regular Variation of Measures in $\mathbb{M}(\mathbb{S} \setminus F)$

A flexible theory of multivariate heavy tails requires careful definition of concepts. For instance, we must specify carefully what we mean by *remote* or *extreme* and this relies on the choice of scaling function and forbidden zone. Since there are many examples where a particular random element possesses different heavy tail properties depending on choice of scaling function and forbidden zone, the theory must accommodate these different properties within the same model.

Applications of multivariate heavy tails are most often phrased in terms of measures derived from distributions of random elements. The concept of regular variation of measures requires:

1. A notion of convergence of a family of measures. This is given by the convergence concept in $\mathbb{M}(\mathbb{S} \setminus F)$ discussed in Sect. 1.4
2. A notion of scaling. Closed sets $F \in \mathcal{F}(\mathbb{S})$ serving as forbidden zones should be closed under scaling which requires F to be a *cone*.

2.1.1 Measures on $\mathbb{R}_+$

As motivation for why it is natural to frame a definition of heavy tails in terms of convergence of measures, we consider the standard definition in one dimension. If you have not already done so, some browsing in the Appendix, particularly Sect. A.3, would be useful.

Suppose X is a non-negative random variable with distribution H on $\mathbb{R}_+$ so that $H(x) = P[X \leq x]$. Then X has a heavy tailed distribution or H has a heavy tail at ∞ with index $-\rho$ if $1 - H(x) \in RV_{-\rho}$; that is, $1 - H(x)$ is a regularly varying function at ∞ meaning,

$$\lim_{t \to \infty} \frac{1 - H(tx)}{1 - H(t)} = x^{-\rho}, \quad (x > 0). \tag{2.1}$$

Suppose $1 - H(\infty) = 0$, $\rho > 0$, and set $b(t) = \left(\frac{1}{1-H}\right)^{\leftarrow}(t)$ so that $b(t) \to \infty$. Then (2.1) has the equivalent sequential form

$$\lim_{n \to \infty} n(1 - H(b(n)x)) = x^{-\rho}, \quad (x > 0). \tag{2.2}$$

This can be expressed as convergence of measures in $\mathbb{M}([0, \infty) \setminus \{0\})$,

$$tH(b(t)\cdot) \to \nu_\rho(\cdot), \quad (n \to \infty), \tag{2.3}$$

where $\nu_\rho(\cdot)$ is the measure on $\mathbb{R}_+ \setminus \{0\}$ with density $\rho x^{-\rho-1}dx$, $x > 0$. So if we assume H has a heavy tail we specify the forbidden zone is $\{0\} = F \subset \mathbb{S} = [0, \infty)$ and that regions bounded away from $\{0\}$ should carry finite mass. So for this case it is natural to consider $\mathbb{BA}(\mathbb{R}_+ \setminus \{0\})$ as tail regions; these are sets contained in (ϵ, ∞) for some $\epsilon > 0$ that stay away from the origin. The property that H has a heavy tail can therefore be expressed as convergence of a scaled family of measures in the TABOF space $\mathbb{R}_+ \setminus \{0\}$.

2.1.2 More on Scaling

From (2.3), the concept of a distribution H on $\mathbb{R}_+$ having a heavy tail can be expressed as convergence of a family of scaled measures $\{tH(b(t)\cdot), t > 0\}$ in a TABOF space. In spaces more general than $\mathbb{R}_+$, we still need to require existence of a scaling function.

One of the advantages of the general theory is the ability to use mapping theorems described in Sect. 1.5. In the context of heavy tails, mapping from one space to another can be thought of as shifting coordinate systems. If we have a random element X of $\mathbb{S}$ with a heavy tail and we apply a map h satisfying conditions of either the First Mapping Theorem 1.2 or the Second Mapping Theorem 1.3, does the random element $h(X)$ still have a heavy tail? Except on the real line, this requires making assumptions about the scaling function in the definition of regular variation of measures so that the definition is robust under change of coordinate systems.

In $\mathbb{S} = \mathbb{R}^p$, the most common scaling (as in (2.2)) is $x \mapsto \lambda x$ for $\lambda > 0$ and $x \in \mathbb{R}^p$. This makes it natural to require a forbidden zone F should be a cone; that is $x \in F$ implies $\lambda x \in F$ for all $x \in F$ and $\lambda > 0$. If F is a cone, then so in $\mathbb{S} \setminus F$.

Since the usual notion of multivariate regular variation involves comparisons along a ray generated by scaling or multiplication, we need a scaling idea in a general complete, separable metric space $\mathbb{S}$. The following is sufficiently flexible.

Assume there exists a mapping from $(0, \infty) \times \mathbb{S}$ into $\mathbb{S}$ called the *scaling function* such that for any real number $\lambda > 0$ and any $x \in \mathbb{S}$, we have $(\lambda, x) \mapsto \lambda x$ satisfying:

2.1 Regular Variation of Measures in $\mathbb{M}(\mathbb{S} \setminus F)$

(S1) the mapping $(\lambda, x) \mapsto \lambda x$ is continuous from $(0, \infty) \times \mathbb{S} \mapsto \mathbb{S}$.
(S2) $1x = x$ and $\lambda_1(\lambda_2 x) = (\lambda_1 \lambda_2)x$.

Assumption (S2) may look obvious, but remember we are defining the scaling function abstractly. Assumptions (S1) and (S2) permit us to define a cone $F \subset \mathbb{S}$ as a set satisfying $x \in F$ implies $\lambda x \in F$ for any $\lambda > 0$. For a closed cone $F \subset \mathbb{S}$, $\mathbb{S} \setminus F$ is an open cone and we assume that

(S3) if $x \in \mathbb{S} \setminus F$, we have for $\lambda > 1$ that $d(x, F) < d(\lambda x, F)$. We interpret this to mean that the scaling function exhibits proper regard for the forbidden zone by pushing a point further away from F when $\lambda > 1$.

2.1.2.1 Examples of Scaling Functions

To show that the assumptions allow flexibility and unification, consider the following circumstances which satisfy (S1)–(S3). The examples are for subsets of $\mathbb{R}_+^2$ but could easily be phrased for higher dimensions.

1. *Standard scaling* used in the definition of standard regular variation: $\mathbb{S} = \mathbb{R}_+^2$, $F = \{\mathbf{0}\}$ and

$$(\lambda, (x_1, x_2)) \mapsto (\lambda x_1, \lambda x_2).$$

2. *Unequal scaling* used in the definition of non-standard regular variation: $\mathbb{S} = \mathbb{R}_+^2$, $F = \{\mathbf{0}\}$ and for positive constants $\alpha_1 > 0, \alpha_2 > 0$

$$\bigl(\lambda, (x_1, x_2)\bigr) \mapsto (\lambda^{\alpha_1} x_1, \lambda^{\alpha_2} x_2).$$

Note:

$$\lambda_1(\lambda_2 \mathbf{x}) = \bigl((\lambda_1\lambda_2)^{\alpha_1} x_1, (\lambda_1\lambda_2)^{\alpha_2} x_2\bigr) = \lambda_1\lambda_2 \mathbf{x}.$$

3. *Delete one ray and scale only one component:* Set $\mathbb{S} = \mathbb{R}^2$ and $F = \mathbb{R}_+ \times \{0\}$. Define $(\lambda, (x_1, x_2)) \mapsto (x_1, \lambda x_2)$.
4. *Polar coordinate transform scaling:* Set $\mathbb{S} = [0, \infty) \times \{\mathbf{a} \in \mathbb{R}_+^2 : \|\mathbf{a}\| = 1\} =: \mathbb{R}_+ \times \aleph_{\mathbf{0}_2}$ and $F = \{0\} \times \aleph_{\mathbf{0}_2}$. For $\lambda > 0$, define $(\lambda, (r, \mathbf{a})) \mapsto (\lambda r, \mathbf{a})$.

2.1.3 Definition of Regular Variation of Measures

Recall from (2.1) and (2.3) that in $\mathbb{R}_+$, the property that a distribution has a heavy tail can be expressed as convergence of measures in a TABOF space. This can be extended to more general measures on metric spaces.

The setup for the definition requires:

1. A complete, separable metric space $\mathbb{S}$, with metric $d(x, y)$ giving the notion of distance.
2. A scaling function $(0, \infty) \times \mathbb{S} \mapsto \mathbb{S}$ mapping $(\lambda, x) \mapsto \lambda x$. Extend the notion of scaling from elements of $\mathbb{S}$ to scaling subsets of $\mathbb{S}$ by writing for a set $A \subset \mathbb{S}$,

$$tA = \{tx : x \in A\}.$$

3. A forbidden zone $F \in \mathcal{F}(\mathbb{S})$ that is assumed to be a cone with respect to the scaling.
4. The class $\mathbb{BA}(\mathbb{S} \setminus F)$ defines regions regarded as remote or extreme; these are the regions at positive distance from F.

Definition 2.1 A measure $\mu \in \mathbb{M}(\mathbb{S} \setminus F)$ is a *regularly varying measure* with forbidden zone F if there exist a *limit measure* $\eta \in \mathbb{M}(\mathbb{S} \setminus F)$, $\eta \not\equiv 0$ and a scaling function $c(t) \in RV_\alpha, \alpha \in \mathbb{R}$, such that as $t \to \infty$,

$$c(t)\mu(t\cdot) \to \eta(\cdot), \qquad \text{in } \mathbb{M}(\mathbb{S} \setminus F). \tag{2.4}$$

Remark 2.1 Some comments:

1. The typical case is when $\mu(\cdot)$ is the distribution of a random element of $\mathbb{S}$ and $c(t) \in RV_\alpha$ with $\alpha > 0$ so $c(t) \to \infty$ as $t \to \infty$. This divergence to ∞ compensates for the fact for a set $A \in \mathbb{BA}(\mathbb{S} \setminus F)$, the scaled set tA gets driven further and further away from the forbidden zone F by Property (S3), page 33. Set $b(t) = c^{\leftarrow}(t) \in RV_{1/\alpha}$ and (2.4) is equivalent to

$$t\mu(b(t)\cdot) \to \eta(\cdot) \quad \text{in } \mathbb{M}(\mathbb{S} \setminus F). \tag{2.5}$$

To emphasize that this convergence depends on choice of scaling function $b(t)$ and forbidden zone F and to keep track of the index α and limit measure $\eta(\cdot)$, we introduce the *mouthful notation*:

$$\mu \in \text{MRV}(\alpha, b(t), \eta(\cdot), \mathbb{S} \setminus F), \tag{2.6}$$

to indicate the measure μ is multivariate regularly varying.

2. From (2.4), the limit measure $\eta(\cdot)$ has the scaling property

$$\eta(c\cdot) = c^{-\alpha}\eta(\cdot).$$

3. If $c(t) \in RV_\alpha$ with $\alpha < 0$, $1/c(t) \in RV_{|\alpha|}$ and $1/c(t) \to \infty$ since the index $|\alpha|$ is positive. So (2.4) is equivalent to

$$\frac{\mu(b(t)\cdot)}{t} \to \eta(\cdot) \quad \text{in } \mathbb{M}(\mathbb{S} \setminus F) \tag{2.7}$$

2.1 Regular Variation of Measures in $\mathbb{M}(\mathbb{S} \setminus F)$

at $t \to \infty$, where $b(t) = (1/c)^{\leftarrow}(t) \in RV_{1/|\alpha|}$. Uses of this case are in Sects. 2.3.2.5 and 2.3.2.6, page 60 and page 62.

Keep in mind the distinction between $\alpha > 0$ and $\alpha < 0$ described by (2.5) and (2.7). We mostly deal with (2.5) as summarized by (2.6). Without belaboring the point, we always assume limit measures are not identically zero.

If a random element X in $\mathbb{S} \setminus F$ has a regularly varying distribution μ as described in (2.6), only the uncouth say X is regularly varying.[1]

2.1.3.1 Examples of Regular Variation

In $\mathbb{R}_+^p$, there are numerous possible forbidden zones depending on the application. Possibilities include

(i) the origin;
(ii) one axis;
(iii) multiple axes;
(iv) the diagonal $\{(x, x, \ldots, x) : x \in \mathbb{R}_+\}$;
(v) some other cone.

The most common forbidden zone is $F = \{\mathbf{0}_p\}$ and a random vector $X = (X_1, \ldots, X_p)$ has a regularly varying distribution $\mu(\cdot)$ on $\mathbb{R}_+^p \setminus \{\mathbf{0}_p\}$ if there exists $b(t) \in RV_{1/\alpha}$, for some $\alpha > 0$, such that in $\mathbb{M}(\mathbb{R}_+^p \setminus \{\mathbf{0}_p\})$,

$$tP\left[\frac{X}{b(t)} \in \cdot\right] = t\mu(b(t)\cdot) \to \eta(\cdot). \tag{2.8}$$

With conventional scaling

$$(\lambda, \mathbf{x}) \mapsto \lambda \mathbf{x} = (\lambda x_1, \ldots, \lambda x_p)$$

it is straightforward to verify that $\eta(\cdot)$, when expressed in polar coordinates, is a product measure. Suppose A is a measurable subset of the unit sphere $\aleph_{\mathbf{0}_p}$ and $r > 0$. Then

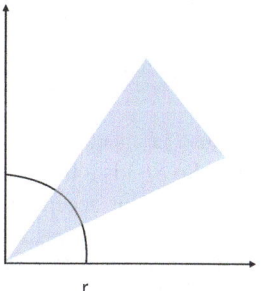

---

[1] Really now, a random element in a metric space is a fixed function so how could it vary regularly or irregularly?

$$\eta\{x \in \mathbb{R}_+^p \setminus \{\mathbf{0}_p\} : \|x\| > r, \frac{x}{\|x\|} \in A\}$$

$$= \eta\{x \in \mathbb{R}_+^p \setminus \{\mathbf{0}_p\} : \|\frac{x}{r}\| > 1, \frac{x/r}{\|x/r\|} \in A\}$$

and changing variable $y = x/r$

$$= \eta\{r y \in \mathbb{R}_+^p \setminus \{\mathbf{0}_p\} : \|y\| > 1, \frac{y}{\|y\|} \in A\}$$

which by the scaling property of $\eta(\cdot)$ is

$$= r^{-\alpha} \eta\{y \in \mathbb{R}_+^p \setminus \{\mathbf{0}_p\} : \|y\| > 1, \frac{y}{\|y\|} \in A\} =: r^{-\alpha} S(A), \quad (2.9)$$

where S is a measure on $\aleph_+$.

Suppose $h : \mathbb{R}_+^p \setminus \{\mathbf{0}_p\} \mapsto (0, \infty) \times \aleph_{\mathbf{0}_p}$ is the *polar coordinate transformation*,

$$h(x) = \left(\|x\|, \frac{x}{\|x\|}\right) = (r, a). \quad (2.10)$$

Define, as usual, the Pareto measure $\nu_\alpha(\cdot)$ on $\mathbb{R}_+ \setminus \{0\}$ by $\nu_\alpha(r, \infty) = r^{-\alpha}$, $r > 0$ and then using h we write the result in (2.9) as

$$\eta \circ h^{-1} = \nu_\alpha \times S \quad (2.11)$$

which is a measure in $\mathbb{M}\big(([0, \infty) \times \aleph_{\mathbf{0}_p}) \setminus (\{0\} \times \aleph_{\mathbf{0}_p})\big)$. Since $\eta(\cdot) \in \mathbb{M}(\mathbb{R}_+^p \setminus \{\mathbf{0}_p\})$, $\eta(\cdot)$ satisfies

$$\eta\{x \in \mathbb{R}_+^p \setminus \{\mathbf{0}_p\} : \|x\| > 1\} < \infty$$

because $\{x : \|x\| > 1\} \in \mathbb{BA}(\mathbb{R}_+^p \setminus \{\mathbf{0}_p\})$ is bounded away from the forbidden zone $\{\mathbf{0}_p\}$. So $S(\cdot)$ is a finite measure on $\aleph_{\mathbf{0}_p}$ and at the possible expense of defining $b(t)$ differently but acceptably, we may as well assume $S(\cdot)$ is a probability measure. This probability measure $S(\cdot)$ is the *angular measure*.

The takeaway is that if X has a distribution $\mu(\cdot)$ on $\mathbb{R}_+^p$ which is regularly varying, the limit measure $\eta(\cdot)$ has a particular structure. Section 2.2.1 uses this takeaway to construct examples of regularly varying distributions.

2.2 Building Regularly Varying Distributions

The definition of regular variation of measures is all well and good but is there a ready way to build models containing such measures? If we combine the First Binding Lemma on page 47, polar and generalized polar coordinates as well as mapping theorems, we get a straightforward method for building such distributions.

Master builder

We begin with the somewhat traditional method using conventional polar coordinates hinted at in Sect. 2.1.3.1.

2.2.1 Constructing Regularly Varying Distributions in $\mathbb{R}_+^p$ and Moving Between Cartesian and Polar Coordinates

We begin by proceeding to construct regularly varying distributions in $\mathbb{R}_+^p$ using the mapping theorems and conventional polar coordinates. Moving comfortably between Cartesian and polar coordinate systems is key.

We develop the construction in several stages. Recall $\aleph_{0_p} = \{x \in \mathbb{R}_+^p : \|x\| = 1\}$ where $\|x\|$ is any conveniently chosen norm.

Stage 1. Define the two TABOF spaces $\mathbb{S}_i \setminus F_i, i = 1, 2$, by

$$\mathbb{S}_1 = \mathbb{R}_+^p \qquad\qquad F_1 = \{\mathbf{0}\}$$
$$\mathbb{S}_2 = \mathbb{R}_+ \times \aleph_{0_p} \qquad\qquad F_2 = \{\mathbf{0}\} \times \aleph_{0_p}.$$

On $\mathbb{S}_1$ use the usual Euclidean metric specified by, say, the L_2 norm and on $\mathbb{S}_2$

$$d_2\big((r_1, a_1), (r_2, a_2)\big) = |r_1 - r_2| \vee \|a_1 - a_2\|,$$

is a convenient choice.

Now specify scaling functions on $\mathbb{S}_1$ and $\mathbb{S}_2$:

- On $\mathbb{S}_1$: $(\lambda, x) \mapsto \lambda x = (\lambda x_1, \ldots, \lambda x_p)$;
- On $\mathbb{S}_2$: $\big(\lambda, (r, a)\big) \mapsto (\lambda r, a)$.

Define $h : \mathbb{S}_1 \setminus F_1 \mapsto \mathbb{S}_2 \setminus F_2$ as in (2.10), page 36 by

$$h(x) = \big(\|x\|, x/\|x\|\big) = (r, a).$$

Then $h^\leftarrow : \mathbb{S}_2 \setminus F_2 \mapsto \mathbb{S}_1 \setminus F_1$ is defined by

$$h^\leftarrow(r, a) = ra. \qquad\qquad (2.12)$$

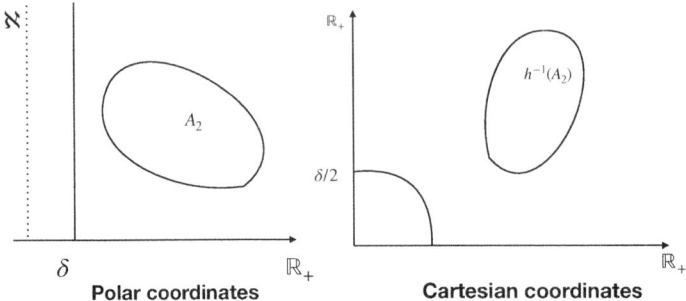

Stage 2. Background. Let X be a random vector with distribution in $\mathrm{MRV}(\alpha, b(t), \eta, \mathbb{R}_+^p \setminus \{\mathbf{0}\})$ so (2.8) page 35 holds. The map h is (bi-)continuous from $\mathbb{S}_1 \setminus F_1$ to $\mathbb{S}_2 \setminus F_2$ and we seek to apply the First Mapping Theorem 1.2, page 16. This requires that we confirm h respects forbidden zones as in Definition 1.2. Suppose A_2 is bounded away from $F_2 = \{0\} \times \aleph_{\mathbf{0}_p}$. Then for some $\delta > 0$,

$$\delta = d_2(A_2, \{0\} \times \aleph_{\mathbf{0}_p}) = \inf\{d_2\big((r_2, \mathbf{a}_2), (0, \mathbf{a})\big) : (r_2, \mathbf{a}_2) \in A_2, \mathbf{a} \in \aleph_{\mathbf{0}_p}\}$$

and because we can minimize by taking $\mathbf{a} = \mathbf{a}_2$, this becomes

$$= \inf\{r_2 > 0 : (r_2, \mathbf{a}_2) \in A_2\} = \delta.$$

Therefore, for the set inverse h^{-1} we have

$$h^{-1}(A_2) = \{\mathbf{x} \in \mathbb{R}_+^p \setminus \{\mathbf{0}\} : h(\mathbf{x}) \in A_2\} = \{\mathbf{x} \in \mathbb{R}_+^p \setminus \{\mathbf{0}\} : \big(\|\mathbf{x}\|, \frac{\mathbf{x}}{\|\mathbf{x}\|}\big) \in A_2\}$$

$$\subset \{\mathbf{x} \in \mathbb{R}_+^p \setminus \{\mathbf{0}\} : \|\mathbf{x}\| > \delta/2\}.$$

Thus $h^{-1}(A_2) \in \mathbb{BA}(\mathbb{S}_1 \setminus F_1)$ when $A_2 \in \mathbb{BA}(\mathbb{S}_2 \setminus F_2)$. If the regular variation in (2.8) holds, then the First Mapping Theorem 1.2, page 16 gives in $\mathbb{M}(\mathbb{S}_2 \setminus F_2)$,

$$tP\Big[\frac{X}{b(t)} \in \cdot\Big] \circ h^{-1} = tP\Big[h\Big(\frac{X}{b(t)}\Big) \in \cdot\Big]$$
$$= tP\Big[\Big(\frac{\|X\|}{b(t)}, \frac{X}{\|X\|}\Big) \in \cdot\Big] \to \eta \circ h^{-1}, \quad (2.13)$$

as $t \to \infty$. Note the convergence in (2.13) is regular variation expressed for polar coordinates if scaling is defined as we have done on $\mathbb{S}_2$. If we set

$$R := \|X\| \quad \text{and} \quad \Theta := X/\|X\|$$

then using (2.11), the polar coordinate version of regular variation is

2.2 Building Regularly Varying Distributions

$$tP\big[(R/b(t), \Theta) \in \cdot\big] \to \eta \circ h^{-1} = \nu_\alpha \times S(\cdot). \tag{2.14}$$

Stage 3 construction. Suppose R is a non-negative random variable with $P[R > r] = r^{-\alpha}$, $r > 1$ (or suppose R has a distribution with a regularly varying tail) and Θ is a random element of $\aleph_{0_p}$ independent of R. Then (R, Θ) satisfy multivariate regular variation in the form (2.14) and if we define $X = R\Theta$, then the mapping theorem implies that Cartesian coordinate regular variation (2.8) holds for this X. The simplest case is $p = 2$ with the L_1-norm and $\aleph_+ = \{(\theta, 1 - \theta) : 0 \le \theta \le 1\}$ and we just have to specify a distribution on $[0, 1]$.

2.2.2 Construction of Regular Variation on $\mathbb{S} \setminus F$ Using the Generalized Polar Coordinate Transform

The polar coordinate transformation in $\mathbb{R}_+^p$ is heavily relied upon when making inferences about the limit measure of regular variation. Transforming from Cartesian to polar coordinates disintegrates the transformed limit measure into a product measure, one of whose factors concentrates on the unit sphere and is called the *angular measure*. Estimating the angular measure and then transforming back to Cartesian coordinates provides a common inference technique for tail probability estimation in $\mathbb{R}_+^p$ using heavy-tail asymptotics.

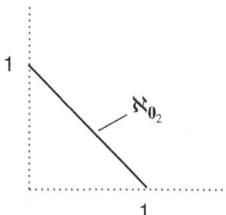

Fig. 2.1 $\aleph_{0_2}$ not BA

(Review Sect. 2.1.3.1, (2.9), (2.10), Sect. 2.2 and [156, pages 173ff, 313].) When removing more than $\{0\}$ from $\mathbb{R}_+^p$, the conventional unit sphere may no longer be bounded away from what is removed and an alternative technique we call the *generalized polar coordinate transformation* can be used. For example, in Fig. 2.1, if $\mathbb{S} \setminus F = \mathbb{R}_+^2 \setminus$ [axes], $\aleph_{0_2}$, the L_1 unit sphere, is not bounded away from the removed axes. The following generalization [38, 125] can resolve this difficulty, provided the metric on $\mathbb{S}$ satisfies a mild condition.

2.2.2.1 Generalized Polar Coordinates

Temporarily, we proceed generally and assume $\mathbb{S}_1$ is a complete, separable metric space and that scalar multiplication is defined. If F_1 is a cone in $\mathbb{S}_1$, $\theta F_1 = F_1$ for $\theta > 0$. Suppose further that the metric $d_1(\cdot, \cdot)$ on $\mathbb{S}_1$ satisfies

$$d_1(\theta x, \theta y) = \theta d_1(x, y), \quad \theta > 0, \ (x, y) \in \mathbb{S}_1 \times \mathbb{S}_1. \tag{2.15}$$

which holds whenever distance is defined by a norm. (But note, condition (2.15) fails for $\mathbb{R}_+^\infty$.) If we intend to remove the closed cone F_1 from $\mathbb{S}_1$, we can avoid some notational confusion by setting

$$\aleph_{F_1} = \{s \in \mathbb{S}_1 \setminus F_1 : d_1(s, F_1) = 1\}, \tag{2.16}$$

which plays the role of the unit sphere in the TABOF space $\mathbb{S}_1 \setminus F_1$. Note $F_2 := \{0\} \times \aleph_{F_1}$ is closed in the product space $\mathbb{R}_+ \times \mathbb{S}_1$ and

$$\mathbb{S}_2 \setminus F_2 := \big([0, \infty) \times \aleph_{F_1}\big) \setminus \big(\{0\} \times \aleph_{F_1}\big)$$

is a TABOF space. Define the generalized polar coordinate transformation

$$\text{GPOLAR} : \mathbb{S}_1 \setminus F_1 \mapsto \mathbb{S}_2 \setminus F_2$$

by

$$\text{GPOLAR}(s) = \big(d_1(s, F_1), s/d_1(s, F_1)\big), \quad s \in \mathbb{S}_1 \setminus F_1. \tag{2.17}$$

Since F_1 is a cone, $\theta F_1 = F_1$, and because $d_1(\cdot, \cdot)$ has property (2.15), we have for any $s \in \mathbb{S}_1 \setminus F_1$ that

$$d\left(\frac{s}{d_1(s, F_1)}, F_1\right) = d\left(\frac{s}{d_1(s, F_1)}, \frac{1}{d_1(s, F_1)} F_1\right) = \frac{1}{d_1(s, F_1)} d_1(s, F_1) = 1,$$

so the second coordinate of GPOLAR belongs to $\aleph_{F_1}$. We maintain the conceptual distinction that GPOLAR^{-1} is the set inverse and $\text{GPOLAR}^{\leftarrow}$ is the pointwise inverse but the GPOLAR map is a bijection and the inverse map $\text{GPOLAR}^{\leftarrow} = \text{GPOLAR}^{-1} : \mathbb{S}_2 \setminus F_2 = \big([0, \infty) \times \aleph_{F_1}\big) \setminus \big(\{0\} \times \aleph_{F_1}\big) \mapsto \mathbb{S}_1 \setminus F_1$ is

$$\text{GPOLAR}^{-1}(r, \boldsymbol{a}) = r\boldsymbol{a}, \quad r \in (0, \infty), \boldsymbol{a} \in \aleph_{F_1}.$$

Example 2.1 Here are four examples of forbidden zones F_1 and the sets $\aleph_{F_1}$ they generate.

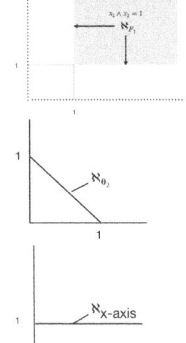

1. If $\mathbb{S}_1 = \mathbb{R}_+^2$, remove the cone consisting of the axes through the origin **0**, so that $F_1 = \{0\} \times \mathbb{R}_+ \cup \mathbb{R}_+ \times \{0\} =$ [axes]. The generalized unit sphere is $\aleph_{F_1} = \{x \in \mathbb{R}_+^2 : x_1 \wedge x_2 = 1\}$.
2. If $\mathbb{S}_1 = \mathbb{R}_+^2$, $F_1 = \{\mathbf{0}_2\}$, then $\aleph_{\{\mathbf{0}_2\}} = \{x \in \mathbb{R}_+^2 : \|x\| = 1\}$. In the figure, distance is L_1. Unlike Fig. 2.1, the conventional unit sphere is bounded away from the forbidden zone.
3. If $\mathbb{S}_1 = \mathbb{R}_+^2$, and $F_1 = [x\text{-axis}]$, then $\mathbb{S}_1 \setminus F_1 = \mathbb{D}_\sqcap = [0, \infty) \times (0, \infty)$. The unit sphere $\aleph_{F_1}$ is the horizontal line at height 1.

4. If $\mathbb{S}_1 = \mathbb{R}_+^2$ and $F_1 = [\text{diag}] = \{(x,x) : x \geq 0\}$ then $\mathbb{S}_1 \setminus F_1 = \{\boldsymbol{x} \in \mathbb{R}_+^2 : x_1 \neq x_2\}$. The unit sphere $\aleph_{F_1}$ is two lines parallel to [diag].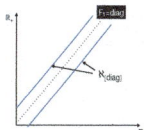

The transformation GPOLAR respects forbidden zones which mimics the discussion in Sect. 2.2 which used the First Mapping Theorem 1.2. So we get the following:

Corollary 2.1 *Assume $\mathbb{S}_1$ is a complete, separable metric space such that (2.15) holds, scalar multiplication is defined and suppose F_1 is a closed cone. Then $\mu_n \to \mu_0$ in $\mathbb{M}(\mathbb{S}_1 \setminus F_1)$ iff*

$$\mu_n \circ GPOLAR^{-1} \to \mu_0 \circ GPOLAR^{-1}$$

in $\mathbb{M}(\mathbb{S}_2 \setminus F_2) = \mathbb{M}\bigl(([0,\infty) \times \aleph_{F_1}) \setminus (\{0\} \times \aleph_{F_1})\bigr)$.

The converse is proven in a similar way.

2.2.2.2 Construction of Regular Variation on $\mathbb{S} \setminus F$ Using GPOLAR

We now follow a path familiar from Sect. 2.2.1. Suppose Θ is a random element of $\aleph_F$ with probability distribution $S(\cdot)$ and assume R is a random element of $\mathbb{R}_+$ independent of Θ and $P[R > x] \in RV_{-\alpha}$ with $\alpha > 0$ and scaling function $b(t) = (1/P[R > \cdot])^{\leftarrow}(t)$. Define $X := R\Theta$. Then from the First Binding Lemma page 47),

$$tP\left[\left(\frac{R}{b(t)}, \Theta\right) \in \cdot\right] \to \nu_\alpha \times S(\cdot), \quad \text{in } \mathbb{M}\bigl((\mathbb{R}_+ \setminus \{0\}) \times \aleph_F\bigr) \qquad (2.18)$$

and hence in $\mathbb{M}(\mathbb{S} \setminus F)$,

$$tP\left[\frac{X}{b(t)} \in \cdot\right] = tP\left[\frac{R \cdot \Theta}{b(t)} \in \cdot\right] \to \nu_\alpha \times S \circ (GPOLAR^{\leftarrow})^{-1}(\cdot). \qquad (2.19)$$

This provides a way to construct regular variation on $\mathbb{S} \setminus F$ provided (2.15) holds.

Remark 2.2 We review why we can always assume $S(\cdot)$ in (2.18) is a probability measure. Since F is a cone and because we assume the metric $d(\cdot,\cdot)$ on $\mathbb{S}$ satisfies (2.15), for any $x \in \mathbb{S} \setminus F$,

$$d(\frac{x}{d(x,F)}, F) = d(\frac{x}{d(x,F)}, \frac{1}{d(x,F)}F) = \frac{1}{d(x,F)}d(x,F) = 1,$$

and $x/d(x, F) \in \aleph_F$. Also

$$d(\aleph_F, F) = \inf\{d(x, F) : x \in \aleph_F\} = 1,$$

and $\aleph_F \in \mathbb{BA}(\mathbb{S} \setminus F)$. If the limit measure of regular variation on $\mathbb{S} \setminus F$ is $\eta(\cdot)$, this means, for any Borel $\Lambda \subset \aleph_F$,

$$\eta\{x \in \mathbb{S} \setminus F : d(x, F) \geq 1, \frac{x}{d(x, F)} \in \Lambda\}$$

is finite. So $\aleph_F$ is bounded away from the forbidden zone and $S(\cdot)$ is always finite on $\aleph_F$; with the right choice of scaling function we can always make S a probability.

2.2.2.3 Examples

We now consider illustrative examples. Several examples use (a) Breiman's Theorem 2.1, page 57 and (b) binding lemmas pages 47, 49, 13, and to assuage my conscience about forward referencing, here is what you need to know: (a) If X, Y are two independent non-negative random variables with $P[Y > x] \in RV_{-\alpha}$ and X has a lighter tail than Y expressed by assuming that $E(X^\beta) < \infty$, $\beta > \alpha$ then $P[XY > x] \sim E(X^\alpha) P[Y > x]$, $x \to \infty$. (b) If R is a non-negative random variable with regularly varying tail and $\Theta \perp\!\!\!\perp R$ then (R, Θ) has a regularly varying distribution.

Example 2.2 Suppose in $\mathbb{R}_+^2$ we have a random element Θ with a distribution H concentrating on

$$\aleph_{[\text{axes}]} = \{x \in \mathbb{R}_+^2 \setminus [\text{axes}] : x_1 \wedge x_2 = 1\}.$$

Suppose $R > 0$ is a random variable with $P[R > x] = x^{-2}$, $x \geq 1$ and $R \perp\!\!\!\perp \Theta$. Application of the First Binding Lemma (page 47) yields the distribution of (R, Θ) is in MRV($\alpha = 2, b(t) = \sqrt{t}, \nu_2 \times H(\cdot), (\mathbb{R}_+ \setminus \{0\}) \times \aleph_{[\text{axes}]})$ and

$$tP\left[\left(\frac{R}{\sqrt{t}}, \Theta\right) \in \cdot\right] \to \nu_2 \times H(\cdot), \qquad (2.20)$$

in $\mathbb{M}\big((\mathbb{R}_+ \setminus \{0\}) \times \aleph_{[\text{axes}]}\big)$. We also have the Cartesian version of (2.20)

$$tP\left[\frac{R\Theta}{\sqrt{t}} \in \cdot\right] \to \nu_2 \times H \circ (\text{GPOLAR}^{\leftarrow})^{-1}(\cdot), \quad \text{in } \mathbb{M}(\mathbb{R}_+^2 \setminus [\text{axes}]) \qquad (2.21)$$

How can we construct H? Take a Bernoulli switching variable B^* satisfying

$$P[B^* = 1] = \frac{1}{2} = P[B^* = 0]$$

and a probability distribution G concentrating on $(1, \infty)$. Suppose Z_1, Z_2 are two iid random variables with distribution G satisfying $(Z_1, Z_2) \perp\!\!\!\perp B^*$. Define

2.2 Building Regularly Varying Distributions

$$\Theta = B^*(1, Z_2) + (1 - B^*)(Z_1, 1)$$

which lays 1/2 the mass of G on $\{(x, 1) : x > 1\}$ and the other half the mass on $\{(1, y) : y > 1\}$.

Marginal regular variation is not guaranteed by (2.21) since, for instance, the set $[0, \infty) \times (y, \infty)$ $y > 0$ is not bounded away from the forbidden zone [axes] and hence cannot just be plugged into (2.21). For marginal regular variation, more must be assumed such as condition (2.22). This allows us to apply Breiman's Theorem 2.1.

For this example, to get marginal regular variation of distribution tails for individual components, assume

$$\int_1^\infty x^{2+\delta} G(dx) < \infty, \quad \delta > 0. \tag{2.22}$$

Since $(R\Theta)_1 = B^* R + (1 - B^*) R Z_1$ we have

$$P[(R\Theta)_1 > x] = \frac{1}{2} P[R > x] + \frac{1}{2} P[R Z_1 > x]$$

$$\sim \frac{1}{2} x^{-2} + \frac{1}{2} E Z_1^2 \cdot x^{-2} \quad \text{(Breiman)}$$

$$= x^{-2}(1 + E Z_1^2)/2. \tag{2.23}$$

Assumption (2.22) guarantees that tail behavior is carried by R. Similarly,

$$P[(R\Theta)_2 > x] \sim (const) P[R > x] = (const) x^{-2}, \quad (x \to \infty). \tag{2.24}$$

So (2.23) and (2.24) confirm marginal regular variation and (2.21) confirms multivariate regular variation on $\mathbb{R}_+^2 \setminus [\text{axes}]$.

Notes on Example 2.2

1. We emphasize that marginal regular variation is not implied by just (2.21) without an extra condition such as (2.22).
2. However, regular variation on $\mathbb{R}_+^2 \setminus [\text{axes}]$ *plus* marginal regular variation implies regular variation on $\mathbb{R}_+^2 \setminus \{\mathbf{0}_2\}$.
3. There are examples where the limit measure on $\mathbb{R}_+^2 \setminus [\text{axes}]$ cannot be extended to $\mathbb{R}_+^2 \setminus \{\mathbf{0}\}$ and a regularly varying measure on $\mathbb{R}_+^2 \setminus [\text{axes}]$ does NOT have marginal regular variation at the given scale. In Example 3.1.1.3 of Sect. 3.1.1.3, page 98 with scaling $b(\sqrt{t})$, we have regular variation on $\mathbb{R}_+^2 \setminus [\text{axes}]$ with limit measure

$$\eta_2(\mathbf{x}, \infty) = \frac{1}{x_1 x_2}, \quad \mathbf{x} > \mathbf{0}_2$$

which cannot be extended to $\mathbb{R}_+^2 \setminus \{\mathbf{0}_2\}$ nor does marginal regular variation exist with this scaling.

Example 2.3 ([130, Theorem 4.1]) This example builds a distribution that is in two MRV classes simultaneously. First, consider the random vector

$$V = B(Y_1, 0) + (1-B)(0, Y_2),$$

where $B \perp\!\!\!\perp (Y_1, Y_2)$ and (Y_1, Y_2) are iid with $P[Y_i > x] = x^{-1}$, $x > 1$. It is a quick verification that

$$P[V \in \cdot] \in \text{MRV}\Big(\alpha = 1, b(t) = t, \eta(\cdot) = \frac{1}{2}\nu_1 \times \epsilon_0 + \frac{1}{2}\epsilon_0 \times \nu_1, \mathbb{R}_+^2 \setminus \{\mathbf{0}_2\}\Big).$$

Let $f \in \mathbb{C}(\mathbb{R}_+^2 \setminus \{\mathbf{0}_2\})$ so that both $f(0, \cdot)$ and $f(\cdot, 0)$ are in $\mathbb{C}(\mathbb{R}_+ \setminus \{0\})$. (See Problem 1.18, page 29.) Then since Y_i are random variables with regularly varying tails, as $t \to \infty$,

$$\begin{aligned} tEf(V/t) &= \frac{t}{2}Ef(\frac{Y_1}{t}, 0) + \frac{t}{2}Ef(0, \frac{Y_2}{t}) \\ &\to \frac{1}{2}\nu_1(f(\cdot, 0) + \frac{1}{2}\nu_1(f(0, \cdot) \\ &= \frac{1}{2}\nu_1 \times \epsilon_0(f) + \frac{1}{2}\epsilon_0 \times \nu_1(f). \end{aligned} \quad (2.25)$$

The limit measure concentrates on the axes.

Also recall the elements in Example 2.2 giving regular variation on $\mathbb{R}_+^2 \setminus [\text{axes}]$. We had

$$\Theta = B^*(1, Z_2) + (1-B^*)(Z_1, 1)$$

where Z_1, Z_2 were iid with common distribution G and independent of the Bernoulli switching variable B^* and the Pareto variable R with $P[R > x] = x^{-2}$, $x > 1$. We found that

$$P[R\Theta \in \cdot] \in \text{MRV}\big(\alpha = 2, b(t) = \sqrt{t}, \eta(\cdot), \mathbb{R}_+^2 \setminus [\text{axes}]\big).$$

Now mix $R\Theta$ and V together: Take another Bernoulli switching variable B^{**}, independent of $B, Y_1, Y_2, B^*, Z_1, Z_2, R$ and define

$$\xi = B^{**}V + (1-B^{**})R\Theta. \quad (2.26)$$

Properties of ξ:

2.2 Building Regularly Varying Distributions

- With scale t, $P[\boldsymbol{\xi} \in \cdot]$ is regularly varying as a measure in $\mathbb{M}(\mathbb{R}_+^2 \setminus \{\boldsymbol{0}_2\})$ with limit measure $\frac{1}{2}\nu_1 \times \epsilon_0 + \frac{1}{2}\epsilon_0 \times \nu_1$. To verify this fact, as above, let $f \in \mathbb{C}(\mathbb{R}_+^2 \setminus \{\boldsymbol{0}_2\})$ and there is some $\delta > 0$ such that $f(\boldsymbol{x}) = 0$ if $x_1 + x_2 \leq \delta$. Then,

$$tEf(\boldsymbol{\xi}/t) = \frac{t}{2}Ef(\boldsymbol{V}/t) + \frac{t}{2}Ef(R\boldsymbol{\Theta}/t) = A + B$$

$$= \left[o(1) + \left(\frac{1}{2}\nu_1 \times \epsilon_0(f) + \frac{1}{2}\epsilon_0 \times \nu_1(f)\right)\right] + o(1).$$

The limit for A is given in (2.25) and we must show $B \to 0$. For B, remember that for some $\delta > 0$, the support of f is in $\{(x_1, x_2) : x_1 + x_2 > \delta\}$ and f being bounded, there is an upper bound $\|f\| < \infty$. Therefore,

$$B \leq \frac{t}{2}\|f\| P[(R\boldsymbol{\Theta})_1/t + (R\boldsymbol{\Theta})_2/t > \delta]$$

$$\leq \frac{t}{2}\|f\| \left(P[(R\boldsymbol{\Theta})_1/t > \delta/2] + P[(R\boldsymbol{\Theta})_2/t > \delta/2]\right)$$

and from (2.23) and (2.24) both terms asymptotically look like

$$= (const) t \cdot t^{-2} \to 0,$$

as $t \to \infty$, so bye bye B.

- With scale $\sqrt{t}$, $P[\boldsymbol{\xi} \in \cdot]$ is regularly varying as a measure in $\mathbb{M}(\mathbb{R}_+^2 \setminus [\text{axes}])$ with limit measure given in (2.21). To verify this fact, let $g \in \mathbb{C}(\mathbb{R}_+^2 \setminus [\text{axes}])$ so there exists $\delta > 0$ such that $g(\boldsymbol{x}) = 0$ if $x_1 \wedge x_2 \leq \delta$. Then write

$$tEg(\boldsymbol{\xi}/\sqrt{t}) = \frac{t}{2}Eg(\boldsymbol{V}/\sqrt{t}) + \frac{t}{2}Eg(R\boldsymbol{\Theta}/\sqrt{t}) = A + B.$$

This time $A = 0$ since $\boldsymbol{V}$ was designed so its distribution concentrated on $[\text{axes}]$ and thus $V_1 \wedge V_2 = 0$ which is outside the support of g. Convergence of B is handled by (2.21).

So depending on the scale and state space, $\boldsymbol{\xi}$ has a distribution satisfying two distinct regular variation properties. We conclude $P[\boldsymbol{\xi} \in \cdot]$ is in

$$\text{MRV}(\alpha_1 = 1, b_1(t) = t, \mathbb{R}_+^2 \setminus \{\boldsymbol{0}_2\}) \bigcap \text{MRV}(\alpha_2 = 2, b_2(t) = \sqrt{t}, \mathbb{R}_+^2 \setminus [\text{axes}]).$$

These ideas will be explored further in Chap. 3 starting on page 93.

Example 2.4 (Strong Dependence) Suppose R is standard Pareto and $\boldsymbol{\Theta} \perp\!\!\!\perp R$ is concentrated on $[a, b] \subset [0, 1]$ and $[a, b] \subsetneq [0, 1]$. Do the construction of Sect. 2.2.2.2 as in (2.18) with $\boldsymbol{\Theta} = (\Theta, 1 - \Theta)$. Take $\mathbb{S} = \mathbb{R}_+^2$, $\mathbb{F} = \{\boldsymbol{0}_2\}$ and $\aleph_{\{\boldsymbol{0}_2\}} = \{(u, v) : u + v = 1\}$. Let S_1 be the distribution of Θ (not $\boldsymbol{\Theta}$) on $[0, 1]$ so that with $T_1(x, 1 - x) = x$ we have $S_1 = S \circ T_1^{-1}$. We have

$$tP\left[\left(\frac{R}{t},\Theta\right)\in\cdot\right]\to\nu_1\times S_1(\cdot),\quad\text{in }\mathbb{M}\big((\mathbb{R}_+\setminus\{0\})\times[0,1]\big). \qquad (2.27)$$

Set $X=R(\Theta,1-\Theta)$ and

$$tP\left[\frac{X}{t}\in\cdot\right]\to\nu_1\times S\circ(\text{POLAR}^{\leftarrow})^{-1}(\cdot). \qquad (2.28)$$

When Θ has distribution with support a proper subset of $[0,1]$ we say the model has *strong dependence*. Two dimensional data from this model will produce a Cartesian scatterplot that puts points in a narrow wedge. The scatterplot on the right are 100,000 simulated points from $R(\Theta,1-\Theta)$ where $R\perp\!\!\!\perp\Theta$, R is standard Pareto and $\Theta\sim U(0.25,0.55)$.

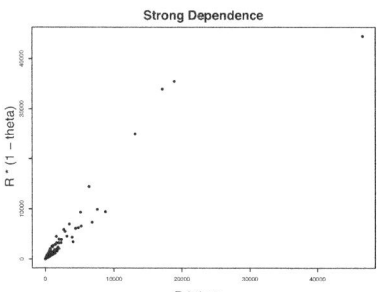

2.3 How to Prove Regular Variation of Measures?

This section meanders through some methods for verifying regular variation. In $\mathbb{R}_+$ we often prove regular variation by proving convergence on a distinguished class of sets sufficient to imply convergence such as (x,∞), $x>0$. However, in higher, even infinite, dimensions the choice of a distinguished class of sets is not always obvious and is considered in Sect. 2.3.4.

2.3.1 Proof Method 1: Great Oaks From Little Acorns Grow

Sometimes we prove regular variation in more complex spaces by combining regular variation properties in simple spaces.[2] Proposition 1.1, page 13, can be adapted to produce one such result but it is for a particular forbidden zone in the product space. There are variants.

For $i=1,2$, let $\mathbb{S}_i\setminus F_i$ be two TABOF spaces and because we will discuss regular variation, each F_i should also be a cone. The metric space $\mathbb{S}_i$ comes with a metric $d_i(x_i,y_i)$ for

A little acorn

---

[2] The Oxford Dictionary of Quotations states that 'great oaks from little acorns grow' is a fourteenth century proverb. You're welcome.

2.3 How to Prove Regular Variation of Measures?

$i = 1, 2$ and define the metric on the product $\mathbb{S}_1 \times \mathbb{S}_2$ by

$$d\big((x_1, x_2), (y_1, y_2)\big) = d_1(x_1, y_1) + d_2(x_2, y_2).$$

(Other equivalent choices are possible.) There are multiple possibilities for the forbidden zone of $\mathbb{S}_1 \times \mathbb{S}_2$ such as $F_1 \times F_2$ or, as used in Proposition 1.1, $(F_1 \times \mathbb{S}_2 \cup \mathbb{S}_1 \times F_2)$; the latter is the forbidden zone obtained in Sect. 1.2.2 (see (1.3) and (1.4)), from taking the product of the two TABOF spaces

$$(\mathbb{S}_1 \setminus F_1) \times (\mathbb{S}_2 \setminus F_2) = \mathbb{S}_1 \times \mathbb{S}_2 \setminus (F_1 \times \mathbb{S}_2 \cup \mathbb{S}_1 \times F_2).$$

Note if $F_2 = \emptyset$, then the product becomes the TABOF space

$$(\mathbb{S}_1 \setminus F_1) \times (\mathbb{S}_2 \setminus \emptyset) = \mathbb{S}_1 \times \mathbb{S}_2 \setminus (F_1 \times \mathbb{S}_2). \tag{2.29}$$

The next section uses this forbidden zone and considers a special case of Proposition 1.1.

2.3.1.1 Combining Regular Variation Properties; The First Binding Lemma

Suppose $\mu_1 \in \mathbb{M}(\mathbb{S}_1 \setminus F_1)$ and $\mu_2 \in \mathbb{M}(\mathbb{S}_2 \setminus \emptyset)$ and further assume

$$\mu_1 \in \text{MRV}(\alpha_1, b_1(t), \eta_1(\cdot), \mathbb{S}_1 \setminus F_1). \tag{2.30}$$

Recall that $\mu_2 \in \mathbb{M}(\mathbb{S}_2 \setminus \emptyset)$ means the forbidden zone in $\mathbb{S}_2$ is $\emptyset$ and therefore μ_2 must be a finite measure.

Proposition 2.1 (First Binding Lemma) *Suppose μ_1 satisfies the little acorn regular variation condition* (2.30) *and $\mu_2 \in \mathbb{M}(\mathbb{S}_2 \setminus \emptyset)$. Then $\mu_1 \times \mu_2$ is regularly varying on the TABOF space $\mathbb{S}_1 \times \mathbb{S}_2 \setminus (F_1 \times \mathbb{S}_2)$ (see* (2.29)*) with scaling satisfying*

$$(\lambda, (x, y)) \mapsto (\lambda x, y)$$

from $\mathbb{R}_+ \times (\mathbb{S}_1 \times \mathbb{S}_2) \mapsto \mathbb{S}_1 \times \mathbb{S}_2$, and assuming the regular variation of $\mu_1(\cdot)$ is of the form (2.5)*, page 34,*

$$t\mu_1 \times \mu_2(b_1(t)dx, dy) \to \eta_1(dx) \times \mu_2(dy).$$

The connection to Proposition 1.1: If $\mu_n^{(1)}(\cdot) \in \mathbb{M}(\mathbb{S}_1 \setminus F_1)$, $n \geq 0$ and $\mu_n^{(1)} \to \mu_0^{(1)}$ in $\mathbb{M}(\mathbb{S}_1 \setminus F_1)$, and $\mu_0^{(2)}(\cdot) \in \mathbb{M}(\mathbb{S}_2 \setminus \emptyset)$, then $\mu_n^{(1)} \times \mu_0^{(2)} \to \mu_0^{(1)} \times \mu_0^{(2)}$ in $\mathbb{M}(\mathbb{S}_1 \times \mathbb{S}_2 \setminus (F_1 \times \mathbb{S}_2))$. (*Cf. Problem 1.8, page 27.*)

Random element language: If X is a random element of $\mathbb{S}_1$ with a regularly varying distribution with limit measure $\eta_1(\cdot)$ and if Y is a random element of $\mathbb{S}_2$ which is independent of X, then (X, Y) is regularly varying:

$$tP\left[\left(\frac{X}{b_1(t)}, Y\right) \in \cdot\right] \to \eta_1 \times P[Y \in \cdot] \qquad (2.31)$$

in $\mathbb{M}(\mathbb{S}_1 \times \mathbb{S}_2 \setminus (F_1 \times \mathbb{S}_2))$. Special case: Suppose

$$\mathbb{S}_1 = \mathbb{R}_+ \qquad F_1 = \{0\}$$
$$\mathbb{S}_2 = \mathbb{R}_+ \qquad F_2 = \emptyset.$$

Then the convergence in (2.31) is convergence in $\mathbb{M}(\mathbb{R}_+^2 \setminus (\{0\} \times \mathbb{R}_+)) = \mathbb{M}(\mathbb{D}_\sqcap)$, according to (1.1), page 3.

Proof Giving a proof is redundant because of Proposition 1.1. We give the proof in the regular variation context but even the most conscientious can feel free to skip ahead. Use the definition of convergence in $\mathbb{M}(\mathbb{S}_1 \times \mathbb{S}_2 \setminus (F_1 \times \mathbb{S}_2))$ and suppose

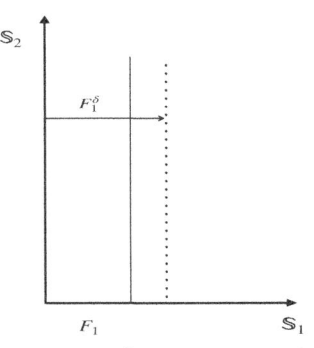

$$f \in \mathbb{C}(\mathbb{S}_1 \times \mathbb{S}_2 \setminus (F_1 \times \mathbb{S}_2))$$

and $d((x, y), F_1 \times \mathbb{S}_2) = d_1(x, F_1)$ and there exists $\delta > 0$ such that

$$\mathrm{supp}(f) \subset \mathbb{S}_1 \times \mathbb{S}_2 \setminus (F_1^\delta \times \mathbb{S}_2)).$$

Rely on Fubini's theorem to iterate the integrals and as $t \to \infty$, we have to show convergence of

$$\int_{\mathbb{S}_2} \int_{\mathbb{S}_1 \setminus F_1} f(x, y) t \mu_1(b_1(t)dx) \mu_2(dy) \to \int_{\mathbb{S}_2} \int_{\mathbb{S}_1 \setminus F_1} f(x, y) t \eta_1(b_1(t)dx) \mu_2(dy).$$

The absolute value of the difference of the integrals bounded by

$$\int_{\mathbb{S}_2} \mu_2(dy) \left| \int_{\mathbb{S}_1 \setminus F_1} f(x, y) t \mu_1(b_1(t)dx) - \int_{\mathbb{S}_1 \setminus F_1} f(x, y) \eta_1(dx) \right|. \qquad (2.32)$$

For any fixed $y \in \mathbb{S}_2$, $f(\cdot, y) \in \mathbb{C}(\mathbb{S}_1 \setminus F_1)$ since $f(\cdot, y)$ is bounded and continuous and if $d_1(x, F_1) = d((x, y), F_1 \times \mathbb{S}_2) < \delta$, $f(x, y) = 0$. Therefore, as $t \to \infty$,

$$\left| \int_{\mathbb{S}_1 \setminus F_1} f(x, y) t \mu_1(b_1(t)dx) - \int_{\mathbb{S}_1 \setminus F_1} f(x, y) \eta_1(dx) \right| \to 0.$$

2.3 How to Prove Regular Variation of Measures?

Remembering that μ_2 is a finite measure, to show the difference in (2.32) converges to 0 is a dominated convergence argument. We must show the absolute value in (2.32) is bounded in t. The difference is bounded by

$$\|f\|\Big(t\mu_1\big(b_1(t)(\mathbb{S}_1 \setminus F_1^\delta)\big) + \eta_1(\mathbb{S}_1 \setminus F_1^\delta)\Big).$$

Since from (1.12) and $\mathbb{S}_1 \setminus F_1^\delta \in \mathcal{F}(\mathbb{S}_1)$,

$$\limsup_{t\to\infty} t\mu_1\big(b_1(t)(\mathbb{S}_1 \setminus F_1^\delta)\big) \le \eta_1(\mathbb{S}_1 \setminus F_1^\delta),$$

the boundedness is clear. □

2.3.1.2 Combining Regular Variation Properties; The Second Binding Lemma

The second binding lemma has some of the flavor of Proposition 1.1, page 13 but is not covered by Proposition 1.1, which proves regular variation on a smaller set. This result is best expressed in Euclidean spaces where the forbidden zone is the closed cone consisting of the origin. Problem 2.4, page 83, discusses the difficulties of creating a potentially more general version of this result.

For $i = 1, 2$, suppose the random vector X_i in $\mathbb{R}_+^{p_i}$ has a distribution in

$$\mathrm{MRV}\big(\alpha_i, b_i(t), \eta_i(\cdot), \mathbb{R}_+^{p_i} \setminus \{\mathbf{0}_{p_i}\}\big),$$

where $b_i(t)$ is a regularly varying function with positive index $1/\alpha_i$. So as $t \to \infty$,

$$tP\left[\frac{X_i}{b_i(t)} \in \cdot\right] \to \eta_i(\cdot), \quad \text{in } \mathbb{M}(\mathbb{R}_+^{p_i} \setminus \{\mathbf{0}_{p_i}\}), \ i = 1, 2. \qquad (2.33)$$

Proposition 2.2 (Second Binding Lemma) *In addition to the regularly varying little acorns (2.33), assume that $X_1 \perp\!\!\!\perp X_2$. Then (X_1, X_2), the random vector in $\mathbb{R}_+^{p_1+p_2}$, has a non-standard regularly varying distribution and explicitly,*

$$tP\left[\left(\frac{X_1}{b_1(t)}, \frac{X_2}{b_2(t)}\right) \in \cdot\right] \to \eta_1 \times \epsilon_{\mathbf{0}_{p_2}} + \epsilon_{\mathbf{0}_{p_1}} \times \eta_2, \qquad (2.34)$$

in $\mathbb{M}(\mathbb{R}_+^{p_1+p_2} \setminus \{\mathbf{0}_{p_1+p_2}\})$. *The limit measure concentrates on* $(\mathbb{R}_+^{p_1} \times \{\mathbf{0}_{p_2}\}) \cup \{\mathbf{0}_{p_1}\} \times \mathbb{R}_+^{p_2}$. *If $p_1 = p_2 = 1$, the limit measure concentrates on* [axes].

More generally, the independence assumption can be weakened. Assume,

(i) Marginal regular variations (2.33) hold.
(ii) It is unlikely that both X_1, X_2 are remote from the origin in the sense that for any $\delta_i > 0$, $i = 1, 2$,

$$tP[X_1/b_1(t) \in (\{\mathbf{0}_{p_1}\}^{\delta_1})^c, X_2/b_2(t) \in (\{\mathbf{0}_{p_2}\}^{\delta_2})^c] \to 0, \quad (t \to \infty). \tag{2.35}$$

Then (2.34) holds.

Furthermore, if we assume $b_1(t) = b_2(t) = b(t) \in RV_{1/\alpha}$, *we have that* (X_1, X_2) *has a joint distribution which is standard regularly varying and*

$$P[(X_1, X_2) \in \cdot] \in MRV(\alpha, b(t), \eta_1 \times \epsilon_{\mathbf{0}_{p_2}} + \epsilon_{\mathbf{0}_{p_1}} \times \eta_2, \mathbb{R}_+^{p_1+p_2} \setminus \{\mathbf{0}_{p_1+p_2}\}). \tag{2.36}$$

Remark 2.3

1. If (X_1, X_2) are jointly regularly varying with limit measure of the form given by the right side of (2.34), then X_1 and X_2 are called *asymptotically independent.*[3] This is implied by marginal regular variation and the assumption that not both of $X_1.X_2$ can be simultaneously large as in (2.35). *Asymptotic independence* gets systematic treatment in its own Sect. 5.6, starting page 187.
2. If we apply Proposition 1.1 to (2.33), we get convergence

$$tP\left[\frac{X_1}{b_1(t)} \in \cdot\right] \times tP\left[\frac{X_2}{b_2(t)} \in \cdot\right] = t^2 P\left[\left(\frac{X_1}{b_1(t)}, \frac{X_2}{b_2(t)}\right) \in \cdot\right] \to \eta_1 \times \eta_2(\cdot)$$

in $\mathbb{M}\left(\mathbb{R}_+^{p_1+p_2} \setminus (\{\mathbf{0}_{p_1}\} \times \mathbb{R}_+^{p_2} \cup \mathbb{R}_+^{p_1} \times \{\mathbf{0}_{p_2}\})\right)$, which is different from (2.34).
3. Condition (2.35) is implied by $X_1 \perp\!\!\!\perp X_2$ since then the joint probability equals

$$tP\left[X_1/b_1(t) \in (\{\mathbf{0}_{p_1}\}^{\delta_1})^c\right] P\left[X_2/b_2(t) \in (\{\mathbf{0}_{p_2}\}^{\delta_2})^c\right] \to 0.$$

The reason for convergence to 0: The first probability multiplied by t stays bounded by (2.33). Assuming distance is defined by L_1 norm, the second probability (without being multiplied by t) is bounded by

$$P[\sum_{l=1}^{p_2} X_2(l) > \delta_2 b_2(t)] \leq \sum_{l=1}^{p_2} P[X_2(l) > \delta_2 b_2(t)/p_2] \to 0,$$

as $t \to \infty$ since $b_2(t) \to \infty$.
4. *Special case.* If for $i = 1, 2$, if $p_1 = p_2 = 1$ and $\{0\}$ is the forbidden zone in each factor space, the limit measure concentrates on $\mathbb{R}_+ \times \{0\} \cup \{0\} \times \mathbb{R}_+ =$ **[axes]**. The limit measure spreads mass on each axis according to ν_{α_i}. This is the case if each X_i has a regularly varying distribution with index α_i, scaling function $b_i(t)$ and as $t \to \infty$,

---

[3] This name is endlessly confusing for newbies and for those with a good feel for language. For more justification of the name, see Problem 2.2, page 82.

2.3 How to Prove Regular Variation of Measures?

$$tP[\frac{X_1}{b_1(t)} > x, \frac{X_2}{b_2(t)} > y] \to 0, \quad \forall x > 0, y > 0.$$

Proof Proceed assuming (2.35). Let $f \in \mathbb{C}(\mathbb{R}_+^{p_1+p_2} \setminus \{\mathbf{0}_{p_1+p_2}\}$ be positive, uniformly continuous and bounded with upper bound $\|f\|$ and for convenience, assume $\|f\| = 1$. We must show

$$tEf(X_1/b_1(t), X_2/b_2(t)) \to \eta_1(f(\cdot, \mathbf{0}_{p_2})) + \eta_2(f(\mathbf{0}_{p_1}, \cdot)).$$

Since $f(\mathbf{0}_{p_1}, \cdot) \in \mathbb{C}(\mathbb{R}_+^{p_2} \setminus \{\mathbf{0}_{p_2}\})$ and $f(\cdot, \mathbf{0}_{p_2}) \in \mathbb{C}(\mathbb{R}_+^{p_1} \setminus \{\mathbf{0}_{p_1}\})$, we get from marginal regular variation that

$$tEf(X_1/b_1(t), \mathbf{0}_{p_2}) \to \eta_1(f(\cdot, \mathbf{0}_{p_2})), \quad \text{in } \mathbb{M}(\mathbb{R}_+^{p_1} \setminus \{\mathbf{0}_{p_1}\}) \tag{2.37}$$

$$tEf(\mathbf{0}_{p_1}, X_2/b_2(t)) \to \eta_2(f(\mathbf{0}_{p_1}, \cdot)), \quad \text{in } \mathbb{M}(\mathbb{R}_+^{p_2} \setminus \{\mathbf{0}_{p_2}\}). \tag{2.38}$$

Write $d_i = d_i(X_i/b_i(t), \mathbf{0}_{p_i})$, $i = 1, 2$. Since f has support bounded away from $\{\mathbf{0}_{p_1+p_2}\}$, there exists $\delta' > 0$ such that $f(x, y) > 0$ implies $d((x, y), \{\mathbf{0}_{p_1+p_2}\}) > 2\delta'$. Then,

$$tEf(X_1/b_1(t), X_2/b_2(t)) = tEf \cdot 1_{[d_1(\frac{X_1}{b_1(t)}, \mathbf{0}_{p_1}) \vee d_2(\frac{X_2}{b_2(t)}, \mathbf{0}_{p_2}) > \delta']}$$

$$= tEf \cdot \left(1_{[d_1 > \delta', d_2 > \delta']} + 1_{[d_1 < \delta', d_2 > \delta']} + 1_{[d_1 > \delta', d_2 < \delta']}\right)$$

$$= A + B + C.$$

Term A is easily killed since it is bounded by $t\|f\|P[d_1 > \delta', d_2 > \delta'] \to 0$ using (2.35). For the rest we need the uniform continuity of f. Let the modulus of continuity of f be

$$\omega_f(\delta) = \sup_{d((x_1, y_1), (x_2, y_2)) < \delta} |f(x_1, y_1) - f(x_2, y_2)|$$

and uniform continuity means $\omega_f(\delta) \to 0$ as $\delta \to 0$.

Compare B, where d_1 is small with $Ef(\mathbf{0}_{p_1}, X_2/b_2(t))$. For $\delta < \delta'$,

$$B = tEf(X_1/b_1(t), X_2/b_2(t))1_{[d_1 < \delta', d_2 > \delta']}$$

$$= tEf(X_1/b_1(t), X_2/b_2(t))1_{[\delta < d_1 < \delta', d_2 > \delta']}$$

$$+ tEf(X_1/b_1(t), X_2/b_2(t))1_{[d_1 < \delta, d_2 > \delta']}$$

$$= B1 + B2 = o(1) + B2,$$

since $B1 \to 0$ using the same killer technique which made $A \to 0$. Similarly,

$$Ef(\mathbf{0}_{p_1}, X_2/b_2(t)) = Ef(\mathbf{0}_{p_1}, X_2/b_2(t))1_{[d_2 > \delta']}$$

(by the assumption about the support of f being away from the origin)

$$\begin{aligned}=& Ef(\mathbf{0}_{p_1}, X_2/b_2(t))1_{[d_1 > \delta, d_2 > \delta']} \\ &+ Ef(\mathbf{0}_{p_1}, X_2/b_2(t))1_{[d_1 < \delta, d_2 > \delta']} \\ =& o(1) + Ef(\mathbf{0}_{p_1}, X_2/b_2(t))1_{[d_1 < \delta, d_2 > \delta']}.\end{aligned}$$

So

$$\begin{aligned}&\left| B - t Ef(\mathbf{0}_{p_1}, X_2/b_2(t)) \right| \\ =& o(1) + \left| t Ef(X_1/b_1(t), X_2/b_2(t))1_{[d_1 < \delta, d_2 > \delta']} \right. \\ & \left. - t Ef(\mathbf{0}_{p_1}, X_2/b_2(t)1_{[d_1 < \delta, d_2 > \delta']} \right|\end{aligned}$$

and therefore

$$\limsup_{t \to \infty} \left| B - t Ef(\mathbf{0}_{p_1}, X_2/b_2(t)) \right|$$
$$\leq \omega_f(\delta) \limsup_{t \to \infty} t P[d_2 > \delta'].$$

The second term stays bounded by marginal regular variation and the modulus of continuity contains the free variable δ which may be sent to 0 so the lim sup is 0.

Using similar turkey carving techniques, we compare

$$C - t Ef(X_1/b_1(t), \mathbf{0}_{p_2})$$

so the meal can be served. □

But wait! There's more: In the Binding Lemma 2.2, the forbidden zone was $\mathbf{0}_{p_1+p_2}$ but the limit measure concentrated on the subset of the state space $(\mathbb{R}_+^{p_1} \times \{\mathbf{0}_{p_2}\}) \cup (\{\mathbf{0}_{p_1}\} \times \mathbb{R}_+^{p_2})$. If we use this support as a forbidden zone, might there be another regular variation property lurking about?[4]

Proposition 2.3 *Suppose* $X_1, X_2,$ *satisfy* (2.33) *and are independent. Then in* $\mathbb{M}(\mathbb{R}_+^{p_1+p_2} \setminus (\mathbb{R}_+^{p_1} \times \{\mathbf{0}_{p_2}\} \cup \{\mathbf{0}_{p_1}\} \times \mathbb{R}_+^{p_2}))$ *we have*

$$tP\left[\left(\frac{X_1}{b_1(\sqrt{t})}, \frac{X_2}{b_2(\sqrt{t})}\right) \in \cdot\right] \to \eta_1 \times \eta_2(\cdot). \qquad (2.39)$$

---

[4] Answer: Yes. This is an example of hidden regular variation (HRV) discussed in Chap. 3 starting on page 93.

2.3 How to Prove Regular Variation of Measures?

In particular, if $p_1 = p_2 = 1$ and $b_1(t) = b_2(t) = b(t)$ and the tail index is α, then

$$tP\left[\left(\frac{X_1}{b(\sqrt{t})}, \frac{X_2}{b(\sqrt{t})}\right) \in \cdot\right] \to \nu_\alpha \times \nu_\alpha(\cdot) \text{ in } \mathbb{M}(\mathbb{R}_+^2 \setminus [\text{axes}]).$$

If $b_1 = b_2$ and the forbidden zone is the origin, then on the scale of $b(t)$ where the index is α, the limit concentrates on this $F := \left(\mathbb{R}_+^{p_1} \times \{\mathbf{0}_{p_2}\} \cup \{\mathbf{0}_{p_1}\} \times \mathbb{R}_+^{p_2}\right)$. If the forbidden zone becomes F, then the scale is $b(\sqrt{t})$ with index $\alpha/2$ and the limit concentrates on all of $\mathbb{R}_+^{p_1+p_2} \setminus F$.

Proof As pointed out in Remark 2.3, page 50, this is a special case of Proposition 1.1, page 13. □

2.3.2 Proof Method 2: Map Your Way to Happiness by Using the Great(er) Oaks and the Mapping Theorems; Some Examples

Some examples of regular variation can be readily converted to other applications by applying the mapping and binding results.

2.3.2.1 The One-Jump Principle: Lebron vs Peewee (But Who Is Who?)

For Proposition 2.2, page 49, suppose $p_1 = p_2 = 1$ and $X_1 \stackrel{d}{=} X_2$ and the tail index is $\alpha > 0$. Then

$$tP\left[\frac{X_1 + X_2}{b(t)} \in \cdot\right] \to 2\nu_\alpha(\cdot). \qquad (2.40)$$

Equivalently, as $t \to \infty$,

$$tP[X_1 + X_2 > b(t)x] \to 2x^{-\alpha}, \quad \text{for } x > 0,$$

So the sum is large because either X_1 or X_2 is large (but not both). This is called the *one jump principle*.

Reason: The Second Binding Lemma 2.2, page 49 gives,

$$tP\left[\left(\frac{X_1}{b(t)}, \frac{X_2}{b(t)}\right) \in \cdot\right] \to \nu_\alpha \times \epsilon_0 + \epsilon_0 \times \nu_\alpha.$$

Lebron vs Peewee

Define the uniformly continuous map $h : R_+^2 \mapsto R_+$ by $h(x, y) = x + y$ and apply the Second Mapping Theorem 1.3, page 18, to get

$$tP\Big[\Big(\frac{X_1}{b(t)}, \frac{X_2}{b(t)}\Big) \in \cdot\Big] \circ h^{-1} \to \nu_\alpha \times \epsilon_0 \circ h^{-1} + \epsilon_0 \times \nu_\alpha \circ h^{-1}.$$

The left side is $tP[\frac{X_1+X_2}{b(t)} \in \cdot]$ and the right side is $2\nu_\alpha(\cdot)$ since for $z > 0$

$$\nu_\alpha \times \epsilon_0 \circ h^{-1}(z, \infty) = \nu_\alpha \times \epsilon_0\{(x,y) : x+y > z\}$$
$$= \nu_\alpha\{x : x > z\} = z^{-\alpha}$$

since ϵ_0 is a measure that concentrates on the point 0. Similarly

$$\epsilon_0 \times \nu_\alpha \circ h^{-1}(z, \infty) = z^{-\alpha} = \nu_\alpha(z, \infty).$$

□

Remark 2.4 (Variants) Here are variants and remarks about the one-jump principle.

1. Assume $p_1 = p_2 = 1$, $X_1 \stackrel{d}{=} X_2$ and $P[X_1 > x] \in RV_{-\alpha}$. Instead of independence of X_1 and X_2, suppose

$$tP[X_1 > b(t)x_1, X_2 > b(t)x_2] \to 0, \quad x_1 \wedge x_2 > 0. \quad (2.41)$$

This is sufficient (and necessary) for asymptotic independence. It also implies the one jump principle. Condition (2.41) is sometimes phrased as

$$\frac{P[X_1 > b(t)x_1, X_2 > b(t)x_2]}{P[X_1 > b(t)]} \to 0, \quad x_1 > 0, x_2 > 0$$

or

$$P[X_2 > b(t)x_2 | X_1 > b(t)] \to 0, \quad (t \to \infty).$$

2. Suppose for $i = 1, \ldots, p$ that $X_i \in \mathbb{R}_+$ are iid random variables with

$$tP[X_i/b(t) \in \cdot] \to \nu_\alpha(\cdot) \quad \text{in } \mathbb{M}(\mathbb{R}_+ \setminus \{0\}).$$

Then in $\mathbb{M}(\mathbb{R}_+^p \setminus \{\mathbf{0}_p\})$,

$$tP[(X_1, \ldots, X_p)/b(t) \in (dx_1, \ldots, dx_p)]$$

$$\to \sum_{i=1}^{p} \prod_{\substack{j \neq i \\ 1 \leq j \leq p}} \nu_\alpha(dx_i)\epsilon_0(dx_j) \quad (2.42)$$

and as $x \to \infty$,

2.3 How to Prove Regular Variation of Measures?

$$P[\sum_{i=1}^{p} X_i > x] \sim pP[X_1 > x] \in RV_{-\alpha}.$$

So in a sample of size p, the sum is large due to only one large summand and each summand is equally likely to be the large one.

3. Extending the previous item 2, define the uniformly continuous map $\text{CUMSUM}_p : \mathbb{R}_+^p \mapsto \mathbb{R}_+^p$ by

$$\text{CUMSUM}_p(x_1, \ldots, x_p) = (x_1, x_1 + x_2, \ldots, x_1 + x_2 + \cdots + x_p),$$

(refer to Sect. 1.5.4.1, page 18). Apply this map to (2.42) and we get via the Second Mapping Theorem 1.3 that in $\mathbb{M}(\mathbb{R}_+^p \setminus \{\mathbf{0}_p\})$,

$$tP\big[(X_1, X_1 + X_2, \ldots, X_1 + \cdots + X_p)/b(t) \in d\boldsymbol{x}\big] \to$$

$$\sum_{i=1}^{p} 1_{\underbrace{[0, \ldots, 0}_{i-1}, \underbrace{x \ldots, x]}_{p-(i-1)}}(\boldsymbol{x}) \nu_\alpha(dx_i). \tag{2.43}$$

The limit measure concentrates on non-decreasing vectors that have one jump from 0 and the size of the jump is governed by $\nu_\alpha(\cdot)$.

2.3.2.2 Products and the Space $\mathbb{D}_\sqcap$; Spice in the Dish

Consider the setup in Sect. 1.2.2.1, page 5 in which we took the product of two TABOF spaces. Set

$\mathbb{S}_1 = \mathbb{R}_+,\quad F_1 = \emptyset,$
$\mathbb{S}_2 = \mathbb{R}_+,\quad F_2 = \{0\},$
$\mathbb{S}_1 \setminus F_1 = \mathbb{R}_+ \setminus \emptyset,\quad \mathbb{S}_2 \setminus F_2 = \mathbb{R}_+ \setminus \{0\},$

and

$$(\mathbb{S}_1 \setminus F_1) \times (\mathbb{S}_2 \setminus F_2) = [0, \infty) \times (0, \infty) = \mathbb{R}_+^2 \setminus [\text{x-axis}] =: \mathbb{D}_\sqcap.$$

Suppose we have two random variables X and Y with $X \perp\!\!\!\perp Y$ and $P[Y > x] \in RV_{-\alpha}$. So for $b(t) = (1/P[Y > \cdot])^\leftarrow(t) \in RV_{1/\alpha}$ we have

$$tP\left[\frac{Y}{b(t)} \in \cdot\right] \to \nu_\alpha(\cdot) \quad \text{in } \mathbb{M}(\mathbb{R}_+ \setminus \{0\}),$$

and independence and regular variation of Y allows use of the First Binding Lemma 2.1, page 47. There is a regular variation for the pair with scaling defined by $(\lambda, (x, y)) \mapsto (x, \lambda y)$ and

$$tP\left[\left(X, \frac{Y}{b(t)}\right) \in \cdot\right] \to P[X \in \cdot] \times \nu_\alpha \quad \text{in } \mathbb{M}(\mathbb{D}_\sqcap). \tag{2.44}$$

Define the continuous map $h : \mathbb{D}_\sqcap \mapsto \mathbb{D}_\sqcap$ by

$$h(x, y) = (xy, y)$$

and also the pointwise inverse $h^\leftarrow : \mathbb{D}_\sqcap \mapsto \mathbb{D}_\sqcap$ by

$$h^\leftarrow(x, y) = \left(\frac{x}{y}, y\right).$$

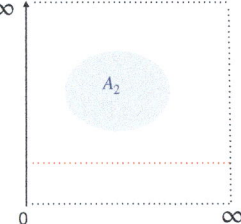

Both h and $h^\leftarrow$ respect the forbidden zone [x-axis] of $\mathbb{D}_\sqcap$. Suppose A_2 is bounded away from the x-axis and $d(A_2, [\text{x-axis}]) > 0$ so there exists $\delta > 0$ such that $\inf\{y : (x, y) \in A_2\} = \delta$. Then the set inverse h^{-1} satisfies

$$h^{-1}(A_2) = \{(x, y) \in \mathbb{D}_\sqcap : h(x, y) = (xy, y) \in A_2\}$$

which is a set whose second coordinates are bounded away from the x-axis.

So we may apply the First Mapping Theorem 1.2 to the convergence in (2.44) to get in $\mathbb{M}(\mathbb{D}_\sqcap)$,

$$tP\left[\left(X, \frac{Y}{b(t)}\right) \in \cdot\right] \circ h^{-1} = tP\left[\left(\frac{XY}{b(t)}, \frac{Y}{b(t)}\right) \in \cdot\right] \to P[X \in \cdot] \times \nu_\alpha \circ h^{-1}(\cdot). \tag{2.45}$$

Since both h and $h^\leftarrow$ respect forbidden zones, in fact, the two conditions (2.45) and (2.44) are equivalent as observed in [131].

It is tempting to try to conclude from (2.45) that XY has a regularly varying tail but this requires an extra condition. The reason is that if we try to apply the projection $h_1 : \mathbb{D}_\sqcap \mapsto \mathbb{R}_+$ defined by $h_1(x, y) = x$, we discover it does not respect forbidden zones. For instance take $[0, 1] \subset \mathbb{R}_+$, a perfectly nice set in $\mathbb{R}_+$, but its pre-image

$$h_1^{-1}([0, 1]) = \{(u, v) \in \mathbb{D}_\sqcap : h_1(u, v) \in [0, 1]\}$$
$$= \{(u, v) \in \mathbb{D}_\sqcap : u \in [0, 1]\} = [0, 1] \times (0, \infty)$$

is not bounded away from the x-axis and hence the pre-image is not so nice. What should we do?

2.3.2.3 Breiman's Theorem

This charming 1965 result by L. Breiman [17] has a pride of place in heavy tail theory and has spawned a gazillion generalizations, several of which are explored in the exercises at the end of this chapter. Some extensions are matrix generalizations and others provide clever and useful applications motivated by risk calculations. Here is a partial list: [3, 4, 20, 21, 25, 37, 47, 48, 100, 102, 130, 131, 149, 156, 186].

There are many subtleties but the simplest case is where there is a product of independent random variables, one of which has a heavy tail and the other has a lighter tail.

Theorem 2.1 *Suppose X, Y satisfy the assumptions of Sect. 2.3.2.2 and in addition*

$$E(X^\beta) < \infty, \qquad \beta > \alpha. \tag{2.46}$$

Then as $t \to \infty$,

$$tP\left[\frac{XY}{b(t)} > x\right] \to E(X^\alpha) x^{-\alpha}, \quad x > 0,$$

or as $x \to \infty$,

$$P[XY > x] \sim E(X^\alpha) P[Y > x].$$

The moment condition (2.46) on X means X has a lighter tail than Y and resolves the issue in (2.45) which prevented us from marginalizing the regular variation on $\mathbb{M}(\mathbb{D}_\sqcap)$ to $\mathbb{R}_+ \setminus \{0\}$.

Proof To begin with, recall that Karamata's theorem on integration, reviewed in Theorem A.1, page 233, gives for any $\epsilon > 0$, as $t \to \infty$,

$$tE\left(\frac{Y}{b(t)}\right)^\beta 1_{[Y/b(t) \leq \epsilon]} \to \frac{\alpha}{\beta - \alpha} \epsilon^{\beta - \alpha}. \tag{2.47}$$

In (2.45) we are permitted to insert any continuity set bounded away from the x-axis so we put in $(x, \infty) \times (\epsilon, \infty)$ and with $G(\cdot) = P[X \in \cdot]$ we get, as $t \to \infty$,

$$v_\epsilon(t) := tP\left[\frac{XY}{b(t)} > x, \frac{Y}{b(t)} > \epsilon\right] \to v_\epsilon(\infty) := G \times v_\alpha \circ h^{-1}\big((x, \infty) \times (\epsilon, \infty)\big)$$

$$= G \times v_\alpha \{(u, v) \in \mathbb{D}_\sqcap : uv > x, v > \epsilon\}$$

$$= \int_0^\infty G(du) v_\alpha(\frac{x}{u} \vee \epsilon, \infty) = \int_0^\infty \left(\frac{x}{u} \vee \epsilon\right)^{-\alpha} G(du).$$

The goal is to replace ϵ by 0. Note $x/u < \epsilon$ iff $x/\epsilon < u$ so the integral becomes

$$= \int_{u=x/\epsilon}^{\infty} \epsilon^{-\alpha} G(du) + \int_{0}^{x/\epsilon} \left(\frac{x}{u}\right)^{-\alpha} G(du)$$

$$= \epsilon^{-\alpha} \bar{G}(x/\epsilon) + x^{-\alpha} \int_{0}^{x/\epsilon} u^{\alpha} G(du) =: v_{\epsilon}(\infty).$$

With x fixed,

$$x^{\alpha} \lim_{\epsilon \to 0} \epsilon^{-\alpha} \bar{G}(x/\epsilon) \leq \lim_{\epsilon \to 0} \int_{x/\epsilon}^{\infty} u^{\alpha} G(du) = 0$$

since $E(X^{\alpha}) = \int_{0}^{\infty} u^{\alpha} G(du) < \infty$ owing to the assumption $E(X^{\beta}) < \infty$. Further,

$$\lim_{\epsilon \to 0} \int_{0}^{x/\epsilon} u^{\alpha} G(du) = E X^{\alpha}.$$

We conclude as $t \to \infty$, $v_{\epsilon}(t) \to v_{\epsilon}(\infty)$ and as $\epsilon \to 0$, $v_{\epsilon}(\infty) \to v_0(\infty) = x^{-\alpha} E X^{\alpha}$.

To show $v_0(t) \to v_0(\infty)$, we finish with a Slutsky style argument ([13, page 28], [153, page 269]) after recognizing $|v_0(t) - v_0(\infty)| \leq |v_0(t) - v_{\epsilon}(t)| + |v_{\epsilon}(t) - v_{\epsilon}(\infty)| + |v_{\epsilon}(\infty) - v_0(\infty)|$ and it suffices to show

$$0 = \lim_{\epsilon \downarrow 0} \limsup_{t \to \infty} |v_{\epsilon}(t) - v_0(\infty)|$$

$$= \lim_{\epsilon \downarrow 0} \limsup_{t \to \infty} \left| tP\left[\frac{XY}{b(t)} > x, \frac{Y}{b(t)} > \epsilon\right] - tP\left[\frac{XY}{b(t)} > x\right]\right|. \quad (2.48)$$

The difference between the two probabilities is

$$tP\left[\frac{XY}{b(t)} > x, \frac{Y}{b(t)} \leq \epsilon\right] = tP\left[\frac{XY}{b(t)} 1_{[\frac{Y}{b(t)} \leq \epsilon]} > x\right]$$

and by Markov's inequality and remembering the moment assumption (2.46) this is bounded by

$$\leq \frac{t}{x^{\beta}} E\left(\frac{XY}{b(t)}\right)^{\beta} 1_{[Y/b(t) \leq \epsilon]} = \frac{E(X^{\beta})}{x^{\beta}} t E\left(\frac{Y}{b(t)}\right)^{\beta} 1_{[Y/b(t) \leq \epsilon]},$$

and applying Karamata's theorem given in (2.47), as $t \to \infty$

$$\to \frac{E(X^{\beta})}{x^{\beta}} \frac{\alpha}{\beta - \alpha} \epsilon^{\beta - \alpha} \to 0,$$

as $\epsilon \to 0$ since $\beta > \alpha$. □

2.3.2.4 A Special Case of Breiman's Theorem Applied to a Tauberian Theorem

Suppose $Y \geq 0$ has distribution H. The Laplace transform of Y or H is the transform

$$\hat{H}(\lambda) = E(e^{-\lambda Y}) = \int_0^\infty e^{-\lambda u} H(du), \quad \lambda > 0,$$

and a convenience formula (integration by parts or Fubini's theorem in disguise) is

$$\frac{1 - \hat{H}(\lambda)}{\lambda} = \int_0^\infty e^{-\lambda u} \bar{H}(u) du, \quad \lambda > 0. \tag{2.49}$$

Theorem 2.2 (Tauberian Special Case) *If $\bar{H} \in RV_{-\alpha}$, $0 < \alpha < 1$, then we have,*

$$1 - \hat{H}(\frac{1}{x}) \sim \bar{H}(x)\Gamma(1 - \alpha), \quad (x \to \infty). \tag{2.50}$$

(The converse is true too but takes us a bit far afield.)

Proof We apply Breiman's theorem and set $X = 1/\mathcal{E}$ where $P[\mathcal{E} > x] = e^{-x}$, $x > 0$, is standard exponential. Then for $x > 0$,

$$P[X > x] = P[\mathcal{E} \leq \frac{1}{x}] = 1 - e^{-1/x} \sim \frac{1}{x}, \quad x \to \infty.$$

So the distribution of X has asymptotically a standard Pareto tail and for $1 > \beta > \alpha$, $EX^\beta < \infty$. Furthermore

$$EX^\beta = E(\frac{1}{\mathcal{E}^\beta}) = \int_0^\infty u^{-\beta} e^{-u} du = \int_0^\infty e^{-u} u^{(1-\beta)-1} du = \Gamma(1-\beta) < \infty.$$

Now apply Breiman's theorem and as $x \to \infty$,

$$P[XY > x] = P[\frac{1}{\mathcal{E}} Y > x] \sim E(1/\mathcal{E})^\alpha P[Y > x].$$

Unpack the left side which by Fubini is

$$\int_0^\infty \bar{H}(sx) e^{-s} ds = \int_0^\infty \bar{H}(u) \frac{1}{x} e^{-u/x} du$$

which by (2.49) is

$$= 1 - \hat{H}(1/x).$$

□

There are extensions [15, 151, 160] of Theorem 2.2 giving behavior of infinite measures which are regularly varying at ∞ to behavior of the transform near 0. For instance, consider the renewal function [76, 152] $U(x)$ associated to H,

$$U(x) = \sum_{n=0}^{\infty} H^{n\star}(x)$$

where $H^{n\star}(x)$ is the n-th convolution power of H. Then if $\bar{H} \in RV_{-\alpha}$, $0 < \alpha < 1$,

$$\hat{U}(\lambda) = \sum_{n=0}^{\infty} \hat{H}^n(\lambda) = \frac{1}{1 - \hat{H}(\lambda)},$$

and from (2.50) we can read off the behavior of $\hat{U}$ near 0 since as $x \to \infty$,

$$\hat{U}(1/x) = \frac{1}{1 - \hat{H}(1/x)} \sim \frac{1}{\Gamma(1-\alpha)\bar{H}(x)}.$$

But what is the behavior of U near ∞? The start of enlightenment comes in Sect. 2.3.2.6, page 62.

2.3.2.5 Binding Plus Mapping Gives a Version of Karamata's Theorem on Integration

Using the first binding lemma and the mapping theorem, we recover a case of Theorem A.1, page 233 that says the indefinite integral of a regularly varying function is regularly varying: If $U(x) \in RV_\alpha$, $\alpha > 0$, then $\int_0^x U(s) ds \sim (const) x U(x) \in RV_{1+\alpha}$, as $x \to \infty$. In addition to binding and mapping, this version of Karamata's theorem just requires the definition of a regularly varying function.

Suppose $U : \mathbb{R}_+ \mapsto \mathbb{R}_+$ is a non-decreasing and regularly varying function at ∞, so $U \in RV_\alpha$, $\alpha > 0$. (With more effort, we could include the case $-1 < \alpha \le 0$ but the forbidden zone given below would change.) Let $\mathbb{S} = [0, \infty]$ with metric

$$d(x, y) = \left| \frac{x}{1+x} - \frac{y}{1+y} \right|,$$

so $[0, \infty]$ is homeomorphic to $[0, 1]$. This requires the understanding that $\infty/(1+\infty) = 1$. Set $F = \{\infty\}$ so the sets bounded away from F are contained

2.3 How to Prove Regular Variation of Measures?

in $[0, x]$ for some $0 \leq x < \infty$. Then with $b(t) = U^{\leftarrow}(t) \in RV_{1/\alpha}$, we have an example of convergence of measures in $\mathbb{M}(\mathbb{S} \setminus \mathbb{F})$ discussed in (2.7),

$$U_t(\cdot) = \frac{U(b(t)\cdot)}{t} \to U_\infty(\cdot), \qquad (2.51)$$

where $U_\infty(dx) = \alpha x^{\alpha-1} dx$, $x \geq 0$.

Let $\text{Leb}(\cdot) \in \mathbb{M}([0, 1] \setminus \{0\})$ be Lebesgue measure. Apply Proposition 1.1, page 13 to (2.51) and $\text{Leb}(\cdot)$ to get in $\mathbb{M}\big(([0, 1]\setminus\{0\}) \times ([0, \infty]\setminus\{\infty\})\big) = \mathbb{M}\big([0, 1] \times [0, \infty] \setminus (\{0\} \times [0, \infty] \cup [0, 1] \times \{\infty\})\big)$,

Fig. 2.2 Red=Forbidden Zone

$$\text{Leb} \times U_t \to \text{Leb} \times U_\infty, \qquad (2.52)$$

(see Fig. 2.2). From the experience in Sect. 2.3.2.2, page 55, define $h : ([0, 1] \setminus \{0\}) \times ([0, \infty] \setminus \{\infty\}) \mapsto ([0, 1] \setminus \{0\}) \times ([0, \infty] \setminus \{\infty\})$ by $h(s, x) = (s, x/s)$ and h respects the forbidden zone $\{0\} \times [0, \infty] \cup [0, 1] \times \{\infty\}$; recall (1.3) and (1.4), page 4. Using Theorem 1.2, page 16 and (2.52), we get

$$\text{Leb} \times U_t \circ h^{-1} \to \text{Leb} \times U_\infty \circ h^{-1},$$

in $\mathbb{M}\big(([0, 1] \setminus \{0\}) \times ([0, \infty] \setminus \{\infty\})\big)$ and since we are allowed to insert (cf. Theorem 1.1, page 9) continuity sets of the limit that are bounded away from the forbidden zone, we insert $[\eta, 1] \times [0, 1]$ for $0 < \eta < 1$, to get

$$\text{Leb} \times U_t \circ h^{-1}([\eta, 1] \times [0, 1]) \to \text{Leb} \times U_\infty \circ h^{-1}([\eta, 1] \times [0, 1]), \qquad (2.53)$$

Unpacking the right side gives

$$\text{Leb} \times U_\infty\{(s, x) \in (0, 1] \times [0, \infty) : \eta \leq s \leq 1, 0 \leq x/s \leq 1\}$$
$$= \int_\eta^1 ds \int_{0 \leq x/s \leq 1} U_\infty(dx) = \int_\eta^1 U_\infty(s) ds = \frac{1 - \eta^{\alpha+1}}{1+\alpha}.$$

Unpacking the left side similarly yields

$$\int_\eta^1 \frac{U(b(t)s)}{t} ds = \int_{b(t)\eta}^{b(t)} \frac{U(y)}{tb(t)} dy.$$

So after a change of variable, $v = b(t) = U^{\leftarrow}(t)$, the unpacted version of (2.53) is

$$\lim_{v \to \infty} \int_{v\eta}^v \frac{U(y)}{vU(v)} dy = \frac{1 - \eta^{\alpha+1}}{1+\alpha}, \qquad (2.54)$$

and it remains to replace η by 0. First of all, for any $\eta > 0$,

$$\liminf_{v \to \infty} \int_0^v \frac{U(y)}{vU(v)} dy \geq \lim_{v \to \infty} \int_{v\eta}^v \frac{U(y)}{vU(v)} dy = \frac{1 - \eta^{\alpha+1}}{1 + \alpha},$$

so letting $\eta \to 0$ on the right yields the lower bound $1/(1+\alpha)$. For the upper bound, since $U(y)$ is non-decreasing and knowing (2.54) we get,

$$\int_0^v \frac{U(y)}{vU(v)} dy = \int_0^{\eta v} + \int_{\eta v}^v \frac{U(y)}{vU(v)} dy$$

$$\leq \frac{\eta v U(\eta v)}{U(v)v} + \frac{1 - \eta^{\alpha+1}}{1 + \alpha} + o(1).$$

Using the definition of regular variation

$$\limsup_{v \to \infty} \int_0^v \frac{U(y)}{vU(v)} dy \leq \eta^{\alpha+1} + \frac{1 - \eta^{\alpha+1}}{1 + \alpha}.$$

Let $\eta \to 0$ to get the required upper bound $1/(1 + \alpha)$.

For a multivariate version of this integration result, try Problem 2.6, page 84.

2.3.2.6 Binding Plus Mapping Gives a Multivariate Version of Karamata's Tauberian Theorem

Continuing the theme of Sect. 2.3.2.4, page 59, we relate the regular variation of an infinite measure near infinity to the behavior of the Laplace transform near 0. We adapt the methods of the previous Sect. 2.3.2.5, page 60. For antecedents see [15, 151, 160, 176, 177, 194, 195] and note the philosophical approach is indebted to the *method of collective marks* [90, 111, 145, 168].

For this section, let

$$\mathbb{S}_1 = \mathbb{R}_+^p = [0, \infty)^p, \quad F_1 = \{x \in \mathbb{S}_1 : \wedge_{i=1}^p x_i = 0\}, \quad \mathbb{S}_1 \setminus F_1 = (0, \infty)^p$$

$$\mathbb{S}_2 = [0, \infty]^p, \quad F_2 = \{x \in \mathbb{S}_2 : \vee_{i=1}^p x_i = \infty\}, \quad \mathbb{S}_2 \setminus F_2 = \mathbb{R}_+^p,$$

and on $\mathbb{S}_2$, use the metric

$$d(x, y) = \sum_{i=1}^p \left| \frac{x_i}{1 + x_i} - \frac{y_i}{1 + y_i} \right|,$$

with the understanding $\infty/(1 + \infty) = 1$. Suppose $U(\cdot) \in \mathbb{M}(\mathbb{S}_2 \setminus F_2)$, that U is an infinite measure concentrating on $\mathbb{R}_+^p$ and that

$$U \in \text{MRV}(\alpha, b(t), U_\infty(\cdot), \mathbb{S}_2 \setminus F_2) \tag{2.55}$$

2.3 How to Prove Regular Variation of Measures?

so that (2.7) holds

$$U_t(\cdot) := \frac{U(b(t)\cdot)}{t} \to U_\infty(\cdot),$$

in $\mathbb{M}(\mathbb{S}_2 \setminus F_2)$. Additionally assume the Laplace transform $\hat{U}$ of $U(\cdot)$ exists,

$$\hat{U}(\boldsymbol{\lambda}) := \int_{\mathbb{R}_+^p} e^{-\sum_{i=1}^p \lambda_i y_i} U(d\boldsymbol{y}) < \infty, \quad \boldsymbol{\lambda} > \boldsymbol{0}.$$

Then also the Laplace transform $\hat{U}_\infty(\boldsymbol{\lambda})$ of U_∞, exists finite for $\boldsymbol{\lambda} > \boldsymbol{0}$ as well. To verify this, note $U_\infty(\cdot)$ has a scaling property so for any $a > 0$,

$$U_\infty\{\boldsymbol{x} \in \mathbb{S}_2 \setminus F_2 : \sum_{i=1}^p x_i \leq a\} = a^\alpha U_\infty\{\boldsymbol{x} : \sum_{i=1}^p x_i \leq 1\} < \infty.$$

Switching to polar coordinates $\boldsymbol{x} \mapsto (\sum_{i=1}^p x_i, \boldsymbol{x}/\sum_{i=1}^p x_i) =: (r, \boldsymbol{a})$ as in Sect. 2.2.1 we get for $\boldsymbol{\lambda} > \boldsymbol{0}$,

$$\hat{U}_\infty(\boldsymbol{\lambda}) = \int_{\mathbb{R}_+^p} e^{-\sum_{i=1}^p \lambda_i x_i} U_\infty(d\boldsymbol{x}) \leq \int_{\mathbb{R}_+^p} e^{-(\wedge_{i=1}^p \lambda_i) \sum_{i=1}^p x_i} U_\infty(d\boldsymbol{x})$$

$$= U_\infty\{\boldsymbol{x} : \sum_{i=1}^p x_i \leq 1\} \int_{r>0} e^{-(\wedge_{i=1}^p \lambda_i)r} \alpha r^{\alpha-1} dr < \infty.$$

Let $(E_1, \ldots, E_p)$ be iid unit exponentially distributed random variables and set $\mathcal{E}(\cdot) = P[(E_1, \ldots, E_p) \in \cdot] \in \mathbb{M}(\mathbb{S}_1 \setminus F_1)$. From the binding result Proposition 1.1, page 13,

$$\mathcal{E} \times U_t(\cdot) \to \mathcal{E} \times U_\infty(\cdot),$$

in $\mathbb{M}\big((\mathbb{S}_1 \setminus F_1) \times (\mathbb{S}_2 \setminus F_2)\big)$. Define

$$h : (\mathbb{S}_1 \setminus F_1) \times (\mathbb{S}_2 \setminus F_2) \mapsto (\mathbb{S}_1 \setminus F_1) \times (\mathbb{S}_2 \setminus F_2)$$

by $h(s, \boldsymbol{x}) = (s, \boldsymbol{x}/s)$ and h respects the forbidden zone in the product space. Respectfulness is helped by defining $\mathcal{E}$ on $(0, \infty)^p$ and you can puzzle the rest out using the visual aid Fig. 2.3 for $p = 1$ which depicts a set $A_2 \in \mathbb{BA}\big((\mathbb{S}_1 \setminus F_1) \times (\mathbb{S}_2 \setminus F_2)\big)$. From the first mapping Proposition 1.2, page 16, we get

$$\mathcal{E} \times U_t \circ h^{-1}(\cdot) \to \mathcal{E} \times U_\infty \circ h^{-1}(\cdot). \quad (2.56)$$

Fig. 2.3 Red=Forbidden Zone

Theorem 1.1, page 9) allows us to insert a continuity

set bounded away from the forbidden zone into convergence (2.56), so for $\epsilon > 0$, $\mathbf{1} = (1, \ldots, 1)$ and $\mathbb{R}_+^p \ni \mathbf{x} > \mathbf{0}$,

$$\mathcal{E} \times U_t \circ h^{-1}\big((\epsilon \mathbf{1}, \infty) \times [\mathbf{0}, \mathbf{x}]\big) \to \mathcal{E} \times U_\infty \circ h^{-1}\big((\epsilon \mathbf{1}, \infty) \times [\mathbf{0}, \mathbf{x}]\big). \quad (2.57)$$

Unpack the left side of (2.57), and we get,

$$\int_{\{(\mathbf{s},\mathbf{y}): \mathbf{s} > \epsilon \mathbf{1}, \mathbf{y} \le \mathbf{x} \mathbf{s}\}} \mathcal{E}(d\mathbf{s}) U_t(d\mathbf{y}) = \int_{\mathbb{R}_+^p} U_t(d\mathbf{y}) \prod_{i=1}^p e^{-(\frac{y_i}{x_i} \vee \epsilon)}$$

and applying the convergence in (2.57), as $t \to \infty$, the right side converges to,

$$\int_{\mathbb{R}_+^p} U_\infty(d\mathbf{y}) \prod_{i=1}^p e^{-(\frac{y_i}{x_i} \vee \epsilon)}, \quad (2.58)$$

so as $\epsilon \to 0$, this converges to

$$\int_{\mathbb{R}_+^p} U_\infty(d\mathbf{y}) \prod_{i=1}^p e^{-(\frac{y_i}{x_i})} = \hat{U}_\infty(\frac{\mathbf{1}}{\mathbf{x}}). \quad (2.59)$$

The last convergence as $\epsilon \to 0$ follows from monotonicity in ϵ and dominated convergence since we know $\hat{U}_\infty(\boldsymbol{\lambda}) < \infty$ for $\boldsymbol{\lambda} > \mathbf{0}$. Remembering that $U_t(\cdot) = U(b(t)\cdot)/t$, the goal is to replace ϵ with 0 in (2.57) which would give a regular variation property for the transform,

$$\frac{1}{t} \hat{U}\Big(\frac{\mathbf{1}}{b(t)\mathbf{x}}\Big) \to \hat{U}_\infty(\frac{\mathbf{1}}{\mathbf{x}}) \quad (t \to \infty). \quad (2.60)$$

To replace ϵ with 0 in convergence (2.58), it suffices to show

$$0 = \lim_{\epsilon \to 0} \limsup_{t \to \infty} \Big| \int U_t(d\mathbf{y}) e^{-\sum_{i=1}^p y_i/x_i \vee \epsilon} - \int U_t(d\mathbf{y}) e^{-\sum_{i=1}^p y_i/x_i} \Big|.$$

Referring to (2.57), the difference of the integrals is

$$\Big| \mathcal{E} \times U_t \circ h^{-1}(\epsilon \mathbf{1}, \infty) \times [\mathbf{0}, \mathbf{x}]) - \mathcal{E} \times U_t \circ h^{-1}(\mathbf{0}, \infty) \times [\mathbf{0}, \mathbf{x}]) \Big|$$

$$= \mathcal{E} \times U_t \circ h^{-1}\Big((\mathbf{0}, \epsilon \mathbf{1}] \times [\mathbf{0}, \mathbf{x}]\Big)$$

$$= \iint_{\mathbf{0} < \mathbf{s} \le \epsilon \mathbf{1}, \mathbf{y} \le \mathbf{x}\mathbf{s}} \prod_{i=1}^p e^{-s_i} ds_i U_t(d\mathbf{y}) \le \int_{\mathbf{0} < \mathbf{s} \le \epsilon \mathbf{1}} U_t([\mathbf{0}, \mathbf{x}\mathbf{s}]) \prod_{i=1}^p e^{-s_i} ds_i$$

$$\le (U_t([\mathbf{0}, \epsilon \mathbf{x}])(1 - e^{-\epsilon})^p \le U_t([\mathbf{0}, \epsilon \mathbf{x}]).$$

2.3 How to Prove Regular Variation of Measures? 65

and remembering U_∞ has a scaling property,

$$\limsup_{t\to\infty} U_t([\mathbf{0}, \epsilon \mathbf{x}]) \leq U_\infty([\mathbf{0}, \epsilon \mathbf{x}]) = \epsilon^\alpha U_\infty([\mathbf{0}, \mathbf{x}])$$

which tends to 0 as $\epsilon \to 0$. This is what was required and we now summarize.

Theorem 2.3 *Assume that U is a regularly varying measure satisfying* (2.55) *whose Laplace transform exists. Then the Laplace transforms $\hat{U}(1/\mathbf{x})$ and $\hat{U}_\infty(1/\mathbf{x})$ are distribution functions of measures on $\mathbb{R}_+^p$ satisfying* (2.60).

The statement about the Laplace transforms being distribution function of measures follows from the representation in (2.57) with $\epsilon = 0$.

Of course, we have not quite nailed down the fact that (2.60) means regular variation of the measure since we have not commented on why convergence on square-ish regions is sufficient for convergence of measures. Information on this for random vectors is in Sect. 2.3.4.1.

2.3.3 Proof Method 3: Reduce Problems in $\mathbb{R}^\infty$ to Finite Dimensions

The standard method of proving weak convergence of a sequence of probability measures in $\mathbb{R}^\infty$ is to verify weak convergence of the finite dimensional distributions of the sequence; see [13, page 19]. Something similar should be true when we try to prove regular variation in $\mathbb{S} = \mathbb{R}^\infty$ and we follow the principle that it is best to reduce a hard convergence problem to an easier convergence problem.

For $\mathbf{x} = (x_1, x_2, \dots) \in \mathbb{R}_+^\infty$, write

$$\mathrm{PROJ}_p(\mathbf{x}) = \mathbf{x}_{|p} = (x_1, \dots, x_p)$$

so $\mathrm{PROJ}_p : \mathbb{R}_+^\infty \mapsto \mathbb{R}_+^p$. The function PROJ_p takes a sequence and retains the first p components and discards the rest. If $\mathbb{R}_+^\infty$ is metrized by $d_\infty(\mathbf{x}, \mathbf{y})$ in (1.2), page 3, the fact that PROJ_p is continuous and even uniformly continuous was discussed in Sect. 1.5.4.2, page 20. The continuity properties are not a surprise and allow application of the Second Mapping Theorem 1.3. However, going from $\mathbb{R}_+^\infty \mapsto \mathbb{R}_+^p$ is not the point as it is the reverse direction that is often significant for applications.

Example 2.5 (Special Case of Corollary 1.4, page 21) Suppose $F = \{\mathbf{0}_\infty\}$ so that $\mathrm{PROJ}_p(F) = \{\mathbf{0}_p\}$. Then if $\mu_n \to \mu_0$ in $\mathbb{M}(\mathbb{R}_+^\infty \setminus \{\mathbf{0}_\infty\})$ we also have

$$\mu_n \circ \mathrm{PROJ}_p^{-1} \to \mu_0 \circ \mathrm{PROJ}_p^{-1} \quad \text{in } \mathbb{M}(\mathbb{R}_+^p \setminus \{\mathbf{0}_p\}).$$

Now we consider when the more useful converse is true.

Theorem 2.4 *Suppose for every $p \geq 1$, that the closed set $F \in \mathcal{F}(\mathbb{R}_+^\infty)$ satisfies*

1. *$PROJ_p(F) \in \mathcal{F}(\mathbb{R}_+^p)$; and*
2. *If $z \in F$, so that $z_{|p} = (z_1, \ldots, z_p) \in PROJ_p(F)$, then also it follows $(z_1, \ldots, z_p, \mathbf{0}_\infty) \in F$.*

Then $\mu_n \to \mu_0$ in $\mathbb{M}(\mathbb{R}_+^\infty \setminus F)$ if and only if for all $p \geq 1$ such that $\mathbb{R}_+^p \setminus PROJ_p(F) \neq \emptyset$, we have

$$\mu_n \circ PROJ_p^{-1} \to \mu_0 \circ PROJ_p^{-1} \quad (2.61)$$

in $\mathbb{M}(\mathbb{R}_+^p \setminus PROJ_p(F))$.

As a criteria for convergence in $\mathbb{M}(\mathbb{R}_+^\infty \setminus F)$, this allows proving convergence in $\mathbb{M}(\mathbb{R}_+^p \setminus PROJ_p(F))$ and then identifying μ_0 from knowing $\mu_0 \circ PROJ_p^{-1}$ for any $p \geq 1$.

The Condition 2 says take an infinite sequence z in $F \subset \mathbb{R}_+^\infty$, truncate it to $z_{|p} \in \mathbb{R}_+^p$, and then make it infinite again by filling in zeros for all the components beyond the pth. The result must still be in F. The first application of Theorem 2.4 is in Sect. 2.3.3.1.

Here are two simple examples where Condition 2 holds.

1. $F = \{\mathbf{0}_\infty\}$.
2. Pick an integer $j \geq 1$ and define

$$F_{\leq j} = \{x \in \mathbb{R}_+^\infty : \sum_{i=1}^\infty \epsilon_{x_i}(0, \infty) \leq j\}, \quad (2.62)$$

where recall $\epsilon_x(A) = 1$, if $x \in A$, and 0, if $x \in A^c$. So $F_{\leq j}$ is the set of sequences with at most j positive components. Truncation and then insertion of zeros does not increase the number of positive components so $F_{\leq j}$ is invariant under the operation implied by Condition 2.

Proof of Theorem 2.4 Assume $F \in \mathcal{F}(\mathbb{R}_+^\infty)$ satisfies Conditions 1, 2 and (2.61) holds. Suppose $f \in \mathbb{C}(\mathbb{R}_+^\infty \setminus F)$ so f is bounded, continuous, with a support bounded away from F and without loss of generality suppose f is uniformly continuous with modulus of continuity

$$\omega_f(\eta) = \sup_{\substack{(x,y) \in \mathbb{R}_+^\infty \setminus F \\ d_\infty(x,y) < \eta}} |f(x) - f(y)|.$$

Since the support of f is bounded away from F, there exists $1 > \delta > 0$ such that

$$d_\infty(x, F) < \delta \text{ implies } f(x) = 0. \quad (2.63)$$

2.3 How to Prove Regular Variation of Measures?

Observe,

$$d_\infty\Big((x_{|p}, \mathbf{0}_\infty), x\Big) \leq \sum_{j=p+1}^{\infty} 2^{-j} = 2^{-p}. \tag{2.64}$$

Pick any p so large that $2^{-p} < \delta/2$ and define

$$g(x_1, \ldots, x_p) = f(x_1, \ldots, x_p, \mathbf{0}_\infty).$$

Then we have

(a) From (2.64), g is close to f,

$$|f(x) - g(x_{|p})| = |f(x) - f(x_{|p}, \mathbf{0}_\infty)| \leq \omega_f(2^{-p}). \tag{2.65}$$

(b) Write $d_p\big((x_1 \ldots, x_p), (y_1 \ldots, y_p)\big) = \sum_{i=1}^{p} |x_i - y_i|$ as the usual L_1-metric on $\mathbb{R}_+^p$. With δ from (2.63), the function g satisfies

$$d_p\big(\text{supp}(g), \text{PROJ}_p(F)\big) \geq \delta/2. \tag{2.66}$$

We verify (2.66) below. Assuming its truth, we get $\text{supp}(g) \in \text{BA}(\mathbb{R}_+^p \setminus \text{PROJ}_p(F))$ and $g \in \mathbb{C}(\mathbb{R}_+^p \setminus \text{PROJ}_p(F))$.

(c) The function g is uniformly continuous which is inherited from the uniform continuity of f.

To verify (2.66), suppose $(x_1, \ldots, x_p)$ is too close to $\text{PROJ}_p(F)$, that is,

$$d_p\big((x_1, \ldots, x_p), \text{PROJ}_p(F)\big) < \delta/2. \tag{2.67}$$

Then there is $(z_1, \ldots, z_p) \in \text{PROJ}_p(F)$ such that

$$d_p\big((x_1, \ldots, x_p), (z_1, \ldots, z_p)\big) < \frac{3}{4}\delta. \tag{2.68}$$

There exists $z \in F$ with $z_{|p} = (z_1, \ldots, z_p)$ and by Condition 2 we have $(z_1, \ldots, z_p, \mathbf{0}_\infty) \in F$ and

$$d_\infty\big((x_1, \ldots, x_p, \mathbf{0}_\infty), (z_1, \ldots, z_p, \mathbf{0}_\infty)\big) = \sum_{i=1}^{p} \frac{|x_i - z_i| \wedge 1}{2^i}$$

$$\leq \sum_{i=1}^{p} |x_i - z_i| \wedge 1 \leq \sum_{i=1}^{p} |x_i - z_i|$$

$$= d_p\big((x_1, \ldots, x_p), (z_1, \ldots, z_p)\big) < \delta,$$

by (2.68). Therefore $(x_1, \ldots, x_p, \mathbf{0}_\infty)$ is too close to F and (2.63) implies

$$g(x_1, \ldots, x_p) = f(x_1, \ldots, x_p, \mathbf{0}_\infty) = 0. \tag{2.69}$$

Summary: If $(x_1, \ldots, x_p)$ satisfies (2.67), then (2.69) holds. Set $[g > 0] = \{(x_1, \ldots, x_p) : g(x_1, \ldots, x_p) > 0\}$ and thus

$$\inf_{[g>0]} d_p((x_1, \ldots, x_p), \mathrm{PROJ}_p(F)) \geq \delta/2.$$

Taking the closure of $[g > 0]$ gives (2.66) as claimed.

Assuming (2.61), we must prove $\mu_n(f) \to \mu_0(f)$. The closeness of f and g in (2.65), and knowing fi-di's converge in (2.61) suggests writing

$$\mu_n(f) - \mu_0(f) = [\mu_n(f) - \mu_n(g \circ \mathrm{PROJ}_p)]$$
$$+ [\mu_n(g \circ \mathrm{PROJ}_p) - \mu_0(g \circ \mathrm{PROJ}_p)] + [\mu_0(g \circ \mathrm{PROJ}_p) - \mu_0(f)]$$
$$= A + B + C. \tag{2.70}$$

Now $g \in \mathbb{C}(\mathbb{R}_+^p \setminus \mathrm{PROJ}_p(F))$ and for $n \geq 0$,

$$\mu_n(g \circ \mathrm{PROJ}_p) = \mu_n \circ \mathrm{PROJ}_p^{-1}(g),$$

so from (2.61), we have that as $n \to \infty$,

$$B = \mu_n(g \circ \mathrm{PROJ}_p) - \mu_0(g \circ \mathrm{PROJ}_p) \to 0$$

How to control A? If $\boldsymbol{x} \in \mathbb{R}_+^\infty \setminus F$ satisfies $d_\infty((x_1, \ldots, x_p, \mathbf{0}_\infty), F) < \delta/2$, then from (2.63), we know

(1) $g(x_1, \ldots, x_p) = f(x_1, \ldots, x_p, \mathbf{0}_\infty) = 0$; and

(2) $f(\boldsymbol{x}) = 0$,

where (2) follows from (2.63), since from (2.64),

$$d_\infty(\boldsymbol{x}, F) \leq d_\infty(\boldsymbol{x}, (\boldsymbol{x}_{|p}, \mathbf{0}_\infty)) + d_\infty((\boldsymbol{x}_{|p}, \mathbf{0}_\infty), F) < 2^{-p} + \delta/2 < \delta.$$

Therefore, on

$$\Lambda(p) := \{\boldsymbol{x} \in \mathbb{R}_+^\infty \setminus F : d_\infty\big((\boldsymbol{x}_{|p}, \mathbf{0}_\infty), F\big) < \delta/2\}$$

both f and $g \circ \mathrm{PROJ}_p$ are zero. The complement of $\Lambda(p)$ is

$$\Lambda^c(p) = (\mathbb{R}_+^\infty \setminus F) \setminus \Lambda(p) = \{\boldsymbol{x} \in \mathbb{R}_+^\infty \setminus F : d_\infty\big((\boldsymbol{x}_{|p}, \mathbf{0}_\infty), F\big) \geq \delta/2\}.$$

2.3 How to Prove Regular Variation of Measures?

We call attention to the fact that $\Lambda^c(p)$ may be expressed as

$$\Lambda^c(p) = \text{PROJ}_p^{-1}\{(x_1, \ldots, x_p)$$
$$\in \mathbb{R}_+^p \setminus \text{PROJ}_p(F) : d_\infty\big((x_1, \ldots, x_p), \mathbf{0}_\infty\big), F\big) \geq \delta/2\}$$
$$=: \text{PROJ}_p^{-1}(\Lambda_{\text{fidi}}^c(p)).$$

The dubious can note that if $x \in \mathbb{R}_+^\infty$ with $d_\infty\big((x_{|p}, \mathbf{0}_\infty), F\big) \geq \delta/2$, then also $x \in \mathbb{R}_+^\infty \setminus F$ since if $x \in F$, we have by assumption, $(x_{|p}, \mathbf{0}_\infty) \in F$ implying $d_\infty\big((x_{|p}, \mathbf{0}_\infty), F\big) = 0$, a contradiction. Critically $\Lambda_{\text{fidi}}^c(p)$ is a closed, p-dimensional set so (2.61) becomes applicable and we have

$$|\mu_n(f) - \mu_n(g \circ \text{PROJ}_p)| \leq \int |f - g \circ \text{PROJ}_p| d\mu_n$$
$$= \int_{\Lambda^c(p)} |f - g \circ \text{PROJ}_p| d\mu_n \leq \mu_n(\Lambda^c(p)) \omega_f(2^{-p})$$
$$= \mu_n \circ \text{PROJ}_p^{-1}(\Lambda_{\text{fidi}}^c(p)) \omega_f(2^{-p}),$$

owing to (2.65). For term C, we have $|\mu_0(f) - \mu_0(g \circ \text{PROJ}_p)| \leq \mu_0(\Lambda^c(p)) \omega_f(2^{-p})$.

Apply (1.12), page 9 of the Portmanteau theorem 1.1 to the finite-dimensional convergence (2.61) and we have

$$\limsup_{n \to \infty} |\mu_n(f) - \mu_0(f)| \leq 2\omega_f(2^{-p}) \mu_0 \circ \text{PROJ}_p^{-1}(\Lambda_{\text{fidi}}^c(p)) + 0$$
$$= 2\omega_f(2^{-p}) \mu_0(\Lambda^c(p)). \tag{2.71}$$

Since $\Lambda^c(p)$ is bounded away from F, $\mu_0(\Lambda^c(p)) < \infty$. If $x \in \Lambda^c(p)$ so that $d_\infty((x_{|p}, \mathbf{0}_\infty), F) \geq \delta/2$, we also have

$$d_\infty((x_{|p}, \mathbf{0}_\infty), F) \leq d_\infty((x_{|p}, \mathbf{0}_\infty), x) + d_\infty(x, F)$$

implies

$$d_\infty(x, F) \geq \delta/2 - 2^{-p} \geq \delta/4$$

for all large p. Therefore, for such p,

$$\Lambda^c(p) = \{x \in \mathbb{R}_+^\infty \setminus F : d_\infty((x_{|p}, \mathbf{0}_\infty), F) \geq \delta/2\}$$
$$\subset \{x \in \mathbb{R}_+^\infty \setminus F : d_\infty(x, F) \geq \delta/4\}$$

which is bounded away from F and free of p. Thus the bound in (2.71) is strengthened to the form $(const)\omega_f(2^{-p})$ and now we may let $p \to \infty$ and get $\mu_n(f) \to \mu_0(f)$. □

2.3.3.1 Regular Variation in R_+^∞; What Happens in $\mathbb{R}_+^\infty$ Stays in $\mathbb{R}_+^\infty$

This section concentrates on the case $\mathbb{S} = \mathbb{R}_+^\infty$ and assumes F is a closed cone, so $\mathbb{S} \setminus F$ is still a cone. Using Definition 2.1, a random element X of $\mathbb{S} \setminus F$ has a regularly varying distribution if for some regularly varying function $b(t) \to \infty$, as $t \to \infty$,

$$tP[X/b(t) \in \cdot] \to \nu(\cdot) \quad \text{in } \mathbb{M}(\mathbb{S} \setminus F),$$

for some limit measure $\eta(\cdot) \in \mathbb{M}(\mathbb{S} \setminus F)$. We illustrate Theorem 2.4 by considering the $\mathbb{R}_+^\infty$ valued random element generated by iid components.

The iid Case: Remove $\{0_\infty\}$
Combining Theorem 2.4 and the Binding Lemma 2.2, page 49 yields the following result in $\mathbb{R}_+^\infty$.

Proposition 2.4 *Suppose $X = (X_1, X_2, \dots)$ is iid having non-negative components with common distribution $H(\cdot)$ where $\bar{H} \in RV_{-\alpha}$ so that*

$$P[X_1 > tx]/P[X_1 > t] \to x^{-\alpha}, \quad \text{as } t \to \infty, x > 0, \alpha > 0.$$

Equivalently, as $t \to \infty$,

$$tH(b(t)\cdot) := tP[X_1/b(t) \in \cdot] \to \nu_\alpha(\cdot) \quad \text{in } \mathbb{M}(\mathbb{R}_+ \setminus \{0\}), \tag{2.72}$$

where $\nu_\alpha(x, \infty) = x^{-\alpha}$, $\alpha > 0$ and $b(t) = \left(1/\bar{H}(\cdot)\right)^{\leftarrow}(t)$. Then in $\mathbb{M}(\mathbb{R}_+^\infty \setminus \{0_\infty\})$, we have

$$\mu_t((dx_1, dx_2, \dots)) := tP[X/b(t) \in (dx_1, dx_2, \dots)]$$

$$\to \sum_{l=1}^\infty \Big(\prod_{i \neq l} \epsilon_0(dx_i)\Big) \nu_\alpha(dx_l) =: \mu^{(0)}(dx_1, dx_2, \dots),$$

(2.73)

and the limit measure $\mu^{(0)}(\cdot)$ concentrates on

$$F_{=1} = \{\boldsymbol{x} \in \mathbb{R}_+^\infty : \sum_{i=1}^\infty \epsilon_{x_i}\big((0, \infty)\big) = 1\},$$

the set of sequences with exactly one component positive.

2.3 How to Prove Regular Variation of Measures?

Note $\{\mathbf{0}_\infty\} \cup F_{=1} =: F_{\leq 1}$, the set of sequences with at most one component positive, is closed.

Proof To verify (2.73), note from Theorem 2.4, it suffices to verify finite-dimensional convergence since $\{\mathbf{0}_\infty\}$ satisfies Condition (2), page 66. So we must prove as $t \to \infty$, for $p \geq 1$,

$$\mu_t \circ \mathrm{PROJ}_p^{-1}((dx_1,\ldots,dx_p)) := t P[(X_1,\ldots,X_p)/b(t) \in (dx_1,\ldots,dx_p)]$$

$$\to \mu^{(0)} \circ \mathrm{PROJ}_p^{-1}((dx_1,\ldots,dx_p)) = \sum_{l=1}^{p} \prod_{i \neq l} \epsilon_0(dx_i) \nu_\alpha(dx_l), \quad (2.74)$$

in $\mathbb{M}(\mathbb{R}_+^p \setminus \{\mathbf{0}_p\})$. But (2.74) follows from elaborating the Second Binding Lemma 2.2, page 49. □

The CUMSUM Operator

Owing to Corollary 1.3, page 20, the uniformly continuous operator CUMSUM and may be applied to (2.73). This gives

$$t P[\mathrm{CUMSUM}(X)/b(t) \in (dx_1, dx_2, \ldots)] \to \mu^{(0)} \circ \mathrm{CUMSUM}^{-1}((dx_1, dx_2, \ldots))$$

$$= \sum_{l=1}^{\infty} \prod_{i=1}^{l-1} \epsilon_0(dx_i) \nu_\alpha(dx_l) \prod_{i=l+1}^{\infty} \epsilon_{x_l}(dx_i) \quad \text{in } \mathbb{M}(\mathbb{R}_+^\infty \setminus \{\mathbf{0}_\infty\})), \quad (2.75)$$

where the limit concentrates on non-decreasing sequences with one jump and the size of the jump is governed by ν_α. Then applying the operator PROJ_p we get, by Corollary 1.4, page 21,

$$t P\Big[(X_1, X_1+X_2, \ldots, \sum_{i=1}^{p} X_k)/b(t) \in (dx_1, dx_2, \ldots, dx_p)\Big]$$

$$\to \mu^{(0)} \circ \mathrm{CUMSUM}^{-1} \circ \mathrm{PROJ}_p^{-1}((dx_1, dx_2, \ldots, dx_p))$$

$$= \sum_{l=1}^{p} \prod_{i=1}^{l-1} \epsilon_0(dx_i) \nu_\alpha(dx_l) \prod_{i=l+1}^{p} \epsilon_{x_l}(dx_i) \quad \text{in } \mathbb{M}(\mathbb{R}_+^p \setminus \{\mathbf{0}_p\})), \quad (2.76)$$

giving an elaboration of the one-big-jump heuristic in Sect. 2.3.2.1 which said that summing independent risks which have the same heavy tail results in a tail risk which is the number of summands times the individual tail risk; see also [156, p. 230]. We recover the big jump heuristic by applying the projection from $\mathbb{R}_+^p \setminus \{\mathbf{0}_p\} \mapsto \mathbb{R}_+ \setminus \{0\}$ defined by $T_p : (x_1, \ldots, x_p) \mapsto x_p$ to (2.76). This gives

$$t P[\sum_{i=1}^{p} X_i > b(t)x] \to \mu^{(0)} \circ \mathrm{CUMSUM}^{-1} \circ \mathrm{PROJ}_p^{-1} \circ T_p^{-1}(x, \infty) \quad (2.77)$$

$$= p\nu_\alpha(x, \infty) = px^{-\alpha}.$$

The projection T_p is uniformly continuous so the Second Mapping Theorem 1.3 applies. Also the First Mapping Theorem 1.2 applies to T_p since for $y > 0$, $T_p^{-1}(y, \infty) = \{(x_1, \ldots, x_p) : x_p > y\}$ is at positive distance from $\{\mathbf{0}_p\}$.

To unpack the right side of (2.77) and see why it resolves to $p\nu_\alpha(x, \infty)$, observe for $x > 0$, that $T_p(x, \infty) = \{(y_1, \ldots, y_p) \in \mathbb{R}_+^p \setminus \mathbf{0}_p : y_p > y\}$ and thus from (2.76),

$$\mu^{(0)} \circ \text{CUMSUM}^{-1} \circ \text{PROJ}_p^{-1} \circ T_p^{-1}(x, \infty)$$

$$= \sum_{l=1}^{p} \prod_{i=1}^{l-1} \epsilon_0(dx_i) \nu_\alpha(dx_l) \prod_{i=l+1}^{p} \epsilon_{x_l}(dx_i) T_p^{-1}(y, \infty)$$

$$= \sum_{l=1}^{p} \prod_{i=1}^{l-1} \epsilon_0(dx_i) \nu_\alpha(dx_l) \prod_{i=l+1}^{p} \epsilon_{x_l}(dx_i) \big(\{(y_1, \ldots, y_p) \in \mathbb{R}_+^p \setminus \mathbf{0}_p : y_p > y\}\big)$$

$$= \sum_{l=1}^{p} \mu_{0,l} \{(y_1, \ldots, y_p) \in \mathbb{R}_+^p \setminus \mathbf{0}_p : y_p > y\},$$

where

$$\mu_{0,l}(dx_1, \ldots, dx_p) = \prod_{i=1}^{l-1} \epsilon_0(dx_i) \nu_\alpha(dx_l) \prod_{i=l+1}^{p} \epsilon_{x_l}(dx_i).$$

Since $\mu_{0,l}(\cdot)$ concentrates on $\{(0, \ldots, 0, y, \ldots, y) : y \in \mathbb{R}_+ \setminus \{0\}\}$ with the y first appearing in the lth spot,

$$\mu_{0,l}\{(x_1, \ldots, x_p : x_p > x\} = x^{-\alpha}, \quad l = 1, \ldots, p$$

and the result follows.

With minor modifications, this discussion could be carried out without the iid assumption by assuming independence, marginal regularly varying tails and

$$P[X_j > x]/P[X_1 > x] \to c_j > 0, \quad j \geq 2.$$

2.3.3.2 The Tail Process of a Random Sequence Whose Distribution Is Regularly Varying

Basrak and Segers [6] define the *tail process* of a stationary sequence of the form

$$\mathbf{X} = \{X_j, j \in \mathbb{Z}\},$$

2.3 How to Prove Regular Variation of Measures?

where $\mathbb{Z}$ are the natural numbers $\{\ldots, -1, 0, 1, \ldots\}$, each X_j is $\mathbb{R}^p$-valued and the distribution of X is regularly varying on $(\mathbb{R}^p)^{\mathbb{Z}} \setminus \{\mathbf{0}_{\mathbb{Z}}\}$. Here $\{\mathbf{0}_{\mathbb{Z}}\}$ is the closed cone consisting of the sequence of $\mathbf{0}$'s in $(\mathbb{R}^p)^{\mathbb{Z}}$. We proceed in a reduced context keeping the focus on showing implications of the regular variation of the distribution. Further information is in the following incomplete list of references: [16, 57, 99, 101, 114, 134, 142, 165, 196].

We only assume $X = \{X_0, X_1, \ldots\}$ is $\mathbb{R}_+^\infty$-valued and its distribution satisfies,

$$P[X \in \cdot] \in \mathrm{MRV}(\alpha, b(t), \eta(\cdot), \mathbb{S} \setminus F = \mathbb{R}_+^\infty \setminus \{\mathbf{0}_\infty\})$$

so that as $t \to \infty$,

$$tP\left[\frac{X}{b(t)} \in \cdot\right] \to \eta(\cdot), \quad \text{in } \mathbb{M}(\mathbb{R}_+^\infty \setminus \{\mathbf{0}_\infty\}).$$

Recall the limit measure $\eta(\cdot)$ has the scaling property $\eta(c\cdot) = c^{-\alpha}\eta(\cdot)$, for $c > 0$. For a measurable set $A \subset \mathbb{R}_+^\infty$ such that $A \cap \{x \in \mathbb{R}_+^\infty : x_0 > 1\}$ has zero η-mass on the boundary, we have

$$tP\left[\frac{X}{b(t)} \in A, \frac{X_0}{b(t)} > 1\right] \to \eta\{x \in \mathbb{R}_+^\infty \setminus \{\mathbf{0}_\infty\} : x \in A, x_0 > 1\}$$

and therefore,

$$tP\left[\frac{X}{b(t)} \in A \,\Big|\, \frac{X_0}{b(t)} > 1\right] \to \frac{\eta\{x \in \mathbb{R}_+^\infty \setminus \{\mathbf{0}_\infty\} : x \in A, x_0 > 1\}}{\eta\{x : x_0 > 1\}} =: \mathbb{Q}(A),$$

and the ratio on the right of the arrow is a probability measure $\mathbb{Q}(\cdot)$ in A; that is, a probability measure $\mathbb{Q}$ on $\mathbb{R}_+^\infty$, concentrating on $\mathbb{R}_+^\infty \cap [x_0 > 1]$.

Thus, from Skorohod representation [12, 13], there exists a process $Y = (Y_0, Y_1, \ldots)$ in $\mathbb{R}_+^\infty$ with $Y_0 > 1$ such that in some probability space $(\Omega, \mathcal{B}, P)$ the P-distribution $P \circ Y^{-1}$ of Y is $\mathbb{Q}$. Following [6], call Y the *tail process* and for $l \geq 1$, set $\Theta_l = Y_l/Y_0 \in \mathbb{R}_+$ with $\Theta_0 = 1$. Set $\Theta = (\Theta_0, \Theta_1, \Theta_2, \ldots) \in \{1\} \times \mathbb{R}_+^\infty$ and call Θ the *spectral* or *angular* process. We have the following result [6, Theorem 3.1].

Proposition 2.5 *We have Y_0 is a Pareto random variable, $P[Y_0 > x] = x^{-\alpha}$, $x > 1$ and $Y_0 \perp\!\!\!\perp \Theta$. Therefore the tail process Y has the representation*

$$Y = Y_0 \Theta$$

as an independent product of a Pareto random variable and the spectral process.

Proof For $y > 1$, the distribution of Y_0 and Θ is specified by

$$\frac{\eta\{x \in \mathbb{R}_+^\infty : x_0 > y, \frac{x}{x_0} \in \cdot\}}{\eta\{x : x_0 > 1\}} = \frac{\eta\{x : \frac{x_0}{y} > 1, \frac{x/y}{x_0/y} \in \cdot\}}{\eta\{x : x_0 > 1\}}$$

and applying the scaling property of $\eta(\cdot)$ after setting $z = x/y$ gives

$$= y^{-\alpha} \frac{\eta\{z : z_0 > 1, \frac{z}{z_0} \in \cdot\}}{\eta\{x : x_0 > 1\}}$$

$$= P[Y_0 > y] P[\Theta \in \cdot].$$

□

2.3.4 Proof Method 4: Prove Convergence on a Sufficiently Rich Class of Sets

Theoretically, I often avoid this method[5] but sometimes it makes short work of verifications. The idea is to find a class $\mathcal{A}$ of subsets of $\mathbb{S}$ such that

$$\forall A \in \mathcal{A}: \quad \mu_n(A) \to \mu_0(A) \qquad (n \to \infty)$$

implies $\mu_n \to \mu_0$ in $\mathbb{M}(\mathbb{S}_F)$. We present some important examples.

2.3.4.1 Example: $\mathbb{S} \setminus F = \mathbb{R}_+^p \setminus \{\mathbf{0}_p\}$ for Some $p \geq 1$

Here the convergence determining class is

$$\mathcal{A} = \{[\mathbf{0}_p, x]^c, x \geq \mathbf{0}_p, x \neq \mathbf{0}_p\}.$$

We have the following classical criterion for heavy tails for distributions of vectors in the positive quadrant.

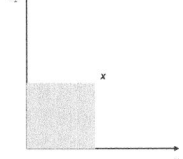

Proposition 2.6 *Suppose $X = (X_1, \ldots, X_p)$ is a random element of $\mathbb{R}_+^p$ that has joint distribution $G(\cdot) := P[X \in \cdot]$ with tail probabilities defined by*

$$\bar{G}(x) = P\{[X \leq x]^c\} = P[X \in [\mathbf{0}_p, x]^c],$$

---

[5] A traditional emphasis on sets stems from regular variation theory growing from a branch of function theory in analysis. For distributions in one-dimension, this necessitated focus on distribution tails which are probabilities of intervals. In higher dimensions, this led to emphasis on square regions or their complements.

2.3 How to Prove Regular Variation of Measures?

for sets $[\mathbf{0}_p, \mathbf{x}]^c \in \mathrm{BA}(\mathbb{R}_+^p \setminus \{\mathbf{0}_p\})$. Assume $\bar{G}(t\mathbf{1}) \in RV_{-\alpha}$, $\alpha > 0$ and that there exists a measure $\eta(\cdot) \in \mathbb{M}(\mathbb{R}_+^p \setminus \{\mathbf{0}_p\})$ such that as $t \to \infty$,

$$\frac{\bar{G}(t\mathbf{x})}{\bar{G}(t\mathbf{1})} \to \eta\big([\mathbf{0}_p, \mathbf{x}]^c\big) \tag{2.78}$$

for $\mathbf{x} \in \mathbb{R}_+^p \setminus \{\mathbf{0}_p\}$ such that the function $\eta([\mathbf{0}_p, \cdot]^c)$ is continuous at $\mathbf{1}$, $\mathbf{x}$ and $[\mathbf{0}_p, \mathbf{x}]^c \in \mathrm{BA}(\mathbb{R}_+^p \setminus \{\mathbf{0}_p\})$. Then,

$$P[\mathbf{X} \in \cdot] \in \mathrm{MRV}(\alpha, b(t), \eta, \mathbb{R}_+^p \setminus \{\mathbf{0}_p\})$$

and as $t \to \infty$,

$$\eta_t(\cdot) := tP\left[\frac{\mathbf{X}}{b(t)} \in \cdot\right] \to \eta(\cdot), \quad \text{in } \mathbb{M}(\mathbb{R}_+^p \setminus \{\mathbf{0}_p\}),$$

where a choice of $b(t)$ is

$$b(t) = \left(\frac{1}{1 - G(\cdot \mathbf{1})}\right)^{\leftarrow}(t).$$

Notes:

- If $p = 1$, this reassuringly says $G(\cdot)$ is a regularly varying measure on $\mathbb{R}_+ \setminus \{0\}$ iff $1 - G(t)$ is a regularly varying function at ∞
- If $p = 2$ and $x > 0$, the set $[\mathbf{0}_2, (x, 0)]^c \notin \mathrm{BA}(\mathbb{R}_+^2 \setminus \{\mathbf{0}_2\})$.
- If $p > 1$, $[\mathbf{X} \leq \mathbf{x}]^c$ is not an especially natural set, especially compared to the case $p = 1$. Its main claim to attractiveness is it allows you to avoid measures and deal with distribution functions. This is excellent if you are in love with square regions.
- The statement that the function $\eta([\mathbf{0}_p, \cdot]^c)$ is continuous at $\mathbf{x}$ is different than stating $\{\mathbf{x}\}$ is not an atom of $\eta(\cdot)$. Consider [14, page 260] the example that $\eta(\cdot)$ concentrates uniform measure on the line segment $\{(x, 1) : 0 \leq x \leq 1\}$ in $\mathbb{R}_+^2$. Then $\eta(\cdot)$ is atomless but the function $\eta([\mathbf{0}_2, \cdot]^c)$ is not continuous at $(1/2, 1)$.

Proof A simple way to proceed is to use Proposition 1.6, page 24, to reduce convergence in $\mathbb{M}(\mathbb{R}_+^p \setminus \{\mathbf{0}_p\})$ to weak convergence of finite measures where it is known that convergence of distribution functions at continuity points of the limit is sufficient [14, page 378].

For general p, one must strap in for a rough notational ride and we will restrict the proof to $p = 2$. Pick $\delta > 0$ such that $\eta(\cdot) \mapsto \eta(\cdot \cap [\mathbf{0}_2, \delta\mathbf{1}]^c)$ is continuous from $\mathbb{M}(\mathbb{R}_+^2 \setminus \{\mathbf{0}_2\}) \mapsto \mathbb{M}([\mathbf{0}_2, \delta\mathbf{1}]^c \setminus \emptyset)$. It suffices to show (2.78) implies

$$\eta_t(\cdot \cap [\mathbf{0}_2, \delta\mathbf{1}]^c) \to \eta(\cdot \cap [\mathbf{0}_2, \delta\mathbf{1}]^c) \tag{2.79}$$

in $\mathbb{M}\big([\mathbf{0}_2, \delta\mathbf{1}]^c \setminus \emptyset\big)$. Since

$$P_t(\cdot) := \frac{\eta_t(\cdot \cap [\mathbf{0}_2, \delta\mathbf{1}]^c)}{\eta_t([\mathbf{0}_2, \delta\mathbf{1}]^c)}, \quad P(\cdot) := \frac{\eta(\cdot \cap [\mathbf{0}_2, \delta\mathbf{1}]^c)}{\eta([\mathbf{0}_2, \delta\mathbf{1}]^c)}$$

are probability measures concentrating on $[\mathbf{0}_2, \delta\mathbf{1}]^c \subset \mathbb{R}_+^2$, it suffices for (2.79) that distribution functions converge at continuity points of the limit. Therefore, assume $\mathbf{x} \in [\mathbf{0}_2, \delta\mathbf{1}]^c$ and that the function $P([\mathbf{0}_2, \cdot])$ is continuous at $\mathbf{x}$ and we must show that $P_t([\mathbf{0}_2, \mathbf{x}]) \to P([\mathbf{0}_2, \mathbf{x}])$. Of course, $P([\mathbf{0}_2, \cdot])$ is continuous at $\mathbf{x}$ means $\eta([\mathbf{0}_2, \cdot])$ is continuous at $\mathbf{x}$ as well.

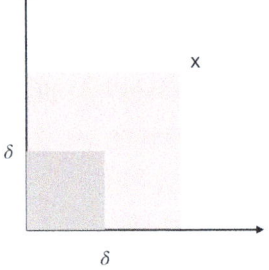

Displayed at right is a straightforward case. Here, $\mathbf{x} \in [\mathbf{0}_2, \delta\mathbf{1}]^c$ and $\mathbf{x}$ is northeast of $\delta\mathbf{1}$, that is, $\mathbf{x} > \delta\mathbf{1}$. Then referring to the light pink area at right, we have as $t \to \infty$,

$$P_t([\mathbf{0}_2, \mathbf{x}]) = \frac{\eta_t([\mathbf{0}_2, \mathbf{x}] \cap [\mathbf{0}_2, \delta\mathbf{1}]^c)}{\eta_t([\mathbf{0}_2, \delta\mathbf{1}]^c)} = \frac{\bar{G}(b(t)\delta\mathbf{1}) - \bar{G}(b(t)\mathbf{x})}{\bar{G}(b(t)\delta\mathbf{1})}$$

$$\to \frac{\eta([\mathbf{0}_2, \delta\mathbf{1}]^c) - \eta([\mathbf{0}_2, \mathbf{x}]^c)}{\eta([\mathbf{0}_2, \delta\mathbf{1}]^c)} = P([\mathbf{0}_2, \mathbf{x}]),$$

and this verifies (2.79).

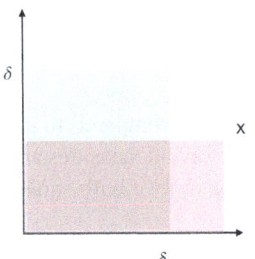

A less straightforward case displayed on the left is when $\mathbf{x} \in [\mathbf{0}_2, \delta\mathbf{1}]^c$ but it is not the case that $\mathbf{x} > \delta\mathbf{1}$. In the graphic, $x_1 > \delta$ but $x_2 < \delta$; the reverse could be true but this is not illustrated. (This case also illustrates why explaining $p > 2$ is tedious.) For the case illustrated,

$$[\mathbf{0}_2, \mathbf{x}] \cap [\mathbf{0}_2, \delta\mathbf{1}]^c = (\delta, x_1] \times [0, x_2)]$$
$$= [\mathbf{0}_2, \mathbf{x}] \setminus [\mathbf{0}_2, (\delta, x_2],$$

and therefore for this case,

$$P_t([\mathbf{0}_2, \mathbf{x}]) = \frac{\eta_t([\mathbf{0}_2, \mathbf{x}] \cap [\mathbf{0}_2, \delta\mathbf{1}]^c)}{\eta_t([\mathbf{0}_2, \delta\mathbf{1}]^c)} = \frac{G(b(t)\mathbf{x}) - G(b(t)(\delta, x_2))}{\bar{G}(b(t)\delta\mathbf{1})}$$

$$= \frac{\bar{G}(b(t)(\delta, x_2)) - \bar{G}(b(t)\mathbf{x})}{\bar{G}(b(t)\delta\mathbf{1})}$$

and provided convergence takes place as $t \to \infty$, the ratio must converge to

$$\to \frac{\eta([\mathbf{0}_2, (\delta, x_2)]^c) - \eta([\mathbf{0}_2, \mathbf{x}]^c)}{\eta([\mathbf{0}_2, \delta\mathbf{1}]^c)} = P([\mathbf{0}_2, \mathbf{x}]).$$

2.3 How to Prove Regular Variation of Measures?

Convergence holds for the ratios as guaranteed by (2.78) provided the function $\eta([\mathbf{0}_2, \cdot]^c)$ is continuous at x (assumed) and also at (δ, x_2) (hmm!!?). Insisting as we have that $\eta(\cdot) \mapsto \eta(\cdot \cap [\mathbf{0}_2, \delta\mathbf{1}]^c)$ is continuous means $\eta\big(\partial([\mathbf{0}_2, \delta\mathbf{1}]^c)\big) = 0$ and $\eta([\mathbf{0}_2, \cdot]^c)$ is continuous at $\delta\mathbf{1}$. Continuity at x and $\delta\mathbf{1}$ requires [14, page 260] that $\eta(\cdot)$ have no mass on the horizontal and vertical rays from the axes to $\delta\mathbf{1}$ and x and hence the same is true for the horizontal and vertical rays from axes to (δ, x_2). This makes $\eta([\mathbf{0}_2, \cdot]^c)$ continuous at (δ, x_2) and again, this verifies (2.79). □

Some good news: If in $\mathbb{R}_+^p$ we have $x > \mathbf{0}_p$, then $\eta([\mathbf{0}, \cdot]^c)$ is continuous at x and trouble can only occur at a boundary point where a component of x is 0. Reason: If $x > \mathbf{0}_p$, find $\epsilon > 0$ small enough that $(1 - \epsilon)x > \mathbf{0}_p$. Then, because $\eta(\cdot)$ is a regular variation limit measure and has the scaling property,

$$\eta([\mathbf{0}, (1-\epsilon)x]^c) - \eta([\mathbf{0}, x]^c) = \eta((1-\epsilon)[\mathbf{0}, x]^c) - \eta([\mathbf{0}, x]^c)$$
$$= (1-\epsilon)^{-\alpha}\eta([\mathbf{0}, x]^c) - \eta([\mathbf{0}, x]^c) \to 0 \text{ as } \epsilon \to 0,$$

giving continuity at x.

2.3.4.2 Example: $\mathbb{S} \setminus \mathbb{F} = \mathbb{R}_+^\infty \setminus \{\mathbf{0}_\infty\}$

Nothing too exciting here since an obvious strategy is to use Theorem 2.4, page 66 and Proposition 2.6, page 74 to get a finite dimensional criterion. So get out your Legos and click pieces together to get the following masterpiece that is Proposition 2.7. Reminder: If $x = (x_1, x_2, \dots) \in \mathbb{R}_+^\infty$, then $x_{|p} = (x_1, \dots, x_p) = \text{PROJ}_p(x)$.

Proposition 2.7 *Suppose* $\mathbb{S} \setminus \mathbb{F} = \mathbb{R}_+^\infty \setminus \{\mathbf{0}_\infty\}$. *Let* $X = (X_1, X_2, \dots) \in \mathbb{R}_+^\infty$ *be a random sequence in* $\mathbb{R}_+^\infty$ *and* $\eta(\cdot) \in \mathbb{M}(\mathbb{R}_+^\infty \setminus \{\mathbf{0}_\infty\})$. *The following are equivalent:*

1. *The distribution of the random sequence X is regularly varying with limit measure $\eta(\cdot)$,*

$$P[X \in \cdot] \in MRV(\alpha, b(t), \eta(\cdot), \mathbb{R}_+^\infty \setminus \{\mathbf{0}_\infty\})$$

and as $t \to \infty$, in $\mathbb{M}(\mathbb{R}_+^\infty \setminus \{\mathbf{0}_\infty\}$,

$$tP\left[\frac{X}{b(t)} \in \cdot\right] \to \eta(\cdot).$$

2. *For any $p \geq 1$, the distribution of $X_{|p} = (X_1, \dots, X_p)$ is regularly varying in $\mathbb{R}_+^p \setminus \{\mathbf{0}_p\}$,*

$$P[X_{|p} \in \cdot] \in MRV(\alpha, b(t), \eta \circ \text{PROJ}_p^{-1}(\cdot), \mathbb{R}_+^p \setminus \{\mathbf{0}_p\})$$

and as $t \to \infty$, in $\mathbb{M}(\mathbb{R}_+^p \setminus \{\mathbf{0}_p\})$,

$$tP[X_{|p}/b(t) \in \cdot\,] \to \eta \circ PROJ_p^{-1}(\cdot).$$

3. For any $p \geq 1$ and $\mathbf{x} \in \mathbb{R}_+^\infty$, the distribution function $G_p(\mathbf{x}_{|p}) = P[X_{|p} \leq \mathbf{x}_{|p}]$ satisfies,

$$\frac{\bar{G}(t\mathbf{x}_{|p})}{\bar{G}(t\mathbf{1}_p)} \to \eta([\mathbf{0}_p, \mathbf{x}_{|p}]^c)$$

provided the function $\eta \circ PROJ_p^{-1}([\mathbf{0}_p, \cdot\,]^c)$ is continuous at $\mathbf{1}_p$, $\mathbf{x}_{|p}$ and $[\mathbf{0}_p, \mathbf{x}_{|p}]^c \in \mathbb{BA}(\mathbb{R}_+^p \setminus \{\mathbf{0}_p\})$.

Example 2.6 Suppose $X = (X_1, X_2, \dots) \in \mathbb{R}_+^\infty$ consists of iid components, each of which has tail regularly varying so that as $t \to \infty$.

$$tP[X_i/b(t) > x] \to \nu_\alpha(x, \infty) = x^{-\alpha}, \quad x > 0.$$

Use Proposition 2.2, page 49, or Part 3 of Proposition 2.7 to get for any $p > 1$, that $X_{|p} = (X_1, \dots, X_p)$ has a regularly varying distribution and hence so does X. The limit measure $\eta(\cdot)$ of the regularly varying measure $P[X \in \cdot\,]$ concentrates on [axes] where only one component is different from 0 and

$$\eta(dx_1, dx_2, \dots) = \sum_{i=1}^\infty \prod_{j \neq i} \epsilon_{\{0\}}(dx_j) \nu_\alpha(dx_i).$$

More discussion of this example is in Sect. 3.2, page 99.

2.3.4.3 Example: $\mathbb{D}_\sqcap = (\mathbb{R}_+ \setminus \emptyset) \times (\mathbb{R}_+ \setminus \{0\}) = \mathbb{R}_+^2 \setminus \text{[x-axis]}$

Here,

$$\mathcal{A} = \{[0, x] \times (y, \infty) : x \geq 0, y > 0\}$$

is a rich enough and natural class. How to verify this? By now, we know the drill: Use restriction and Proposition 1.5, page 22 and Proposition 1.6, page 24 to convert convergence in $\mathbb{M}(\mathbb{D}_\sqcap)$ to weak convergence and then use equivalence of weak convergence to convergence of distribution functions at points of continuity. We get the following unsurprising result.

Proposition 2.8 Suppose $X = (X, Y)$ is a random element of $\mathbb{R}_+^2$ and scaling is defined by $(\lambda, (x, y)) \mapsto (x, \lambda y)$. Let $\eta(\cdot) \in \mathbb{M}(\mathbb{D}_\sqcap)$. The following are equivalent.

2.3 How to Prove Regular Variation of Measures?

1. *The measure $P[X \in \cdot]$ is regularly varying on $\mathbb{D}_\sqcap$,*

$$P[X \in \cdot] \in MRV(\alpha, b(t), \eta(\cdot), \mathbb{D}_\sqcap),$$

so that, as $t \to \infty$, in $\mathbb{M}(\mathbb{D}_\sqcap)$,

$$tP\left[\left(X, \frac{Y}{b(t)}\right) \in \cdot\right] \to \eta(\cdot). \tag{2.80}$$

2. *If $(x, y) \in \mathbb{D}_\sqcap$ is a continuity point of the function $\eta\big([0, \cdot] \times (\cdot, \infty)\big)$, then as $t \to \infty$,*

$$tP\left[X \le x, \frac{Y}{b(t)} > y\right] \to \eta\big([0, x] \times (y, \infty)\big). \tag{2.81}$$

Proof We concentrate on showing 2 implies 1. According to Proposition 1.6, to show $\mathbb{M}$-convergence it is enough to show convergence of restrictions where $\mathbb{M}$-convergence is reduced to weak convergence. Pick $\delta > 0$ such that for $B_\delta := \mathbb{R}_+ \times (\delta, \infty)$ we have $\eta(\partial B_\delta) = 0$ which precludes $\eta(\cdot)$ placing any mass on the line $\mathbb{R}_+ \times \{\delta\}$. Applying Proposition 1.6, it is enough to show restrictions to B_δ converge as $t \to \infty$,

$$tP\left[\left(X, \frac{Y}{b(t)}\right) \in (\cdot \cap B_\delta)\right] \to \eta(\cdot \cap B_\delta).$$

For this, it suffices for distribution functions to converge at continuity points of the limit. For $(x, y) \in \mathbb{R}_+ \times (\delta, \infty)$ such that $\eta([0, \cdot] \times (\cdot, \infty))$ is continuous at (x, y) we have from (2.81),

$$tP[X \le x, \delta < Y/b(t) \le y] = tP[X \le x, \delta < Y/b(t)] - tP[X \le x, y < Y/b(t)]$$
$$\to \eta([0, x] \times (\delta, \infty)) - \eta([0, x] \times (y, \infty))$$
$$= \eta([0, x] \times (\delta, y]).$$

Hence, as required, we have convergence of the distribution functions of finite measures at continuity points of the limit. □

Example 2.7 For instance using this class we can readily prove: Suppose X and Y are non-negative, independent random variables and $P[Y > x] \in RV_{-\alpha}$. Then

$$tP\left[\left(X, \frac{Y}{b(t)}\right) \in \cdot\right] \to P[X \in \cdot] \times \nu_\alpha(\cdot), \quad \text{in } \mathbb{M}(\mathbb{D}_\sqcap), \tag{2.82}$$

where $b(t) = (1/P[Y > \cdot])^{\leftarrow}(t)$. This was previously approached using the first binding lemma Proposition 2.1, page 47. Now we verify (2.82) from Proposition 2.8. In (2.82), insert $[0, x] \times (y, \infty)$ and

$$tP\left[\left(X, \frac{Y}{b(t)}\right) \in [0, x] \times (y, \infty)\right] = P[X \le x] \cdot tP\left[\frac{Y}{b(t)} > y\right]$$
$$\to P[X \le x]\nu_\alpha(y, \infty),$$

and, easy peasy, we are done.

The space $\mathbb{D}_\sqcap$ is the basis of the conditional extreme value model (CEV) [40, 88, 89, 166] since if we assume the joint convergence in (2.80) and a heavy tailed Y, we get,

$$P[X \in \cdot | Y > t] \to H(\cdot), \quad (t \to \infty),$$

for some distribution H.

2.3.4.4 Example: $\mathbb{S} \setminus \mathbb{F} = \mathbb{R}_+^2 \setminus$ [axes]

The class

$$\mathcal{A} = \{(x, \infty) = (x_1, \infty) \times (x_2, \infty) : x \in (0, \infty)^2\}$$

is rich enough to describe convergence on $\mathbb{R}_+^2 \setminus$ [axes].

Proposition 2.9 *Suppose $X = (X_1, X_2)$ is a random element of $\mathbb{R}_+^2$ and $\eta(\cdot)$ is a limit measure in $\mathbb{M}(\mathbb{R}_+^2 \setminus$ [axes]$)$. Use conventional scaling, $(\lambda, (x_1, x_2)) \mapsto (\lambda x_1, \lambda x_2)$. The following are equivalent.*

1. *The distribution $P[X \in \cdot]$ is regularly varying on $\mathbb{R}_+^2 \setminus$ [axes]; that is, for some $b(t) \in RV_{1/\alpha}, \alpha > 0$, we have*

$$P[X \in \cdot] \in MRV(\alpha, b(t), \eta(\cdot), \mathbb{R}_+^2 \setminus [\text{axes}]),$$

so that as $t \to \infty$, in $\mathbb{M}(\mathbb{R}_+^2 \setminus$ [axes]$)$,

$$tP[X/b(t) \in \cdot] \to \eta(\cdot).$$

2. *For any $x > \mathbf{0}_2$ such that the function $\eta\big((\cdot, \infty)\big)$ is continuous at x, we have as $t \to \infty$,*

$$tP[X/b(t) > x] \to \eta\big((x, \infty)\big).$$

2.3 How to Prove Regular Variation of Measures?

Proof Assume 2. Second verse same as the first[6] and again apply Proposition 1.6. Pick a $\delta > 0$ such that $\eta\big((\cdot, \infty)\big)$ is continuous at $\delta \mathbf{1}$. This choice means $\eta(\cdot)$ has no mass on the line segments $\{x \in \mathbb{R}_+^2 : x_1 \wedge x_2 = \delta\} = \{(x, \delta) : x \geq \delta\} \cup \{(\delta, x) : x \geq \delta\}$. We must show the finite restricted measures that concentrate on $[x > \delta\mathbf{1}]$ converge as $t \to \infty$, or in symbols, show

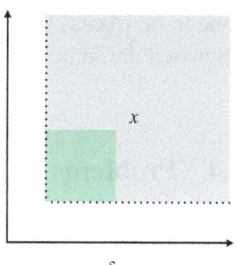

$$tP\big[X/b(t) \in (\cdot) \cap [x > \delta\mathbf{1}]\big] \to \eta\big((\cdot) \cap [x > \delta\mathbf{1}]\big)$$

and this is equivalent to showing convergence of the distribution functions at continuity points of the limit. Assume, as in the figure, that $x > \delta\mathbf{1}$ and that $\eta\big(\cdot, \infty\big)$ is continuous at x. Then the distribution functions of the restricted measures satisfy,

$$tP\Big[\frac{X}{b(t)} \leq x, \frac{X}{b(t)} > \delta\mathbf{1}\Big] = tP\Big[\frac{X}{b(t)} \in (\delta\mathbf{1}, x]\Big]$$

$$= tP\Big[\frac{X}{b(t)} > \delta\mathbf{1}\Big] - P\Big[\frac{X}{b(t)} > x\Big]$$

$$\to \eta\big((\delta\mathbf{1}, \infty)\big) - \eta\big((x, \infty)\big) = \eta\big((\delta\mathbf{1}, x]\big).$$

This suffices. □

Example 2.8 Suppose X_1, X_2 are iid, non-negative with $P[X_i > x] \in RV_{-\alpha}$ and scaling function $b(t)$. Then (cf. Proposition 2.3, page 52),

$$\mu_t(\cdot) := tP\Big[\Big(\frac{X_1}{b(\sqrt{t})}, \frac{X_2}{b(\sqrt{t})}\Big) \in \cdot\Big] \to \eta_2(\cdot) \quad \text{in } \mathbb{M}(\mathbb{R}_+^2 \setminus [\text{axes}]) \quad (2.83)$$

where

$$\eta_2\big((x_1, \infty) \times (x_2, \infty)\big) = (x_1 x_2)^{-\alpha}, \quad x_1 > 0, x_2 > 0.$$

Reason using Proposition 2.9: For $x_1 > 0, x_2 > 0$, write

$$tP\Big[\frac{X_1}{b(\sqrt{t})} > x_1, \frac{X_2}{b(\sqrt{t})} > x_2\Big] = \sqrt{t}P\Big[\frac{X_1}{b(\sqrt{t})} > x_1\Big]\sqrt{t}P\Big[\frac{X_2}{b(\sqrt{t})} > x_2\Big]$$

$$\to (x_1 x_2)^{-\alpha}.$$

According to 2 of Proposition 2.9, this is sufficient since every $x > \mathbf{0}_2$ satisfies $\eta_2\big((\cdot, \infty)\big)$ is continuous at x.

---

[6] Wikipedia: The 1965 British pop group, Herman's Hermits.

We emphasize that although $P[X_i > x] \in RV_{-\alpha}$, $i = 1, 2$, and $b(t) \in RV_{1/\alpha}$, the scale $b(\sqrt{t}) \in RV_{1/(2\alpha)}$ in (2.83) is required once we specify the forbidden zone to be [axes]. Previously the second binding lemma Proposition 2.2, page 49, discussed the same distribution with a different scaling and forbidden zone.

2.4 Problems

2.1 [122] Follow-up to Remark 2.3 (2), page 50: Suppose $X_i \in \mathbb{R}_+^{p_i}$, $i = 1, 2$ and (2.34) holds so that (X_1, X_2) has asymptotic independence. Show this asymptotic independence holds iff as $t \to \infty$,

$$tP\left[\frac{X_1}{b_1(t)} \in B_1, \frac{X_2}{b_2(t)} \in B_2\right] \to 0$$

for all $B_i \in \mathbb{BA}(\mathbb{R}_+^{p_i} \setminus \{\mathbf{0}_{p_i}\})$. (Again, the idea is that two large risks do not happen simultaneously.) Without loss, the sets B_i, $i = 1, 2$ may be assumed to be rectangles in $R_+^{p_i}$ bounded away from $\{\mathbf{0}_{p_i}\}$.

2.2 (More on Asymptotic Independence) Suppose without assuming component independence that $X = (X_1, \ldots, X_p) \in \mathbb{R}_+^p$ satisfies (2.42), page 54, which we call asymptotic independence of $X_1, \ldots, X_p$. Make iid replicates of the vector X and call the copies $\{X^{(1)}, X^{(2)}, \ldots\}$. The following are equivalent to (2.42):

1. For $x > \mathbf{0}_p$, component-wise maxima have a joint product limit distribution,[7]

$$P\left[\bigvee_{l=1}^{n} X_j^{(l)}/b(n) \leq x_j, \, j = 1, \ldots, p\right] \to \prod_{j=1}^{p} e^{-x_j^{-\alpha}}.$$

2. For any $1 \leq i \leq j \leq p$, pair-wise component-wise maxima have a joint product limit distribution,

$$P\left[\bigvee_{l=1}^{n} X_i^{(l)}/b(n) \leq x_i, \bigvee_{l=1}^{n} X_j^{(l)}/b(n) \leq x_j, \, j = 1, \ldots, p\right] \to e^{-x_i^{-\alpha}} e^{-x_j^{-\alpha}}.$$

3. Suppose for every $1 \leq i \leq p$ that marginal regular variation $tP[X_i > b(t)x] \to x^{-\alpha}$ holds and for any $1 \leq i \leq j \leq p$ and for $x_i > 0$, $x_j > 0$,

$$tP[X_i > b(t)x_i, X_j > b(t)x_j] \to 0.$$

4. For any $1 \leq i < j \leq p$,

---

[7] I believe this is the historical reason for the name *asymptotic independence*.

2.4 Problems

$$tP\left[\left(\frac{X_i}{b(t)}, \frac{X_j}{b(t)}\right) \in \cdot\right] \to \epsilon_{\{0\}} \times \nu_\alpha + \nu_\alpha \times \epsilon_{\{0\}},$$

in $\mathbb{M}(\mathbb{R}_+^2 \setminus \{\mathbf{0}_2\})$.

See [157, page 296].

2.3 Suppose $X_1 \perp\!\!\!\perp X_2$ and the distribution of each concentrates on $\mathbb{R}_+$ with $X_1 \stackrel{d}{=} X_2$ and $P[X_i > x] \in RV_{-\alpha}$ for $\alpha > 0$. Let $a > 0$, $b > 0$ and define $X = (aX_1, X_1, bX_2, X_2)$ as a random element of $\mathbb{R}_+^4$. Discuss the regular variation properties of X.

2.4 Suppose for $i = 1, 2$ that X_i is a random element of $\mathbb{S}_i$ that has a distribution in

$$MRV(\alpha_i, b_i(t), \eta_i(\cdot), \mathbb{S}_i \setminus F_i)$$

where F_i is a closed cone in $\mathbb{S}_i$. Further suppose $X_1 \perp\!\!\!\perp X_2$. Proposition 1.1, page 13, informs us that (X_1, X_2) has a distribution satisfying

$$tP\left[\left(\frac{X_1}{b_1(\sqrt{t})}, \frac{X_2}{b_2(\sqrt{t})}\right) \in \cdot\right] \to \eta_1 \times \eta_2,$$

in $\mathbb{M}\big((\mathbb{S}_1 \setminus F_1) \times (\mathbb{S}_2 \setminus F_2)\big) = \mathbb{M}\big(\mathbb{S}_1 \times \mathbb{S}_2 \setminus (F_1 \times \mathbb{S}_2 \cap \mathbb{S}_1 \times F_2)\big)$. What about convergence of

$$tP\left[\left(\frac{X_1}{b_1(t)}, \frac{X_2}{b_2(t)}\right) \in \cdot\right]$$

in $\mathbb{M}\big((\mathbb{S}_1 \times \mathbb{S}_2 \setminus F_1 \times F_2)\big)$? This is difficult to analyze in this generality.

1. Where would any limit measure concentrate?
2. If we have Borel sets $A_1 \subset F_1^\delta$ and $A_2 \subset F_2$, then ideally, if $\eta_1(\partial A_1) = 0$,

$$tP\left[\frac{X_1}{b_1(t)} \in A_1, \frac{X_2}{b_2(t)} \in A_2\right] \sim \eta_1(A_1) P\left[\frac{X_2}{b_2(t)} \in A_2\right].$$

If A_2 is a sub-cone of F_1 then the last factor is $P[X_1 \in A_2]$ but otherwise, there is no control on the behavior.
3. If for $i = 1, 2$, $\mathbb{S}_i = \mathbb{R}_+^2$ and $F_i = \{(x, x) : x \geq 0\}$, what can be said?

2.5 Suppose in the Second Binding Lemma page 49 we let $p_1 = p_2 = p > 1$, $\eta_1 = \eta_2 = \eta$ and $b_1(t) = b_2(t) = b(t)$. Verify the multivariate one jump principle:

$$tP\left[\left(\frac{X_1 + X_2}{b(t)}\right) \in \cdot\right] \to 2\eta(\cdot)$$

in $\mathbb{M}(\mathbb{R}_+^p \setminus \{\mathbf{0}_p\})$.

2.6 Suppose $\mathbb{S}_2 = [0, \infty]^p$ with metric

$$d_2(\boldsymbol{x}, \boldsymbol{y}) = \sum_{i=1}^{p} \left| \frac{x_i}{1+x_i} - \frac{y_i}{1+y_i} \right|,$$

where we interpret $\infty/(1+\infty) = 1$. Assume the forbidden zone is $F_2 = \{\infty\} = \{(\infty, \ldots, \infty)\}$. Suppose $U(\cdot)$ is a regularly varying measure on $\mathbb{S} \setminus F$ with index $\alpha > 0$, whose support is $[0, \infty)^p$, so that for scaling function $b(t) \in RV_\alpha$ and limit measure U_∞, as $t \to \infty$,

$$U_t := \frac{U(b(t) \cdot)}{t} \to U_\infty(\cdot)$$

in $\mathbb{M}(\mathbb{S}_2 \setminus F_2)$.

1. Take the one-dimensional marginal distributions $U^{(i)}(\cdot)$ defined for $x > 0$ by

$$U^{(i)}([0, x]) = U([0, \infty] \times \cdots \times [0, \infty] \times [0, x] \times [0, \infty] \times \cdots \times [0, \infty]),$$

$$i = 1, \ldots, p.$$

Why is $U^{(i)}[0, x] < \infty$? Why is it a regularly varying function?

2. If we had taken the forbidden zone to be

$$F_2' = \{(x_1, \ldots, x_p) \in \mathbb{S} : \vee_{i=1}^p x_i = \infty, \wedge_{i=1}^p x_i < \infty\},$$

would we be still able to conclude regular variation of the one-dimensional marginals of U?

3. Set $\mathbb{S}_1 = [0, 1]^p$ and $F_1 = \{\boldsymbol{x} \in [0, 1]^p : \wedge_{i=1}^p = 0\}$ and consider the TABOF space $\mathbb{S}_1 \setminus F_1$. Let $\text{Leb}^p = \text{Leb} \times \cdots \times \text{Leb}$ be p-dimensional Lebesgue measure and then

$$\text{Leb}^p \times U_t \to \text{Leb}^p \times U_\infty$$

in $\mathbb{M}\Big((\mathbb{S}_1 \setminus F_1) \times (\mathbb{S}_2 \setminus F_2)\Big)$. What is the product TABOF space and, in particular, what is the forbidden zone of the product?

4. Define $h : (S_1 \setminus F_1) \times (S_2 \setminus F_2) \mapsto (S_1 \setminus F_1) \times (S_2 \setminus F_2)$ by

$$h(\boldsymbol{s}, \boldsymbol{x}) = (\boldsymbol{s}, \boldsymbol{x}/\boldsymbol{s}).$$

Verify h respects the forbidden zone so

$$\text{Leb}^p \times U_t \circ h^{-1} \to \text{Leb}^p \times U_\infty \circ h^{-1},$$

2.4 Problems

in $\mathbb{M}((\mathbb{S}_1 \setminus F_1) \times (\mathbb{S}_2 \setminus F_2))$ and in particular both the left and right sides are measures in $(\mathbb{S}_1 \setminus F_1) \times (\mathbb{S}_2 \setminus F_2)$.

5. Evaluate the convergence in the previous part on the set $[\eta \mathbf{1}, \mathbf{1}] \times [\mathbf{0}, \mathbf{y}]$ ($0 < \eta < 1$, $\mathbf{y} > \mathbf{0}$), to get

$$\int_\eta^1 ds_1 \ldots \int_\eta^1 ds_p U_t([\mathbf{0}, s\mathbf{y}]) \to \int_\eta^1 ds_1 \ldots \int_\eta^1 ds_p U_\infty([\mathbf{0}, s\mathbf{y}])$$

or

$$\int_{b(t)\eta y_1}^{b(t)y_1} \ldots \int_{b(t)\eta y_p}^{b(t)y_p} \frac{U([\mathbf{0}, \mathbf{v}])}{b^p(t) y_1 \ldots y_p} dv_1 \ldots dv_p$$

$$\to \int_{\eta y_1}^{y_1} \ldots \int_{\eta y_p}^{y_p} \frac{U_\infty([\mathbf{0}, \mathbf{v}])}{y_1 \ldots y_p} dv_1 \ldots dv_p.$$

6. Show that η can be replaced by 0.

For variants, see [52–54, 151, 183].

2.7 (The CEV Model and Mass on the Lines Through ∞) [38] If we allow mass on lines through ∞, there can be different limits in (2.82), page 79, under different normalizations.

Suppose Y is Pareto(1) and B is a Bernoulli random variable with $P[B = 1] = P[B = 0] = 1/2$ and $Y \perp\!\!\!\perp B$. Define

$$\mathbf{Z} = (Z_1, Z_2) = B(Y, Y) + (1 - B)(\sqrt{Y}, Y).$$

Verify the following.

1. In $\mathbb{M}(\mathbb{D}_{11}) = \mathbb{M}\big([0, \infty) \times (0, \infty)\big)$

$$tP[\frac{\mathbf{Z}}{t} \in \cdot] \to \eta_1(\cdot),$$

where for $x \geq 0$, $y > 0$,

$$\eta_1([0, x] \times (y, \infty]) = \frac{1}{2}\left(\frac{1}{y} - \frac{1}{x}\right)_+ + \frac{1}{2y}.$$

This convergence also holds in $\mathbb{M}\big([0, \infty] \times (0, \infty]\big)$. Finally check

$$\eta_1\big((x, \infty] \times (y, \infty]\big) = \frac{1}{2(x \vee y)}, \quad x > 0, \ y > 0,$$

and

$$\eta_1(\{\infty\} \times (y, \infty]) = 0.$$

2. In $\mathbb{M}([0, \infty] \times (0, \infty]) = \mathbb{M}(([0, \infty] \setminus \emptyset) \times ([0, \infty] \setminus \{0\}))$, we also have

$$tP[(\frac{Z_1}{\sqrt{t}}, \frac{Z_2}{t}) \in \cdot] \to \eta_2(\cdot),$$

where

$$\eta_2([0, x] \times (y, \infty]) := \lim_{t \to \infty} tP\left[\left(\frac{Z_1}{\sqrt{t}}, \frac{Z_2}{t}\right) \in [0, x] \times (y, \infty]\right]$$

$$= \frac{1}{2}\left(\frac{1}{y} - \frac{1}{x^2}\right)_+$$

and

$$\eta_2(\{\infty\} \times (y, \infty]) = \frac{1}{2y}, \quad y > 0.$$

To verify this last statement check

$$\eta_2((x, \infty] \times (y, \infty]) = \lim_{t \to \infty} \frac{t}{2} P[\frac{Y}{\sqrt{t}} > x, \frac{Y}{t} > y]$$

$$+ \lim_{t \to \infty} \frac{t}{2} P[\frac{\sqrt{Y}}{\sqrt{t}} > x, \frac{Y}{t} > y]$$

$$= \frac{1}{2}\left(\frac{1}{y} + \frac{1}{x^2 \vee y}\right).$$

2.8 [38] Referring to Example 2.1, part 4, page 40, in $\mathbb{R}_+^2$, the forbidden zone was $\mathbb{F} = [\text{diag}] = \{(x, x) : x \in \mathbb{R}_+\}$ and $\aleph_F$ was two lines straddling and parallel to [diag]. Compactification of $\mathbb{R}_+^2$ to $[0, \infty]^2$ with the topology defined by the homeomorphic map $[0, \infty]^2 \mapsto [0, 1]^2$ using $(z_1, z_2) \mapsto (z_1/(1+z_1), z_2/(1+z_2))$ allows use of vague convergence (e.g. see [156]) but compactification can introduce unwanted geometric properties.

1. For example, if we extend [diag] to $\{(x, x) : 0 \le x \le \infty\}$, check that the two parallel lines to [diag] meet at ∞_2 and therefore the two parallel lines are not bounded away from [diag].
2. For $w > 0$, the set $A_{>w} := \{(z_1, z_2) \in [0, \infty]^2 : |z_1 - z_2| > w\}$ is not bounded away from [diag] since the set $\{(z_1, z_2) \in [0, \infty]^2 : z_1 - z_2 = w\}$ is parallel to [diag] and not bounded away.
3. Consider the line $\{(z_1, z_2) \in [0, \infty]^2 : z_2 = 1\}$. This line is bounded away from the horizontal axis.

2.4 Problems

2.9 Verify that the support of the limit measure of regular variation is a closed cone.

2.10 (Extension of the One-Jump Principle) Review Section 2.3.2.1, page 53, and Problem 1.14, page 27. Suppose $\{X_n, n \geq 1\}$ are iid non-negative random variables with common distribution F with $1 - F \in RV_{-\alpha}$. Let N be a bounded integer-valued random variable with range $\{1, \ldots, M\}$ for a fixed integer M. Assume N is independent of $\{X_n\}$. Show

$$P[\sum_{i=1}^{N} X_i > x] \sim E(N) P[X_1 > x], \quad x \to \infty.$$

2.11 Modify Proposition 2.4, page 70 as follows: For fixed n, show as $t \to \infty$,

$$tP[(X_1, \ldots, X_n, \mathbf{0}_\infty)/b(t) \in \cdot] \to \mu^{(0)}(\cdot),$$

in $\mathbb{M}(\mathbb{R}_+^\infty \setminus \{\mathbf{0}_\infty\})$.

2.12 Review Problem 1.17, page 28 and absorb the notation, including the definition of h and $\mathbb{M}_1$. Define a scaling function on $(0, \infty) \times \mathbb{M}_1(\mathbb{R}_+) \mapsto \mathbb{M}_1(\mathbb{R}_+)$ by $(b, \epsilon_x) \mapsto \epsilon_{x/b}$. Verify this definition of scaling satisfies (S1)–(S3), page 33. Suppose X is a random variable in $\mathbb{R}_+$ whose distribution H satisfies (2.1) so that $P[X \in \cdot]$ is a regularly varying measure in $\mathbb{M}(\mathbb{R}_+ \setminus \{0\})$. Then $P[\epsilon_X \in \cdot]$ is a regularly varying measure in $\mathbb{M}(\mathbb{M}_1(\mathbb{R}_+) \setminus \{\epsilon_0\})$.

Define the map $h_2 : \mathbb{R}_+^2 \mapsto \mathbb{M}_1(\mathbb{R}_+) \times \mathbb{M}_1(\mathbb{R}_+)$ by $h_2(x, y) = (\epsilon_x, \epsilon_y)$. Suppose X_1, X_2 are iid with common distribution H satisfying (2.1).

1. Show $P[(\epsilon_{X_1}, \epsilon_{X_2}) \in \cdot]$ is a regularly varying measure in $\mathbb{M}(\mathbb{M}_1(\mathbb{R}_+) \times \mathbb{M}_1(\mathbb{R}_+) \setminus \{(\epsilon_0, \epsilon_0)\})$.
2. What is the scaling function?
3. What is the limit measure? Express it in terms of h_2.
4. Let $\mathbb{M}_2(\mathbb{R}_+)$ is the set of point measures on $\mathbb{R}_+$ with two points:

$$\mathbb{M}_2(\mathbb{R}_+) = \{\epsilon_x + \epsilon_y : (x, y) \in \mathbb{R}_+^2\}.$$

Metrize $\mathbb{M}_2(\mathbb{R}_+)$ with metric d_2; for example, let

$$d_2(\epsilon_x + \epsilon_y, \epsilon_z + \epsilon_w) = |x \vee y - z \vee w| + |x \wedge y - z \wedge w|.$$

Define $h_0 : \mathbb{M}_1(\mathbb{R}_+) \times \mathbb{M}_1(\mathbb{R}_+) \mapsto \mathbb{M}_2(\mathbb{R}_+)$ by $h_0(\epsilon_x, \epsilon_y) = \epsilon_x + \epsilon_y$. Is h_0 continuous? Uniformly continuous? Does it respect forbidden zones as a map $\mathbb{M}_1(\mathbb{R}_+) \times \mathbb{M}_1(\mathbb{R}_+) \setminus \{(\epsilon_0, \epsilon_0)\} \mapsto \mathbb{M}_2(\mathbb{R}_+) \setminus \{\epsilon_0 + \epsilon_0\}$?

5. Is $P[\epsilon_{X_1} + \epsilon_{X_2} \in \cdot]$ a regularly varying measure in $\mathbb{M}(\mathbb{M}_2(\mathbb{R}_+) \setminus \{\epsilon_0 + \epsilon_0\})$? If so, what is the limit measure and where does it concentrate? How would you simulate from this limit measure?

2.13 (Variant 1 of Breiman's Theorem) [186] Suppose $\{X(t), t \geq 0\}$ is an $\mathbb{R}_+^p$-valued stochastic process for some $p \geq 1$. Let Y be a positive random variable with regularly varying distribution satisfying for some scaling function $b(t)$,

$$\lim_{t \to \infty} t P[Y/b(t) > x] = x^{-c} = \nu_c(x, \infty), \quad x > 0, \ c > 0.$$

Further suppose

1. For some finite, positive random vector X_∞ and positive normalizing function $a(t)$,

$$\lim_{t \to \infty} \frac{X(t)}{a(t)} = X_\infty \quad \text{(almost surely)};$$

2. The random variable Y and the process $X(\cdot)$ are independent.

Then,

(i) in $\mathbb{M}(\mathbb{R}_+^p \times (\mathbb{R}_+ \setminus \{0\}))$,

$$t P\left[\left(\frac{X(Y)}{a(Y)}, \frac{Y}{b(t)}\right) \in \cdot\right] \to P[X_\infty \in \cdot] \times \nu_c. \tag{2.84}$$

(ii) If additionally, for some $c' > c$ we have the condition

$$\kappa := \sup_{t \geq 0} \mathbb{E}\left(\frac{\|X(t)\|}{a(t)}\right)^{c'} < \infty, \tag{2.85}$$

for some L_p norm $\|\cdot\|$, then the product $X(Y)Y/a(Y)$ of components in (2.84) has a regularly varying distribution with scaling function $b(t)$ and in $\mathbb{M}(\mathbb{R}_+^p \setminus \{0\})$,

$$t P\left[\left(\frac{X(Y)}{a(Y)} \cdot \frac{Y}{b(t)}\right) \in \cdot\right] \to P[X_\infty \in \cdot] \times \nu_c \circ h^{-1}, \tag{2.86}$$

where $h(\boldsymbol{u}, v) = \boldsymbol{u} \cdot v$.

For the classical Breiman Theorem, $p = 1$ and $X(t) \equiv X_\infty$ and (2.85) is the expected moment condition.

2.14 (Variant 2 of Breiman's Theorem)

1. **(Multiply by Scalar)** [149] Suppose $P[Y \in \cdot] \in \text{MRV}(\alpha, b(t), \eta(\cdot), \mathbb{R}^p \setminus \{0\})$ for some $p \geq 1$ and limit measure $\eta(\cdot)$ with $\alpha > 0$. If $X \geq 0$ is independent of Y with $\mathbb{E}(X^\beta) < \infty$ and $\beta > \alpha$, then $P[XY \in \cdot] \in \text{MRV}(\alpha, b(t), \tilde{\eta}(\cdot), \mathbb{R}^p \setminus \{0\})$. If for some L_p norm $\|\cdot\|$, $b(t)$ satisfies

$$t P[\|Y\| > b(t)] \to 1 \quad (t \to \infty),$$

2.4 Problems

then

$$\tilde{\eta}(\cdot) = \frac{\mathbb{E}(X^\alpha)}{\eta\{y \in \mathbb{R}^p \setminus \{0\} : \|y\| > 1\}} \eta(\cdot).$$

2. **(Multiply by Matrix)** [3] Suppose the random vector $Y \in \mathbb{R}^p$ has a distribution satisfying $P[Y \in \cdot] \in \mathrm{MRV}(\alpha, b(t), \eta(\cdot), \mathbb{R}^p \setminus \{0\})$ for $\alpha > 0$ and scaling function $b(t)$. Let X be a random $q \times p$ matrix, independent of Y. If for $\beta > \alpha$ we have $0 < \mathbb{E}\|X\|^\beta < \infty$, then

$$P[XY \in \cdot] \in \mathrm{MRV}(\alpha, b(t), \tilde{\eta}(\cdot), \mathbb{R}^p \setminus \{0\}),$$

Here $\|X\| = \sup_{\|x\|=1} |Xx|$ is the operator norm of the matrix X and for a set $B \subset \mathbb{R}^p \setminus \{0\}$,

$$\tilde{\eta}(B) = \eta\{x \in \mathbb{R}^p \setminus \{0\} : Xx \in B\}.$$

In particular if X is $1 \times p$, that is a random vector, satisfying $\mathbb{E}|X_i|^\beta < \infty$ for $i = 1, \ldots, p$, then $P[(X_1 Y_1, \ldots, X_p Y_p) \in \cdot]$ is a regularly varying measure with index α.

2.15 (Variant 3 of Breiman's Theorem) [131, 156] The product of two random variables, which are not asymptotically independent, but whose tails satisfy multivariate regular variation, offers contrasting behavior to Breiman's Theorem. Consider (Y, Z) whose joint distribution satisfies non-standard regular variation.

Suppose Y is a non-negative random variable with a regularly varying distribution satisfying for some $\alpha_Y > 0$,

$$P[Y \in \cdot] \in \mathrm{MRV}(\alpha_Y, b_Y(t), \nu_{\alpha_Y}(\cdot), \mathbb{R}_+ \setminus \{0\}).$$

Assume Z be a $\mathbb{R}_+^p$-valued random vector, defined on the same probability space as Y, whose distribution is also regularly varying with index $\alpha_Z > 0$, that is,

$$P[Z \in \cdot] \in \mathrm{MRV}(\alpha_Z, b_Z(t), \nu_Z(\cdot), \mathbb{R}_+^p \setminus \{0_p\}).$$

Suppose further that the joint distribution of (Y, Z) satisfies as $t \to \infty$,

$$tP\left[\left(\frac{Y}{b_Y(t)}, \frac{Z}{b_Z(t)}\right) \in \cdot\right] \to \eta(\cdot) \neq 0,$$

in $\mathbb{M}(\mathbb{R}_+^{p+1} \setminus \{0_{p+1}\})$ and that and there exists $\delta > 0$ such that

$$\eta\{(y, z) \in \mathbb{R}^{p+1} \setminus \{0_{p+1}\} : |y| \wedge \|z\| > \delta\} > 0.$$

This means lack of asymptotic independence. Then the product $Y\mathbf{Z}$ has a regularly varying distribution with index $-\frac{\alpha_Y \alpha_{\mathbf{Z}}}{\alpha_Y + \alpha_{\mathbf{Z}}}$, scaling function $b_Y(\cdot)b_{\mathbf{Z}}(\cdot)$ and limit measure in $\mathbb{M}(\mathbb{R}_+^p \setminus \{\mathbf{0}_p\})$,

$$\eta\{(y, \mathbf{z}) : y\mathbf{z} \in \cdot\}.$$

Therefore,

$$P[Y\mathbf{Z} \in \cdot] \in \text{MRV}(\frac{\alpha_Y \alpha_{\mathbf{Z}}}{\alpha_Y + \alpha_{\mathbf{Z}}}, b_Y(t)b_{\mathbf{Z}}(t), \eta\{(y, \mathbf{z}) : y\mathbf{z} \in \cdot\}, \mathbb{R}_+^p \setminus \{\mathbf{0}_p\}).$$

2.16 (Variant 4 of Breiman's Theorem: iid Factors) Suppose Z_1, Z_2 are iid, non-negative random variables with regularly varying distributions with index $\alpha \in (0, 2)$ and scaling function $b(t)$.

1. **(Finite Mean)** [47] If $\mathbb{E}(Z_i)^\alpha < \infty$ then,

$$\lim_{x \to \infty} \frac{P[Z_1 Z_2 > x]}{P[Z_1 > x]} = 2\mathbb{E}(Z_1)^\alpha$$

iff

$$\lim_{s \to \infty} \limsup_{t \to \infty} \frac{P[Z_1 Z_2 > t, s < Z_1 \leq t/s]}{P[Z_1 > t]} = 0.$$

2. **(Infinite Mean)** [25, 48] If $\mathbb{E}(Z_1)^\alpha = \infty$, then $Z_1 Z_2$ has a regularly varying distribution tail

$$P[Z_1 Z_2 \in \cdot] \in \text{MRV}(\alpha, \tilde{b}(t), \nu_\alpha(\cdot), \mathbb{R}_+ \setminus \{0\})$$

but

$$\lim_{t \to \infty} \tilde{b}(t)/b(t) = \infty,$$

so

$$\lim_{t \to \infty} \frac{P[Z_1 Z_2 > t]}{P[Z_1 > t]} = \infty.$$

Other variants and discussion: See [22, 26, 56, 71, 78, 131].

2.17 (Partial Converse to the Breiman Theorem) [130] Suppose Y and X are two independent, non-negative random variables and Y has a Pareto distribution with parameter 1:

$$P[Y > x] = x^{-1}, \quad x \geq 1.$$

(a) We have
$$P[YX > x] \in RV_{-\alpha}, \quad \alpha < 1,$$
iff
$$P[X > x] \in RV_{-\alpha},$$
and then
$$\frac{P[YX > x]}{P[X > x]} \to \frac{1}{1-\alpha}.$$

(b) If $P[YX > x] \in RV_{-1}$ and YX has a heavier tail than Y meaning
$$\frac{P[YX > x]}{P[Y > x]} = \int_0^x P[X > y]dy \to \infty,$$
i.e., $\mathbb{E}[X] = \infty$, then
$$\int_0^x P[X > s]ds =: L(x) \uparrow \infty$$
is slowly varying.

2.18 (Another Partial Converse of Breiman's Theorem) [4] Let N be a standard normal random variable which is independent of the non-negative random variable X. If $(NX)_+$ has a regularly varying distribution tail with index $\alpha > 0$, then X does also.

2.19 [131] Suppose Y and X are strictly positive, finite random variables and suppose $tP[Y > b(t)] \to 1$ as $t \to \infty$. Further assume
$$\lim_{\epsilon \downarrow 0} \limsup_{t \to \infty} t\mathbb{E}\left[\left(\frac{XY}{b(t)}\right)^\delta 1_{[Y/b(t) \leq \epsilon]}\right] = 0, \quad \text{for some } \delta > 0.$$
Then in $\mathbb{M}(\mathbb{D}_\sqcap)$ we have
$$tP\left[\left(X, \frac{Y}{b(t)}\right) \in \cdot\right] \to G \times \nu_\alpha,$$
for some $\alpha > 0$ and some probability measure G on $\mathbb{R}_+$ with $G(\mathbb{R}_+) = 1$ and G has finite α-moment, iff in $\mathbb{M}(R_+^2 \setminus \{\mathbf{0}\})$,
$$tP\left[\left(\frac{XY}{b(t)}, \frac{Y}{b(t)}\right) \in \cdot\right] \to \tilde{\nu},$$
for some measure $\tilde{\nu}$ satisfying $\tilde{\nu}\{(y, x) : y > u\} > 0$ for some $u > 0$.

Chapter 3
Hidden Regular Variation

If the distribution of a random element possesses hidden regular variation (HRV), multiple regular variation properties coexist. A choice of forbidden zone and scaling function determine each property. An example of hidden regular variation is in Propositions 2.2, page 49 and Proposition 2.3, page 52. An indicator that HRV *might* exist is that regular variation of a measure $\mu(\cdot)$ on $\mathbb{M}(\mathbb{S} \setminus F_1)$ produces a limit measure $\eta_1(\cdot)$ that concentrates on a proper subset $F_2 \subsetneq \mathbb{S} \setminus F_1$. Since $\eta_1(\mathbb{S} \setminus F_2) = 0$, there is a possibility that with a smaller

Hidden under the mat?

normalization, there could be another regular variation property for $\mu(\cdot)$ on $\mathbb{S} \setminus F_2$. For $p_1 = p_2 = 2$ in Propositions 2.2 and 2.3, $\mathbb{S} = \mathbb{R}_+^2$, $F_1 = \{\mathbf{0}\}$ and $F_2 = [\text{axes}]$, the limit measure $\eta_1(\cdot)$ concentrates on [axes] and regular variation exists on $\mathbb{R}_+^2 \setminus [\text{axes}]$ with $\eta_2(\cdot)$ a product measure.

The word *hidden* refers to the fact that convergence on the bigger cone $\mathbb{S} \setminus F_1$ with fast growing scaling may mask a different regular variation property on the smaller cone $\mathbb{S} \setminus F_2$ with slow growing scaling. There are also distributions where there are an infinite number of different regular variation regimes; such distributions possess *steroidal* regular variation. Examples of steroidal regular variation in $\mathbb{S} = \mathbb{R}_+^\infty$ are in Sect. 3.2 and in $\mathbb{S} = \mathbb{R}_+^2$, see Problem 3.7. Chapter 4 considers a function space example of steroidal regular variation where the random element with steroidally regularly varying distribution is a Lévy process with Lévy measure having a regularly varying tail. The random element is in the space $\mathbb{S} = \mathbb{D}[0, 1]$ consisting of finite real functions on [0,1] which are right continuous on [0, 1) and having finite left hand limits on (0, 1].

3.1 Hidden Regular Variation

Suppose as usual that $\mathbb{S}$ is a complete separable metric space and that $F_1 \in \mathcal{F}(\mathbb{S})$ is a closed cone. Let X be a random element of $\mathbb{S}$ with a regularly varying distribution with scaling function $b_1(t)$ and limit measure $\eta_1(\cdot)$; that is, $P[X \in \cdot\,] \in \mathrm{MRV}(\alpha_1, b_1(t), \eta_1(\cdot), \mathbb{S} \setminus F_1)$ or

$$tP\left[\frac{X}{b_1(t)} \in \cdot\right] \to \eta_1(\cdot), \quad \text{in } \mathbb{M}(\mathbb{S} \setminus F_1).$$

where $b_1(t) \in RV_{1/\alpha_1}$. Assume also that there is a second closed cone $F_2 \in \mathcal{F}(\mathbb{S})$ and a second scaling function $b_2(t) \in RV_{1/\alpha_2}$ with

$$b_1(t)/b_2(t) \to \infty, \quad (t \to \infty) \tag{3.1}$$

so that $b_2(t)$ grows more slowly than $b_1(t)$. (Think of b_2 being kinder and gentler than the meat cleaver $b_1(t)$.) Since $F_1 \cup F_2$ is also a closed cone in $\mathcal{F}(\mathbb{S})$ we may think of this as a forbidden zone. The random element X possesses *hidden regular variation* on $\mathbb{S} \setminus (F_1 \cup F_2)$ if for some limit measure $\eta_2(\cdot)$

$$tP\left[\frac{X}{b_2(t)} \in \cdot\right] \to \eta_2(\cdot), \quad \text{in } \mathbb{M}\bigl(\mathbb{S} \setminus (F_1 \cup F_2)\bigr). \tag{3.2}$$

This means

$$P[X \in \cdot\,] \in \mathrm{MRV}(\alpha_1, b_1, \eta_1, \mathbb{S} \setminus F_1) \cap \mathrm{MRV}\bigl(\alpha_2, b_2, \eta_2, \mathbb{S} \setminus (F_1 \cup F_2)\bigr).$$

What makes the second regular variation (3.2) with bigger forbidden zone *hidden* is the condition (3.1).

If $\mathbb{S} = \mathbb{R}_+^2$, $F_1 = \{\mathbf{0}_2\}$ and $F_1 \cup F_2 = [\text{axes}]$, (3.1) implies asymptotic independence since the probability both components are large relative to the scaling $b_1(t)$ is for $x > \mathbf{0}$,

$$tP\left[\frac{X}{b_1(t)} > x\right] = tP\left[\frac{X}{b_2(t)} > \frac{b_1(t)}{b_2(t)}x\right] \to 0,$$

which results from $b_1(t)/b_2(t) \to \infty$ and $tP[X/b_2(t) \in \cdot\,] \to \eta_2(\cdot)$.

We have seen examples of this on page 49 in Lemma 2.2, (2.34) and Proposition 2.3, page 52. If we specialize (2.74), page 71 to $p = 2$ we get in $\mathbb{M}(\mathbb{R}_+^2 \setminus \{\mathbf{0}_2\})$,

$$tP\left[\frac{1}{b(t)}(X_1, X_2) \in (\cdot)\right] \to \eta_1(\cdot) = \epsilon_0 \times \nu_\alpha(\cdot) + \nu_\alpha \times \epsilon_0(\cdot) \tag{3.3}$$

and the limit measure concentrates on [axes]. From Proposition 2.8, page 81 or (2.83) we have a second regular variation property with forbidden zone [axes],

3.1 Hidden Regular Variation

namely, in $\mathbb{M}(\mathbb{R}_+^2 \setminus [\text{axes}])$,

$$tP\left[\frac{1}{b(\sqrt{t})}(X_1, X_2) \in \cdot\right] \to \eta_2(\cdot), \tag{3.4}$$

where

$$\eta_2((x_1, \infty) \times (x_2, \infty)) = (x_1 x_2)^{-\alpha}, \quad x_1 > 0, x_2 > 0.$$

So for this example $P[(X_1, X_2) \in \cdot] \in$

$$\text{MRV}(\alpha, b(t), \eta_1(\cdot), \mathbb{R}_+^2 \setminus \{\mathbf{0}_2\}) \cap \text{MRV}(2\alpha, b(\sqrt{t}), \eta_2(\cdot), \mathbb{R}_+^2 \setminus [\text{axes}]).$$

Using regular variation to estimate risk probabilities means the choice of scale and forbidden zone is critical. Using the asymptotic result (3.3) at scale $b(t)$ with limit measure η_1 to estimate the probability of a region $(x_1, \infty) \times (x_2, \infty)$ requiring both components be simultaneously large gives the estimate 0 since the region is not in the support of the limit measure. Using the asymptotic result (3.4) at scale $b(\sqrt{t})$, the estimate is non-zero.

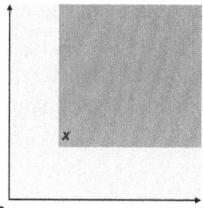

3.1.1 Simple Diagnostics for HRV in $\mathbb{R}_+^p$, $p > 1$

What are simple tools for diagnosing existence of HRV? One method takes a p-dimensional regular variation and finds a one-dimensional reduction. Data diagnostics can then use one-dimensional techniques such as *Hill plots* [156, p. 85, 365] or *QQ-plots* [156, p. 366] to confirm data is consistent with the regular variation assumption in $\mathbb{R}_+^p$.

3.1.1.1 Regular Variation in $\mathbb{R}_+^p \setminus \{\mathbf{0}_p\}$; Reduction to One-Dimension

We leverage the example in Sect. 2.3.4.1 and Proposition 2.6, page 74. Suppose $X = (X_1, \ldots, X_p) \in \mathbb{R}_+^p$ has a regularly varying distribution $P[X \in \cdot] \in \text{MRV}(\alpha_1, b_1(t), \eta_1(\cdot), \mathbb{R}_+^p \setminus \{\mathbf{0}_p\})$ so that

$$tP\left[\frac{X}{b_1(t)} \in \cdot\right] \to \eta_1(\cdot), \quad \text{in } \mathbb{M}(\mathbb{R}_+^p \setminus \{\mathbf{0}_p\}). \tag{3.5}$$

Pick $s \in \mathbb{R}_+^p \setminus \{\mathbf{0}_p\}$ and suppose there exists $c_1(s) > 0$ such that for $x > 0$, as $t \to \infty$,

$$tP\left[\frac{\bigvee_{i=1}^{p} s_i X_i}{b_1(t)} > x\right] \to c_1(s) x^{-\alpha_1}. \tag{3.6}$$

If precisely phrased, the statements (3.5) and (3.6) are equivalent [156, p. 326] and [50] and provide a diagnostic: To check multivariate regular variation of $P[X \in \cdot]$ on $\mathbb{R}_+^p \setminus \{\mathbf{0}_p\}$, we check regular variation of the distribution tail of the one-dimensional max-linear combinations. We cannot check (3.6) for all s but checking for a couple of values, or even just $s = 1$ provides evidence that is consistent with the regular variation assumption.

Consider the two one-dimensional diagnostic plots for regular variation of componentwise maxima of 5000 pairs of standard Pareto random variables. On the left is a Hill plot [156, p. 85, 364], and on the right is a QQ-plot [156, p. 366] of log-data against exponential quantiles. Both plots indicate an estimate of $\alpha_1 \approx 1$.

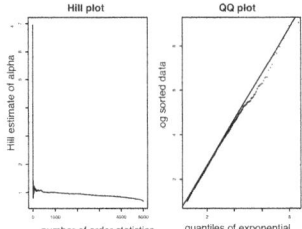

3.1.1.2 Regular Variation in $\mathbb{R}_+^p \setminus$ [axes]

In $\mathbb{R}_+^p$ there are many regular variation properties possible. A simple case is where $X = (X_1, \ldots, X_p)$ has the property that for $i \neq j$, X_i and X_j are asymptotically independent. This means the limit measure for regular variation on $\mathbb{R}_+^p \setminus \{\mathbf{0}_p\}$ is

$$\eta_1(dx_1, \ldots, dx_p) = \sum_{i=1}^{p} \nu_{\alpha_1}(dx_i) \epsilon_{e_i}(x).$$

which concentrates mass on one-dimensional axes and

$$[\text{axes}] = \bigcup_{i=1}^{p} L_i = \bigcup_{i=1}^{p} \{t e_i, t \geq 0\}.$$

A diagnostic for $P[X \in \cdot] \in \text{MRV}(\alpha_2, b_2(t), \eta_2, \mathbb{R}_+^p \setminus [\text{axes}])$ is the following [156, p. 326]: There must exist a function $c_2(a)$, $a \in (0, \infty)^p$ and

$$tP\left[\frac{\bigwedge_{i=1}^{p} a_i X_i}{b_2(t)} > x\right] = c_2(a) x^{-\alpha_2}, \quad a > \mathbf{0}_p, t \to \infty. \tag{3.7}$$

Again, it is not possible to check all $a > \mathbf{0}_p$ but checking a few values or even just $a = 1$ reveals if there is evidence consistent with regular variation on $\mathbb{R}_+^p \setminus [\text{axes}]$.

So if we seek evidence consistent with the presence of hidden regular variation and want evidence that the distribution has membership in

3.1 Hidden Regular Variation

$$\text{MRV}(\alpha_1, b_1(t), \eta_1(\cdot), \mathbb{R}_+^p \setminus \{\mathbf{0}_p\}) \cap \text{MRV}(\alpha_2, b_2(t), \eta_2(\cdot), \mathbb{R}_+^p \setminus \{[\text{axes}]\}).$$

we may accumulate the evidence using the following steps: Apply diagnostics that lead to the belief that,

1. $P[\vee_{i=1}^p X_i > x] \in RV_{-\alpha_1}$ and
2. $P[\wedge_{i=1}^p X_i > x] \in RV_{-\alpha_2}$ and
3. $\alpha_1 < \alpha_2$.

Sometimes data has been standardized in some way so that $\alpha_1 = 1$ and then one must verify $\alpha_2 > 1$.

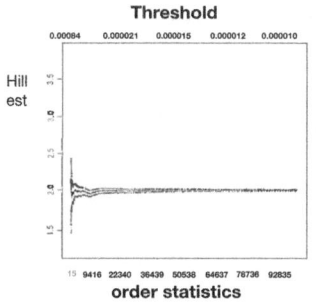

Consider 5000 replicates of $\boldsymbol{X} = (X_1, X_2)$, $X_1 \perp\!\!\!\perp X_2$ and

$$P[X_i > x] = x^{-1}, \quad x > 1, i = 1, 2.$$

Then $\alpha_1 = 1$ and $\alpha_2 = 2$. On the left is the Hill plot with 95% confidence interval for the minima of the components which indicates the (correct) value $\alpha_2 = 2$. Conclusion: Sometimes detection of HRV is possible.

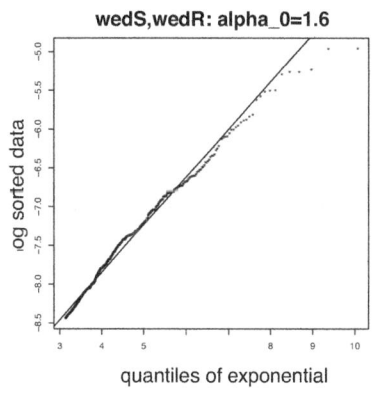

For those skeptical of simulation, here is a real data set UNC Wed (S, R) [91]. This is Internet response data where S is size of response and R is average transmission rate, meaning

$$\frac{\text{size of response}}{\text{download time of response}}.$$

This is old data collected by the CS group at University of North Carolina. Each component is heavy tailed but the indices are not the same as required by the standard model.

So the data was standardized by the ranks method [156, page 310], automatically giving a value $\alpha_1 = 1$ for each component. The QQ plot for the minimum component of rank transformed data using 1000 upper order statistics was plotted and the slope estimation method [156, page 97] yields an estimate of the second index $\hat{\alpha}_2 = 1.6 > 1$. So for $d = 2$, there is evidence supporting the presence of a second regular variation for the transformed pair.

3.1.1.3 Additional HRV Examples

Here are more examples that illustrate subtleties and provide context.

1. Regularly varying but no HRV: Let

$$X = \left(\frac{1}{U}, \frac{1}{(1-U)}\right),$$

where $U \sim U(0, 1)$ is uniformly distributed on $(0, 1)$. The marginal distributions of X are Pareto and we check

$$P[X \in \cdot] \in MRV(\alpha = 1, b(t) = t, \eta(\cdot), \mathbb{R}_+^2 \setminus \{\mathbf{0}_2\})$$

where $\eta([0, \mathbf{x}]^c) = x_1^{-1} + x_2^{-1}$, $\mathbf{x} \neq \mathbf{0}_2$.
We verify multivariate regular variation using Proposition 2.6, page 74, and write

$$tP\left[\frac{X}{t} \in [\mathbf{0}_2, (x_1, x_2)]^c\right] = t\left(1 - P\left[\frac{1}{U} \leq tx_1, \frac{1}{1-U} \leq tx_2\right]\right)$$

$$= t\left(1 - P\left[\frac{1}{tx_1} \leq U \leq 1 - \frac{1}{tx_2}\right]\right).$$

If $1 - \frac{1}{tx_2} > \frac{1}{tx_1}$ (which happens for large enough t), the above is

$$= t\left(1 - \left(1 - \frac{1}{tx_2} - \frac{1}{tx_1}\right)\right)$$

$$= x_1^{-1} + x_2^{-1} =: \eta([\mathbf{0}_2, \mathbf{x}]^c).$$

Note $P[X \in \cdot]$ has asymptotic independence since away from [axes] we have for $\mathbf{x} > \mathbf{0}_2$,

$$tP[X > t\mathbf{x}] = tP\left[\frac{1}{U} > tx_1, \frac{1}{1-U} > tx_2\right]$$

$$= tP\left[1 - \frac{1}{tx_2} \leq U \leq \frac{1}{tx_1}\right] = 0$$

because eventually $1 - \frac{1}{tx_2} > \frac{1}{tx_1}$.
Thus $P[X \in \cdot]$ is regularly varying and has asymptotic independence. We claim this disribution does NOT have HRV. An easy way to see this is to use (3.7) and note

$$X_1 \wedge X_2 = \frac{1}{U} \wedge \frac{1}{1-U} \leq 2$$

and no bounded random variable can have a regularly varying tail.
2. Reminder: Suppose $X = (X_1, X_2)$ where X_1, X_2 are iid with (for concreteness) $P[X_i > x] = x^{-1}$, $i = 1, 2$, $x > 1$. Then X has a distribution which is

   1. regularly varying with asymptotic independence on $\mathbb{R}_+^2 \setminus \{\mathbf{0}_2\}$, index $\alpha_1 = 1$, scaling function $b_1(t) = t$ and limit measure $\eta_1([0, \mathbf{x}]^c) = x_1^{-1} + x_2^{-1}$ for $\mathbf{x} \neq \mathbf{0}_2$.
   2. regularly varying on $\mathbb{R}_+^2 \setminus [\text{axes}]$, index $\alpha_2 = 2$, scaling function $b_2(t) = \sqrt{t}$ with limit measure $\eta_2((\mathbf{x}, \infty)) = (x_1 x_2)^{-1}$, for $\mathbf{x} > \mathbf{0}_2$.

3. HRV absent: Suppose Y_1, Y_2 are iid with $P[Y_i > x] = x^{-1}$, $i = 1, 2$, $x > 1$. Let $B \perp\!\!\!\perp (Y_1, Y_2)$ be a Bernoulli switching variable with $P[B = 1] = 1/2 = P[B = 0]$. Define

$$X = B(Y_1, 0) + (1 - B)(0, Y_2). \tag{3.8}$$

The distribution of X satisfies,

1. $P[X \in \cdot]$ concentrates on [axes].
2. $P[X \in \cdot]$ is regularly varying on $\mathbb{R}_+^2 \setminus \{\mathbf{0}_2\}$ with asymptotic independence.
3. $P[X \in \cdot]$ does NOT have HRV on $\mathbb{R}_+^2 \setminus [\text{axes}]$ since using the criterion (3.7) we see that

$$X_1 \wedge X_2 = B(Y_1 \wedge 0) + (1 - B)(0 \wedge Y_2) = 0.$$

Other examples are in the exercises in Sect. 3.3.

3.2 Steroidal Regular Variation: Roid Rage Strikes

Distributions that exhibit infinitely many coexisting regular variation properties possess *steroidal regular variation*. Each different regular variation property depends on a choice of scaling and forbidden zone. To see how an infinite number of regular variation properties can coexist, we start in $\mathbb{R}_+^\infty$ in Sect. 3.2.1 with the iid model $X = (X_1, X_2, \dots)$.

Roid rage!

In Sect. 3.2.2 we stay in $\mathbb{R}_+^\infty$ and illustrate the steroidal concept with Poisson points. Other examples exist in $\mathbb{R}_+^2$ (see [38] and Problem 3.7, page 120) and $\mathbb{D}[0, 1]$ (functions which are right continuous on $[0, 1)$ with finite left limits on $(0, 1]$). See [125] and Sect. 4.2.

3.2.1 Steroidal Regular Variation in $\mathbb{R}_+^\infty$ for the iid Model

This section picks up threads from Sects. 2.3.3, page 65 and 2.3.3.1 page 70 which discuss reducing regular variation in $\mathbb{R}_+^\infty$ to testing in $\mathbb{R}_+^p$. See also Proposition 2.6, page 74. We review notation introduced in Sect. 2.3.3, Theorem 2.4, page 66.

Assume $X = (X_1, X_2, \dots)$ is a random element of $\mathbb{R}_+^\infty$ consisting of iid random variables with $P[X_i > x] \in \mathrm{RV}_{-\alpha}$ and scaling function $b(t)$,

$$tP[X_1 > b(t)x] \to x^{-\alpha}, \quad x > 0.$$

Recall the following subsets of $\mathbb{R}_+^\infty$,

$$F_{\leq j} = \{x \in \mathbb{R}_+^\infty : \sum_{i=1}^\infty \epsilon_{x_i}(0, \infty) \leq j\},$$

=sequences having at most j components positive,

$$F_{=j} = \{x \in \mathbb{R}_+^\infty : \sum_{i=1}^\infty \epsilon_{x_i}(0, \infty) = j\},$$

=sequences having exactly j components positive.

Properties of $F_{\leq j}$:

(i) $F_{\leq j}$ is closed in $\mathbb{R}_+^\infty$.
(ii) If $x \in F_{\leq j}$, then also $(x_{|p}, \mathbf{0}_\infty) \in F_{\leq j}$ so the criteria Theorem 2.4 for reducing convergence in $\mathbb{R}_+^\infty \setminus F_{\leq j}$ to finite dimensional convergence is available. Review the discussion around (2.62).
(iii) $F_{=0} = F_{\leq 0} = \{\mathbf{0}_\infty\}$.

We now find infinitely many distinct regular variation properties for $P[X \in \cdot]$.

Step 1 Remove $\{\mathbf{0}_\infty\}$ from $\mathbb{R}_+^\infty$. This is the discussion in Sect. 2.3.3.1. Then in $\mathbb{M}(\mathbb{R}_+^\infty \setminus \{\mathbf{0}_\infty\})$, we have

$$\mu_t^{(0)}((dx_1, dx_2, \dots)) := tP[X/b(t) \in (dx_1, dx_2, \dots)]$$
$$\to \sum_{l=1}^\infty \prod_{i \neq l} \epsilon_0(dx_i) \nu_\alpha(dx_l) =: \mu^{(0)}(dx_1, dx_2, \dots), \quad (3.9)$$

and the limit measure concentrates on $F_{=1}$, the set of sequences with exactly one component positive. Also $\{\mathbf{0}_\infty\} \cup F_{=1} =: F_{\leq 1}$, a closed set.

Step 2 Remove $\{\mathbf{0}_\infty\} \cup F_{=1} = F_{=0} \cup F_{=1} = F_{\leq 1}$ from $\mathbb{R}_+^\infty$. Previous discussions, for example, Example 2.3 and Reminder 3.1.1.3, lead us to believe that we should scale with $b(\sqrt{t})$ and that in $\mathbb{M}(\mathbb{R}_+^\infty \setminus F_{\leq 1})$, as $t \to \infty$,

3.2 Steroidal Regular Variation: Roid Rage Strikes

$$\mu_t^{(1)}(dx_1, dx_2, \dots) = tP[X/b(\sqrt{t}) \in (dx_1, dx_2, \dots)] \to \mu^{(1)}((dx_1, dx_2, \dots))$$

$$:= \sum_{l<k} \left(\prod_{j \notin \{l,k\}} \epsilon_0(dx_j) \right) v_\alpha(dx_l) v_\alpha(dx_k), \tag{3.10}$$

and $\mu^{(1)}(\cdot)$ concentrates on $F_{=2}$. So the limit measure $\mu^{(1)}(\cdot)$ concentrates on two-dimensional faces of $\mathbb{R}_+^\infty \setminus F_{\leq 1}$ and apportions mass $v_\alpha \times v_\alpha$ on each two-dimensional face.

The proof uses Theorem 2.4 to reduce infinite dimensional convergence to finite dimensional convergence in $\mathbb{M}(\mathbb{R}_+^p \setminus \text{PROJ}_p(F_{\leq 1}))$ where $\text{PROJ}_p(F_{\leq 1})$ are vectors in $\mathbb{R}_+^p$ with at most one component positive. Staying away from the forbidden zone $\text{PROJ}_p(F_{\leq 1})$ means two or more components are positive so extrapolating from Sect. 2.3.4.4 and Proposition 2.9, page 80, we must verify for $l \neq k$ and $x > 0$, $y > 0$ that

$$tP[X_l > b(\sqrt{t})x, X_k > b(\sqrt{t})y] \to v_\alpha(x, \infty)v_\alpha(y, \infty)$$

and concentration of the limit measure $\mu^{(1)}$ on $F_{=2}$ occurs because for 3 events, l, k, m distinct and x, y, z all positive,

$$tP[X_l > b(\sqrt{t})x, X_k > b(\sqrt{t})y, X_k > b(\sqrt{t})y, X_m > b(\sqrt{t})z] \to 0.$$

Step 3 Remove from $\mathbb{R}_+^\infty \setminus F_{\leq 1}$ the additional set $F_{=2}$ where $\mu^{(1)}$ concentrates in the previous step. Consider convergence on $\mathbb{R}_+^\infty \setminus F_{\leq 2}$.

$\vdots$

Step General In general, we find that in $\mathbb{M}(\mathbb{R}_+^\infty \setminus F_{\leq j})$ as $t \to \infty$,

$$\mu_t^{(j)}(dx_1, dx_2, \dots) = tP[X/b(t^{1/(j+1)}) \in (dx_1, dx_2, \dots)]$$

$$\to \mu^{(j)}((dx_1, dx_2, \dots))$$

$$:= \sum_{i_1 < i_2 < \dots < i_{j+1}} \left(\prod_{j \notin \{i_1, \dots, i_{j+1}\}} \epsilon_0(dx_j) \right) v_\alpha(dx_{i_1}) v_\alpha(dx_{i_2}) \dots v_\alpha(dx_{i_{j+1}}),$$

(3.11)

which concentrates on $F_{=(j+1)}$. This is an elaboration of results in [130, 135, 136]. The result in $\mathbb{R}_+^\infty$ can be proven by reducing to $\mathbb{R}_+^p$ by means of Theorem 2.4 noting that $F_{\leq j}$ satisfies Condition 2.

Table 3.1 An infinite number of coexisting regular-variation properties

j	Remove	Scaling	$\mu^{(j)}$	Support
1	$\{\mathbf{0}_\infty\}$	$b(t)$	$\sum_{l=1}^{\infty} \nu_\alpha(dx_l)\left[\prod_{i\neq l} \epsilon_0(dx_i)\right]$	$F_{=1}$
2	$F_{\leq 1}$	$b(\sqrt{t})$	$\sum_{l,m} \nu_\alpha(dx_l)\nu_\alpha(dx_m)\left[\prod_{i\notin\{l,m\}} \epsilon_0(dx_i)\right]$	2-dim faces $F_{=2}$
$\vdots$	$\vdots$	$\vdots$	$\vdots$	$\vdots$
m	$F_{\leq(m-1)}$	$b(t^{\frac{1}{m}})$	$\sum_{(l_1,\ldots,l_m)}\prod_{p=1}^{m}\nu_\alpha(dx_{l_p})\left[\prod_{i\notin\{l_1,\ldots,l_m\}}\epsilon_0(dx_i)\right]$	$F_{=m}$

Table 3.1 is the Dr. Spock[1] summary in tabular form.

3.2.1.1 What Happens when We Apply CUMSUM?

Remember the map CUMSUM : $\mathbb{R}_+^\infty \mapsto \mathbb{R}_+^\infty$ defined by

$$\text{CUMSUM}(\mathbf{x}) = (x_1, x_1 + x_2, \ldots)?$$

The range of this map is

$$\text{CUMSUM}(\mathbb{R}_+^\infty) =: \mathbb{R}_+^{\infty,\uparrow},$$

the space of non-negative, non-decreasing sequences. Similarly CUMSUM$(F_{\leq j})$ is the set of non-decreasing sequences with at most j positive jumps.

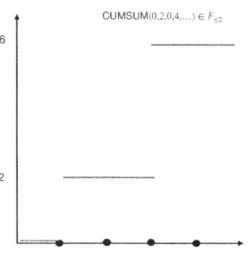

Let

$$\mathbb{R}_+^{\infty,\uparrow} \supset \text{ONE} = \text{CUMSUM}([\text{axes}])$$

which are sequences which are either $\mathbf{0}_\infty$ or sequences which start at 0 until one positive jump after which the sequence is constant.

For $\mathbf{X} = (X_1, X_2, \ldots)$ consisting of iid random variables with regularly varying tails, define

$$\mathbf{S} = \text{CUMSUM}(\mathbf{X}) = (S_1, S_2, \ldots).$$

Proposition 1.2, page 19, cheerfully informs us the map CUMSUM is uniformly continuous and from the Second Mapping Theorem 1.3, page 18, we can therefore apply this map to the regular variation properties of the previous section. So for

---

[1] The Vulcan from Star Trek, not the pediatrician.

3.2 Steroidal Regular Variation: Roid Rage Strikes

Table 3.2 Attempting to marginalize when $j = 1$ vs $j = 2$

	Operation	$j = 1$, $\mathbb{R}_+^\infty \setminus \{0_\infty\}$	$j = 2$, $\mathbb{R}_+^\infty \setminus [\text{axes}]$
1	Apply CUMSUM	$\mathbb{R}_+^\infty \setminus \{0_\infty\} \mapsto \mathbb{R}_+^{\infty,\uparrow} \setminus \{0_\infty\}$	$\mathbb{R}_+^\infty \setminus [\text{axes}] \mapsto \mathbb{R}_+^{\infty,\uparrow} \setminus \text{ONE}$
2	Apply T_2	$\mathbb{R}_+^{2,\uparrow} \setminus \{0_2\} \mapsto \mathbb{R}_+ \setminus \{0\}$	$\mathbb{R}_+^{\infty,\uparrow} \setminus \text{ONE} \mapsto \mathbb{R}_+ \setminus \mathbb{R}_+$
3	Leads to	$tP[S_2/b(t) \in \cdot] \to 2\nu_\alpha$	NADA?

instance, from (3.11), we get for any $j \geq 1$,

$$tP\Big[\frac{S}{b(t^{1/j})} \in \cdot\Big] \to \mu^{(j-1)} \circ \text{CUMSUM}^{-1}(\cdot), \tag{3.12}$$

in $\mathbb{M}(\text{CUMSUM}(\mathbb{R}_+^\infty) \setminus \text{CUMSUM}(F_{\leq(j-1)})) = \mathbb{M}(\mathbb{R}_+^{\infty,\uparrow} \setminus \text{CUMSUM}(F_{\leq(j-1)}))$. Formerly (3.12) looks powerful but it is not clearly useful or informative beyond $j = 1$ as problems arise when attempting to marginalize to get analogues of the one-jump principle. Things work fine for $j = 1$ but not for $j = 2$ as summarized in Table 3.2.

Here is what goes wrong with the $j = 2$ case when attempting to conclude something about one dimensional marginal distributions from (3.12). Let $T_p : \mathbb{R}_+^\infty \mapsto \mathbb{R}_+$ be the one-dimensional projection $T_p((x_1, x_2, \ldots)) = x_p$. Attempt to apply T_2 to (3.12). One possible path tries to apply Proposition 1.4, page 21 but that does not work since the set [axes] is not compact. Another path for applying T_2 is to utilize something like Theorem 1.3, page 18 which suggests the forbidden zone should be $\mathbb{R}_+$ meaning we are attempting to get a result in

$$\mathbb{M}(T_2(\mathbb{R}_+^{\infty,\uparrow}) \setminus T_2(\text{ONE})) = \mathbb{M}(\mathbb{R}_+ \setminus \mathbb{R}_+).$$

Not likely!

Moral: Do not get lost in the notation.

3.2.2 Poisson Points as Random Elements of $\mathbb{R}_+^\infty$

Considering Poisson points provides a variant to the iid case [125] and leads naturally in Chap. 4 to considering regular variation of the distribution of a Lévy process with regularly varying measure. Discussion of Lévy processes starts on page 121.

Suppose $\nu(\cdot) \in \mathbb{M}\big((0, \infty)\big) = \mathbb{M}([0, \infty) \setminus \{0\})$ and $x \mapsto \nu(x, \infty) \in RV_{-\alpha}$ is regularly varying at infinity with index $-\alpha < 0$. Let $Q(x) = \nu([x, \infty))$ and define $Q^\leftarrow(y) = \inf\{t > 0 : \nu([t, \infty)) < y\}$. Then the function $b(t) := Q^\leftarrow(1/t)$ satisfies

$$\lim_{t \to \infty} tQ(b(t)x) = \lim_{t \to \infty} t\nu\big(b(t)x, \infty\big) = x^{-\alpha}, \quad x > 0, \tag{3.13}$$

and b is regularly varying at infinity with index $1/\alpha$.

Let $\{E_n, n \geq 1\}$ be iid standard exponentially distributed random variables so that if $\{\Gamma_n, n \geq 1\} := \text{CUMSUM}\{E_n, n \geq 1\}$, we get points of a homogeneous Poisson process of rate 1. Using the transformation theory for Poisson processes [156, p. 121] or [152, p. 308], we find $\{Q^{\leftarrow}(\Gamma_n), n \geq 1\}$ are points of a Poisson process with mean measure ν written in decreasing order.

Define the following subspaces:

$$\mathbb{R}_+^{\infty\downarrow} = \{\mathbf{x} \in \mathbb{R}_+^\infty : x_1 \geq x_2 \geq \ldots\},$$
$$\mathbb{R}_+^{p,\downarrow} = \{(x_1, \ldots, x_p) \in \mathbb{R}_+^p : x_1 \geq x_2 \geq \cdots \geq x_p\}$$
$$\mathbb{H}_{=j} = \{\mathbf{x} \in \mathbb{R}_+^{\infty\downarrow} : x_j > 0, x_{j+1} = 0\},$$
$$\mathbb{H}_{\leq j} = \{\mathbf{x} \in \mathbb{R}_+^{\infty\downarrow} : x_{j+1} = 0\}. \tag{3.14}$$

So $\mathbb{H}_{\leq 0} = \{\mathbf{0}_\infty\}$, $\mathbb{R}_+^{\infty\downarrow}$ are sequences with decreasing, non-negative components and $\mathbb{H}_{\leq j}$ are decreasing sequences such that components are 0 from the $(j+1)$st component onwards. Furthermore, for each $j \geq 1$, $\mathbb{H}_{\leq j}$ is closed. To verify the closed property, suppose $\{\mathbf{x}(n), n \geq 1\}$ is a sequence in $\mathbb{H}_{\leq j}$ and $\mathbf{x}(n) \to \mathbf{x}(\infty)$ in the $\mathbb{R}_+^\infty$ metric. This means componentwise convergence so for the mth component convergence, where $m > j$, as $n \to \infty$, $0 = x_m(n) \to x_m(\infty)$ and $\mathbf{x}(\infty)$ is 0 beyond the jth component. The monotonicity of the components for each $\mathbf{x}(n)$ is preserved by taking limits. This verifies $\mathbb{H}_{\leq j}$ is closed and for every $j \geq 0$, $\mathbb{R}_+^{\infty\downarrow} \setminus \mathbb{H}_{\leq j}$ is a TABOF space.

Analogous to (3.9), we have the following.

Proposition 3.1 *Assuming (3.13), as $t \to \infty$,*

$$tP[(Q^{\leftarrow}(\Gamma_l)/b(t), l \geq 1) \in \cdot] \to \eta^{(1)}(\cdot), \tag{3.15}$$

in $\mathbb{M}(\mathbb{R}_+^{\infty\downarrow} \setminus \mathbb{H}_{\leq 0})$, *where*

$$\eta^{(1)}(dx_1 \times dx_2 \times \ldots) = \nu_\alpha(dx_1) 1_{[x_1 > 0]} \prod_{l=2}^\infty \epsilon_0(dx_l). \tag{3.16}$$

The limit measure $\eta^{(1)}(\cdot)$ concentrates on

$$(0, \infty) \times \{\mathbf{0}_\infty\} = \mathbb{H}_{=1}.$$

Warmup To verify this, it suffices to prove finite dimensional convergence. For the biggest component and $x > 0$, as $t \to \infty$,

$$tP[Q^{\leftarrow}(\Gamma_1)/b(t) > x] = tP[\Gamma_1 \leq Q(b(t)x)] = t(1 - e^{-Q(b(t)x)})$$
$$\sim tQ(b(t)x) \to x^{-\alpha} = \nu_\alpha(x, \infty) = \eta^{(1)} \circ \text{PROJ}_1^{-1}(x, \infty).$$

3.2 Steroidal Regular Variation: Roid Rage Strikes

For the first two components, let $\text{PRM}(\nu)$ be a Poisson counting function with mean measure $\nu(\cdot)$ on $(0, \infty)$ and for $x > 0$, $y > 0$,

$$tP[Q^{\leftarrow}(\Gamma_1)/b(t) > x,\ Q^{\leftarrow}(\Gamma_2)/b(t) > y] \leq tP[\text{PRM}(\nu)(b(t)(x \wedge y, \infty)) \geq 2]$$

and writing $p(t) = \nu\big(b(t)(x \wedge y, \infty)\big)$, we have

$$tP[\text{PRM}(\nu)(b(t)(x \wedge y, \infty)) \geq 2] = t(1 - e^{-p(t)} - p(t)e^{-p(t)})$$
$$\leq t(p(t) - p(t)e^{-p(t)}) \leq tp^2(t) \to 0$$
$$= \eta^{(1)} \circ \text{PROJ}_2^{-1}\big((x, \infty) \times (y, \infty)\big).$$

So the limit measure cannot charge sequences with 2 components positive and this makes (3.15) at least plausible. A formal proof consists of a maneuver to get things in a state where the iid result Proposition 2.4 and (3.9) can be applied.

Proof To prove (3.15), we apply Theorem 2.4, page 66, and prove as $t \to \infty$,

$$tP\left[\text{PROJ}_p\left(\left(\frac{Q^{\leftarrow}(\Gamma_l)}{b(t)}, l \geq 1\right)\right) \in \cdot\right] = tP\left[\left(\frac{Q^{\leftarrow}(\Gamma_l)}{b(t)}, l = 1, \ldots, p\right) \in \cdot\right]$$
$$\to \eta^{(1)} \circ \text{PROJ}_p^{-1}(\cdot)$$

in $\mathbb{M}(\mathbb{R}_+^{p,\downarrow} \setminus \{\mathbf{0}_p\})$. To prove this convergence, assume $f \in \mathbb{C}(\mathbb{R}_+^{p,\downarrow} \setminus \{\mathbf{0}_p\})$ which means there exists $\delta > 0$ such that if $\mathbf{x} = (x_1, \ldots, x_p) \in \text{supp}(f) \cap \mathbb{R}_+^{p,\downarrow}$, then $d_p(\mathbf{x}, \mathbf{0}_p) \geq \delta$. Because of monotonicity, this means, there exists $\delta_1 > 0$, such that,

$$\mathbf{x} = (x_1, \ldots, x_p) \in \text{supp}(f) \cap \mathbb{R}_+^{p,\downarrow} \text{ implies } x_1 \geq \delta_1. \tag{3.17}$$

So we must show

$$tEf\left(\frac{Q^{\leftarrow}(\Gamma_l)}{b(t)}, l = 1, \ldots, p\right) \to \int_{\mathbb{R}_+^{p,\downarrow} \setminus \{\mathbf{0}_p\}} f\, d\eta^{(1)} \circ \text{PROJ}_p^{-1}$$
$$= \int_{\mathbb{R}_+^{p,\downarrow} \cap [x_1 \geq \delta_1]} f\, d\eta^{(1)} \circ \text{PROJ}_p^{-1}. \tag{3.18}$$

One version of the order statistic property of Poisson points (see, eg., [152, Lemma 4.5.1, page 322]) is the equality in distribution in $\mathbb{R}_+^p$,

$$(\Gamma_1, \ldots, \Gamma_p | \Gamma_{p+1} = s) \stackrel{d}{=} s(U_{1,p}, \ldots, U_{p,p}), \quad s > 0, p > 1,$$

where $U_{1,p} \leq \cdots \leq U_{p,p})$ are uniform order statistics in non-decreasing order from a sample size p on $[0, 1]$. Also define on $\mathbb{R}_+^p$, revsort : $R_+^p \setminus \{\mathbf{0}_p\} \mapsto R_+^{p,\downarrow} \setminus \{\mathbf{0}_p\}$ by

$$\text{revsort}(x_1, \ldots, x_p) = (x_{p,p}, \ldots, x_{1,p}),$$

as the operator taking a vector into the vector with components ordered in a non-increasing way so $x_{p,p} \geq \cdots \geq x_{1,p}$.

In the left most side of (3.18), consider the conditional expectation

$$tE\left(f\left(\frac{Q^{\leftarrow}(\Gamma_l)}{b(t)}, l = 1, \ldots, p\right)\Big|\Gamma_{p+1} = s\right) = tEf\left(\frac{Q^{\leftarrow}(U_{l,p}s)}{b(t)}, l = 1, \ldots, p\right) \tag{3.19}$$

and assuming $U_1, \ldots, U_p$ are iid $U(0, 1)$ random variables this is

$$= tEf \circ \text{revsort}\left(\frac{Q^{\leftarrow}(U_l s)}{b(t)}, l = 1, \ldots, p\right) = tE\tilde{f}\left(\frac{Q^{\leftarrow}(U_l s)}{b(t)}, l = 1, \ldots, p\right) \tag{3.20}$$

where $\tilde{f} = f \circ \text{revsort}$. Observe $\left(Q^{\leftarrow}(U_l s), l = 1, \ldots, p\right)$ are iid and have heavy tailed distributions since for $x > 0$, and large enough t,

$$tP\left[\frac{Q^{\leftarrow}(U_1 s)}{b(t)} > x\right] = tP[U_1 s \leq Q(b(t)x)] = t\left(\frac{Q(b(t)x)}{s}\right) \to \frac{1}{s}x^{-\alpha}.$$

Assume temporarily that we have verified that $\tilde{f} \in \mathbb{C}(R_+^P \setminus \{\mathbf{0}_p\})$. Then apply the Second Binding Lemma and (2.34), page 49 or the $j = 1$ line of Table 3.1, page 102, to get convergence in (3.20) to

$$\to \sum_{l=1}^{p} \int_0^\infty \tilde{f}(0, \ldots, 0, x, 0 \ldots, 0) \nu_\alpha(dx)/s$$

$$= p \int_0^\infty f(x, \mathbf{0}_{p-1}) \nu_\alpha(dx)/s. \tag{3.21}$$

Now undo the conditioning on the left side of (3.19) by integrating,

$$tEf\left(\frac{Q^{\leftarrow}(\Gamma_l)}{b(t)}, 1 \leq l \leq p\right)$$

$$= \int_0^\infty P[\Gamma_{p+1} \in ds] tE\left(f\left(\frac{Q^{\leftarrow}(\Gamma_l)}{b(t)}, 1 \leq l \leq p\right)\Big|\Gamma_{p+1} = s\right) \tag{3.22}$$

3.2 Steroidal Regular Variation: Roid Rage Strikes

and assuming we can interchange integration on s and the limit on t, we should get from (3.21) that as $t \to \infty$, this converges to

$$\to \int_0^\infty \frac{e^{-s}s^p}{p!} \frac{p}{s} \int f(x, \mathbf{0}_{p-1})v_\alpha(dx)ds \qquad (3.23)$$

and for $p \geq 1$, this is

$$= \int_0^\infty \frac{e^{-s}s^{p-1}}{(p-1)!} \int f(x, \mathbf{0}_{p-1})v_\alpha(dx)ds$$

$$= \int f(x, \mathbf{0}_{p-1})v_\alpha(dx) = \int f d\eta^{(1)} \circ \mathrm{PROJ}_p^{-1}.$$

To properly prove the result, here are the remaining tasks.

1. Why is it true that $\tilde{f} \in \mathbb{C}(\mathbb{R}_+^p \setminus \{\mathbf{0}_p\})$?
2. How do we justify the interchange of limit and integration that led to (3.23)?

For task 1, we have been assuming $f \in \mathbb{C}(\mathbb{R}_+^{p,\downarrow} \setminus \{\mathbf{0}_p\})$ so that $d_p(\mathrm{supp}(f), \mathbf{0}_p) = \delta > 0$. If $\tilde{f}(x) = f(\mathrm{revsort}(x)) > 0$, then referring to (3.17) we have $\mathrm{revsort}(x) \in \mathrm{supp}(f)$ implies $\vee_{i=1}^p x_i \geq \delta_1$. Hence $\mathrm{supp}(\tilde{f})$ is bounded away from $\mathbf{0}_p$. This completes Task 1.

Task 2 is resolved by showing dominated convergence is applicable. The right side of (3.22) is of the form

$$\int_0^\infty V_t(s)P[\Gamma_{p+1} \in ds],$$

so we should prove $V_t(s)$ is bounded for all large t by an function integrable in s with respect to $P[\Gamma_{p+1} \in ds]$. Since f is bounded, there exists a number K such that $0 \leq f \leq K$. Then from (3.19),

$$V_t(s) = tEf\left(\frac{Q^\leftarrow(U_{l,p}s)}{b(t)}, l = 1, \ldots, p\right)$$

and from (3.17), this is

$$= tEf\left(\frac{Q^\leftarrow(U_{l,p}s)}{b(t)}, l = 1, \ldots, p\right) 1_{[Q^\leftarrow(U_{1,p}s)/b(t) \geq \delta_1]}$$

$$\leq Kt P[Q^\leftarrow(U_{1,p}s)/b(t) \geq \delta_1]$$

$$\leq Kpt P[U_1 s \leq Q(b(t)\delta_1)] = Kpt \frac{Q(b(t)\delta_1)}{s} \wedge 1$$

$$\leq Kp\frac{tQ(b(t)\delta_1)}{s} \leq Kp2\delta_1^{-\alpha}/s,$$

3.2.2.1 Hidden Regular Variation: The Big Reveal

Poisson points provide another example where there are an infinite number of hidden regular variation regimes. We continue to suppose the mean measure $v(\cdot)$ satisfies the regular variation assumption (3.13) and ease into presenting the general case of hidden regular variation.

In Proposition 3.1, the limit measure $\eta^{(1)}(\cdot)$ concentrates on $\mathbb{H}_{=1}$ and if we add this to the forbidden zone $\mathbb{H}_{\leq 0}$ to get $\mathbb{H}_{\leq 1}$ we claim,

$$tP[(Q^{\leftarrow}(\Gamma_l)/b(t^{1/2}), l \geq 1) \in \cdot] \to \eta^{(2)}(\cdot)$$

Reveal

in $\mathbb{M}(\mathbb{R}_+^{\infty,\downarrow} \setminus \mathbb{H}_{\leq 1})$ as $t \to \infty$, where

$$\eta^{(2)}(dx_1 \times dx_2 \times \ldots) = \nu_\alpha(dx_1)\nu_\alpha(dx_2)1_{[x_1 \geq x_2 > 0]} \prod_{l=3}^{\infty} \epsilon_0(dx_l).$$

Elementary computations give the distribution of $(\Gamma_1, \Gamma_2) = (E_1, E_1 + E_2)$,

$$P(\Gamma_1 \leq z, \Gamma_2 \leq w) = \begin{cases} 1 - e^{-z} - ze^{-w}, & z < w, \\ 1 - e^{-w} - we^{-w}, & z \geq w. \end{cases}$$

From this, for fixed $x > y > 0$,

$$P[Q^{\leftarrow}(\Gamma_1)/b(t^{1/2}) > x, \ Q^{\leftarrow}(\Gamma_2)/b(t^{1/2}) > y]$$
$$= P[\Gamma_1 \leq Q(b(t^{1/2})x), \Gamma_2 \leq Q(b(t^{1/2})y)]$$
$$= 1 - e^{-Q(b(t^{1/2})x)} - Q(b(t^{1/2})x)e^{-Q(b(t^{1/2})y)}$$
$$= Q(b(t^{1/2})x) - \frac{1}{2}\Big(Q(b(t^{1/2})x)\Big)^2 + O(Q(b(t^{1/2})x)^3)$$
$$\quad - Q(b(t^{1/2})x)\Big(1 - Q(b(t^{1/2})y) + O(Q(b(t^{1/2})y)^2)\Big)$$
$$= -\frac{1}{2}\Big(Q(b(t^{1/2})x)\Big)^2 + Q(b(t^{1/2})x)[1 - (1 - Q(b(t^{1/2})y))] + \text{JUNK}.$$

3.2 Steroidal Regular Variation: Roid Rage Strikes

Now lean on (3.13), page 103, to verify that for $x > y > 0$

$$\lim_{t \to \infty} tP[Q^{\leftarrow}(\Gamma_1)/b(t^{1/2}) > x, \ Q^{\leftarrow}(\Gamma_2)/b(t^{1/2}) > y]$$
$$= x^{-\alpha} y^{-\alpha} - x^{-2\alpha}/2$$
$$= \eta^{(2)}(z \in \mathbb{R}^{\infty,\downarrow} : z_1 > x, z_2 > y).$$

Similar computations show that, for $y > x > 0$,

$$\lim_{t \to \infty} tP[Q^{\leftarrow}(\Gamma_1)/b(t^{1/2}) > x, \ Q^{\leftarrow}(\Gamma_2)/b(t^{1/2}) > y]$$
$$= y^{-2\alpha}/2$$
$$= \eta^{(2)}(z \in \mathbb{R}^{\infty,\downarrow} : z_1 > x, z_2 > y).$$

Moreover, for $x > 0$, $y > 0$, $w > 0$,

$$tP[Q^{\leftarrow}(\Gamma_1)/b(t^{1/2}) > x, \ Q^{\leftarrow}(\Gamma_2)/b(t^{1/2}) > y, \ Q^{\leftarrow}(\Gamma_3)/b(t^{1/2}) > w]$$
$$\leq tP[\mathrm{PRM}(\nu)(b(t^{1/2})(x \wedge y \wedge w, \infty) \geq 3]$$

and writing $p(t) = \nu\big(b(t^{1/2})(x \wedge y \wedge w, \infty)\big)$, we have

$$tP[\mathrm{PRM}(\nu)(b(t^{1/2})(x \wedge y \wedge w, \infty) \geq 3]$$
$$= t(1 - e^{-p(t)} - p(t)e^{-p(t)} - p(t)^2 e^{-p(t)}/2)$$
$$\sim t(p(t)^3/3! + o(p(t)^3))$$

as $t \to \infty$. Hence, $\lim_{t \to \infty} tP[\mathrm{PRM}(\nu)(b(t^{1/2})(x \wedge y \wedge w, \infty) \geq 3] = 0$.

Steroidal HRV is present for the sequence of Poisson points as in the iid case described by Table 3.1 and (3.11). This is explained next.

Theorem 3.1 *Suppose Q and $\nu(\cdot)$ satisfy the regular variation condition (3.13). For every $j \geq 1$, the Poisson points $\{Q^{\leftarrow}(\Gamma_l), l \geq 1\}$, considered as a random element of $\mathbb{R}_+^{\infty,\downarrow}$ satisfy,*

$$tP\big[(Q^{\leftarrow}(\Gamma_l)/b(t^{1/j}), l \geq 1) \in \cdot\big] \to \eta^{(j)}(\cdot), \qquad (3.24)$$

in $\mathbb{M}(\mathbb{R}_+^{\infty,\downarrow} \setminus \mathbb{H}_{\leq j-1})$ as $t \to \infty$, where $\eta^{(j)}$ is a measure concentrating on $\mathbb{H}_{=j}$ given by

$$\eta^{(j)}(dx_1, dx_2, \ldots) = \prod_{i=1}^{j} \nu_\alpha(dx_i) 1_{[x_1 \geq x_2 \geq \cdots \geq x_j > 0]} \prod_{i=j+1}^{\infty} \epsilon_0(dx_i). \qquad (3.25)$$

Proof The proof follows the steps needed to prove Proposition 3.1. First reduce the convergence problem to finite dimensions by applying Theorem 2.4, page 66. Then fix $j \geq 1$ and show for any $p \geq 1$, that as $t \to \infty$,

$$tP\left[\left(Q^{\leftarrow}(\Gamma_l)/b(t^{1/j}), l = 1, \ldots, p\right) \in \cdot\right] \to \eta^{(j)} \circ \text{PROJ}_p^{-1}(\cdot),$$

in $\mathbb{M}(\mathbb{R}_+^{p,\downarrow} \setminus \text{PROJ}_p(\mathbb{H}_{\leq j-1}))$. To do this, suppose $f \in \mathbb{C}(\mathbb{R}_+^{p,\downarrow} \setminus \text{PROJ}_p(\mathbb{H}_{\leq j-1}))$. and we must show the analogue of (3.18), namely that

$$tEf\left(\frac{Q^{\leftarrow}(\Gamma_l)}{b(t^{1/j})}, l = 1, \ldots, p\right) \to \int_{\mathbb{R}_+^{p,\downarrow} \setminus \{\text{PROJ}_p(\mathbb{H}_{\leq j-1})\}} f \, d\eta^{(j)} \circ \text{PROJ}_p^{-1}. \tag{3.26}$$

Note

$$\text{PROJ}_p(\mathbb{H}_{\leq j-1}) = \begin{cases} \{x \in \mathbb{R}_+^{p,\downarrow} : x_j = 0\}, & \text{if } j \leq p, \\ \mathbb{R}_+^{p,\downarrow}, & \text{if } j > p, \end{cases}$$

and in the latter case where $j > p$,

$$\mathbb{C}(\mathbb{R}_+^{p,\downarrow} \setminus \text{PROJ}_p(\mathbb{H}_{\leq j-1})) = \emptyset$$

and there is nothing to show. So we proceed assuming $j \leq p$. If $j = 1$ we are back in the known case that $\mathbb{H}_{\leq j-1} = \{\mathbf{0}_\infty\}$ so assume $2 \leq j \leq p$. Further, there exists $\delta > 0$ such that if $x \in \mathbb{R}_+^{p,\downarrow} \cap \text{supp}(f)$, then $x_j \geq \delta$.

Now proceed as in (3.19),

$$tE\left(f\left(\frac{Q^{\leftarrow}(\Gamma_l)}{b(t^{1/j})}, l = 1, \ldots, p\right) \Big| \Gamma_{p+1} = s\right) = tE\tilde{f}\left(\frac{Q^{\leftarrow}(U_l s)}{b(t^{1/j})}, l = 1, \ldots, p\right) \tag{3.27}$$

where $\tilde{f} = f \circ \text{revsort} \in \mathbb{C}(\mathbb{R}_+^p \setminus F_{\leq j-1})$ and $F_{\leq j-1}$ is defined in (2.62), page 66. Remembering that $Q^{\leftarrow}(U_l s), s = 1, \ldots p$ are iid and heavy tailed we apply (3.11) or Table 3.1, page 102 and get as $t \to \infty$, that the above converges to

$$\to \int \sum_{i_1,\ldots,i_j} \tilde{f}(x_1, \ldots, x_p) \prod_{l=1}^{j} \frac{\nu_\alpha(dx_{i_l})}{s} \prod_{k \notin \{i_1,\ldots,i_j\}} \epsilon_0(dx_k).$$

Note $\tilde{f}$ is a symmetric function of its arguments so the previous is

$$= s^{-j} \binom{p}{j} \int \tilde{f}(x_1, \ldots, x_j, \mathbf{0}_{p-j}) \prod_{l=1}^{j} \nu_\alpha(dx_l).$$

3.2 Steroidal Regular Variation: Roid Rage Strikes

Finally, undo the conditioning with respect to the density of Γ_{p+1} which yields,

$$tE\left(f\left(\frac{Q^{\leftarrow}(\Gamma_l)}{b(t^{1/j})}, l=1,\ldots,p\right)\right)$$

$$\to \int_{s=0}^{\infty}\int \frac{e^{-s}s^p}{p!}s^{-j}\binom{p}{j}\tilde{f}(x_1,\ldots,x_j,\mathbf{0}_{p-j})\prod_{l=1}^{j}v_\alpha(dx_l)ds$$

and combining powers of s and cancelling factorials yields

$$=\int_{s=0}^{\infty}\frac{e^{-s}s^{p-j}}{(p-j)!}ds\int \frac{1}{j!}\tilde{f}(x_1,\ldots,x_j,\mathbf{0}_{p-j})\prod_{l=1}^{j}v_\alpha(dx_l)$$

$$=\int \frac{1}{j!}\tilde{f}(x_1,\ldots,x_j,\mathbf{0}_{p-j})\prod_{l=1}^{j}v_\alpha(dx_l)$$

$$=\frac{1}{j!}\int_{\{x_1\geq x_2\geq\cdots\geq x_j\}}j!f(x_1,\ldots,x_j,\mathbf{0}_{p-j})\prod_{l=1}^{j}v_\alpha(dx_l)$$

$$=\int fd\eta^{(j)}\circ\mathrm{PROJ}_p^{-1}.$$

3.2.3 Different Tail Rates in Different Directions

It is possible, but somewhat pathological, that there are uncountably many distinct indices of regular variation along distinct rays in $\mathbb{R}_+^2$. Here is a cunning example from [38] dreamed up by B. Das and A. Mitra.

Suppose (R, U) is a bivariate random vector with probability density function given by:

$$f_{R,U}(r,u) = 4r^{-(u+2)}\left((u+1)\log r - 1\right), \quad r > 2, 0 < u < 1. \tag{3.28}$$

Think of (R, U) as polar coordinates and define a random element $\mathbf{Z}$ of $\mathbb{R}_+^2$ in Cartesian coordinates by $\mathbf{Z} = (RU, R(1-U))$. The best way to investigate the regular variation properties of $P[\mathbf{Z} \in \cdot] \in \mathbb{M}(\mathbb{R}_+^2 \setminus \{\mathbf{0}_2\})$ is to investigate this for the polar co-ordinate version $P[(R,U) \in \cdot] \in \mathbb{M}(\mathbb{R}_+ \setminus \{0\}) \times [0,1])$. The polar coordinate transformation that is convenient for $\mathbb{R}_+^2 \setminus \{\mathbf{0}_2\} \mapsto (\mathbb{R}_+ \setminus \{0\}) \times [0,1]$ is

$$(x,y) \mapsto \left(x+y, \frac{(x,y)}{x+y}\right)$$

and this means
$$Z \mapsto (R, (U, 1-U)).$$

Because $(U, 1-U)$ belongs to the simplex $\{(x, y) \in \mathbb{R}_+^2 : x + y = 1\}$, it suffices to consider $Z \mapsto (R, U)$.

Some calculus applied to (3.28) yields

$$P[R > r, U \in [a, b]] = 4\left(r^{-(a+1)} - r^{-(b+1)}\right), \quad r > 2, \ 0 \le a < b \le 1. \tag{3.29}$$

(As a sanity check, note if $r = 2, a = 0, b = 1$ the right side of (3.29) is 1.) Define $b_a(t) = t^{1/(a+1)}$ for $t > 0$, $0 \le a \le 1$. Then for $r > 2$, $0 \le a < b \le 1$, we have,

$$\lim_{t \to \infty} tP\left[\frac{R}{b_a(t)} > r, U \in [a, b]\right] = 4r^{-(a+1)} = 4\nu_{a+1}(r, \infty) \times \epsilon_a([a, b]),$$

where as usual $\nu_\alpha(r, \infty) = r^{-\alpha}$ for $r > 0, \alpha > 0$. Provided $b > a$, the limit does not depend on b and therefore,

$$tP\left[\left(\frac{R}{b_a(t)}, U\right) \in \cdot\right] \to 4\nu_{a+1} \times \epsilon_a(\cdot)$$

in $\mathbb{M}(\mathbb{R}_+ \setminus \{0\}) \times [0, 1])$ where the limit is the polar coordinate transformation of a measure $\eta_a(\cdot)$ that concentrates Pareto mass on $(\mathbb{R}_+ \setminus \{0\}) \times \{a\}$. Switching back to Cartesian coordinates, we get in mouthful notation,

$$P[Z \in \cdot] \in \mathrm{MRV}(a+1, b_a(t) = t^{1/(a+1)}, \eta_a(\cdot), \mathbb{R}_+^2 \setminus \{\mathbf{0}_2\}),$$

where $\eta_a(\cdot)$ is the measure putting the Pareto measure with index $a+1$ on the single ray through the point $(a, 1-a)$.

This, of course holds for any $a \in [0, 1]$. The bigger the chosen a is, the lighter the tail of the Pareto measure on the ray through $(a, 1-a)$. The heaviest tail is obtained when $a = 0$ and then the Pareto measure ν_1 concentrates on the ray through $(0, 1)$ which is the y-axis.

This example also serves as a caution against always relying on asymptotic structure to infer something about the distribution of Z. If we approximate $P[Z \in \cdot]$ using the heavy tail methodology, which heavy tail regime should be used? The heaviest? The character of the limit measure may be different from that of $P[Z \in \cdot]$.

3.3 Problems

3.1 [135, 136] The thorough reader may want to review Problem 2.3, page 83. Suppose $X \perp\!\!\!\perp Y$ are iid Pareto(1) random variables and set $\mathbf{Z} = (X, 2X, Y) \in \mathbb{R}_+^3$. Verify the following assertions.

1. $P[\mathbf{Z} \in \cdot] \in \text{MRV}(\alpha = 1, b_1(t) = 2t, \eta_1(\cdot), \mathbb{R}_+^3 \setminus \{\mathbf{0}\})$.
2. Where does the limit measure $\eta_1(\cdot)$ concentrate? The measure η_1 gives zero mass to $\mathbb{R}_+^3 \cap \{\mathbf{x} \in \mathbb{R}_+^3 : \wedge_{i=1}^3 x_i > 0\}$ and in fact

$$\text{supp}(\eta_1) = \{\mathbf{x} \in \mathbb{R}_+^3 : x_2 = 2x_1 > 0, x_3 = 0\} \cup \{\mathbf{x} \in \mathbb{R}_+^3 : x_3 > 0, x_1 = x_2 = 0\}.$$

3. The vector $\mathbf{Z}$ cannot possess asymptotic independence since the limit $\eta_1(\cdot)$ does not concentrate on the one-dimensional axes and in particular the components $(X, 2X)$ do not have asymptotic independence. See Problem 2.2(3).
4. The vector $\mathbf{Z}$ cannot possess hidden regular variation on $\mathbb{R}_+^3 \setminus [\text{axes}] = \mathbb{R}_+^3 \setminus \text{PROJ}_3(F_{\le 1})$ (delete the one-dimensional axes) because if it did, then $\mathbf{Z}$ would possess asymptotic independence.
5. Despite no HRV in Part 4 we do have $\mathbf{Z}$ having a regularly varying distribution when the forbidden zone is [axes]:

$$P[\mathbf{Z} \in \cdot] \in \text{MRV}(\alpha = 1, b_1(t) = 2t, \eta_1(\cdot), \mathbb{R}_+^3 \setminus [\text{axes}])$$

since $\eta_1(\mathbb{R}_+^3 \setminus [\text{axes}]) \ne 0$.
6. Despite the story in the previous Part 4, the vector $\mathbf{Z}$ does possess hidden regular variation on $\mathbb{R}_+^3 \setminus [\text{faces}] = \mathbb{R}_+^3 \setminus \text{PROJ}_3(F_{\le 2}) = \{\mathbf{x} \in \mathbb{R}_+^3 : x_1 \wedge x_2 \wedge x_3 > 0\}$ and

$$P[\mathbf{Z} \in \cdot] \in \text{MRV}(\alpha = 2, b_2(t) = \sqrt{t}, \eta_2(\cdot), \mathbb{R}_+^3 \setminus \text{PROJ}_3(F_{\le 2}))$$

and for $z_i > 0$, $i = 1, 2, 3$

$$\eta_2\big((z_1, \infty) \times (z_2, \infty) \times (z_3, \infty)\big) = \big((z_1 \vee (z_2/2)) z_3\big)^{-1}.$$

7. Summary:
   (a) $P[\mathbf{Z} \in \cdot] \in \text{MRV}(\alpha = 1, b_1(t) = 2t, \eta_1, \mathbb{R}_+^3 \setminus \{\mathbf{0}_3\})$.
   (b) $P[\mathbf{Z} \in \cdot] \in \text{MRV}(\alpha = 1, b_1(t) = 2t, \eta_1, \mathbb{R}_+^3 \setminus [\text{axes}])$.
   (c) HRV holds on $\mathbb{R}_+^3 \setminus [\text{faces}]$ but not on $\mathbb{R}_+^3 \setminus [\text{axes}]$).
   (d) $\mathbf{Z}$ does not possess asymptotic independence but HRV is present on $\mathbb{R}_+^3 \setminus [\text{faces}]$. Contrast this with the material in Sect. 3.1.1.2.

Table 3.3 Table giving the values of **Z**

B_1, B_2	Probability	**Z**
0, 0	1/4	$(0, X_2, X_3)$
0, 1	1/4	$(X_1, 0, X_3)$
1, 0	1/4	$(0, X_2, 0)$
1, 1	1/4	$(X_1, 0, 0)$

3.2 [130, 136] Suppose (X_1, X_2, X_3) are iid unit Pareto random variables on $[1, \infty)$ (with index 1) and suppose B_1, B_2 are iid Bernoulli variables independent of the X's and

$$P[B_i = 1] = P[B_i = 0] = \frac{1}{2}, \quad i = 1, 2.$$

Define

$$\mathbf{Z} = \big(B_2 X_1, (1 - B_2) X_2, (1 - B_1) X_3\big).$$

The values of **Z** are given in Table 3.3.

Verify the following:

1. $P[\mathbf{Z} \in \cdot] \in \text{MRV}(\alpha_1 = 1, b_1(t) = t, \eta_1(\cdot), \mathbb{R}_+^3 \setminus \{\mathbf{0}_3\})$. Hint: Let $f_1(x, y, z) \in \mathbb{C}(\mathbb{R}_+^3 \setminus \{\mathbf{0}_3\})$ and consider $t E f_1(\mathbf{Z}/b_1(t))$. This splits into 4 terms according to the 4 lines of Table 3.3.
2. As usual, set $\nu_1(dx) = x^{-2} dx \mathbf{1}_{(0,\infty)}(x)$. Show the limit measure is

$$\eta_1 = \frac{1}{2}\Big((\epsilon_0 \times \epsilon_0 \times \nu_1) + (\nu_1 \times \epsilon_0 \times \epsilon_0) + (\epsilon_0 \times \nu_1 \times \epsilon_0)\Big).$$

3. Hence η_1 concentrates on [**axes**] and **Z** has asymptotic independence.
4. The previous part 3 suggests investigating HRV on $\mathbb{R}_+^3 \setminus [\textbf{axes}]$. Prove,

$$P[\mathbf{Z} \in \cdot] \in \text{MRV}(\alpha_2 = 2, b_2(t) = \sqrt{t}, \eta_2(\cdot), \mathbb{R}_+^3 \setminus [\textbf{axes}])$$

Hint: Let $f_2 \in \mathbb{C}(\mathbb{R}_+^3 \setminus [\textbf{axes}])$ so that $f_2(x, y, z) = 0$ if $(x, y, z) \in [\textbf{axes}]^\delta$ for some $\delta > 0$. Using Table 3.3, the expectation $t E f_2(\mathbf{Z}/\sqrt{t})$ splits into 4 terms two of which are 0 because the argument is in [**axes**].
5. Verify

$$\eta_2 = \frac{1}{4}(\epsilon_0 \times \nu_1 \times \nu_1 + \nu_1 \times \epsilon_0 \times \nu_1)$$

which concentrates on

$$(\{0\} \times \mathbb{R}_+ \times \mathbb{R}_+) \cup (\mathbb{R}_+ \times \{0\} \times \mathbb{R}_+) =: \text{2FACES}$$
$$= \{\mathbf{x} \in \mathbb{R}_+^3 : x_1 \geq 0, x_2 = 0, x_3 \geq 0\} \cup \{\mathbf{x} \in \mathbb{R}_+^3 : x_1 = 0, x_2 \geq 0, x_3 \geq 0\}.$$

3.3 Problems

Note since $(0, \infty)^3 \cap 2\text{FACES} = \emptyset$, we have $\eta_2((0, \infty)^3) = 0$.

6. The conclusion that η_2 concentrates on 2FACES suggests that *perhaps* there is a regular variation property on $\mathbb{R}_+^3 \setminus 2\text{FACES}$. However, show $P[\mathbf{Z} \in \cdot]$ does not have regular variation in this reduced space. Hint: Let $f_3 \in \mathbb{C}(\mathbb{R}_+^3 \setminus 2\text{FACES})$ and let $b_3(t) \to \infty$ be any scaling function.

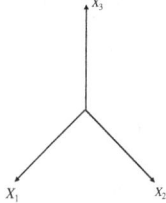

Note there exists some $\delta > 0$ such that $f_3(x, y, z) = 0$ if $(x, y, z) \in \{\mathbf{x} : x_1 = 0, x_2 \geq 0, x_3 \geq 0\}^\delta \cup \{\mathbf{x} : x_1 \geq 0, x_2 = 0, x_3 \geq 0\}^\delta$; that is, if (x, y, z) is too close to two faces; see the minimal visual aid on the right. Evaluate $tEf_3(\mathbf{Z}/b_3(t))$ as 4 terms all of which are 0.

Visual aid

7. Verify

$$P[(Z_1, Z_2) \in \cdot] \in \text{MRV}(\alpha = 1, b(t) = t, \eta_1', \mathbb{R}_+^2 \setminus \{\mathbf{0}_2\})$$

with

$$\eta_1' = \frac{1}{2}(\epsilon_0 \times \nu_1 + \nu_1 \times \epsilon_0),$$

so that (Z_1, Z_2) has asymptotic independence and η_1' concentrates on the two dimensional axes.

From Table 3.3 we see that

$$(Z_1, Z_2) = \begin{cases} (0, X_2), & \text{with probability } \frac{1}{2}, \\ (X_1, 0), & \text{with probability } \frac{1}{2}, \end{cases}$$

and therefore (Z_1, Z_2) cannot have HRV on $\mathbb{R}_+^2 \setminus [\text{axes}] = (0, \infty)^2$.

8. Summary:

   (a) $P[\mathbf{Z} \in \cdot] \in \text{MRV}(\alpha_1 = 1, b_1(t) = t, \eta_1(\cdot), \mathbb{R}_+^3 \setminus \{\mathbf{0}_3\})$ and has asymptotic independence.

   (b) $P[\mathbf{Z} \in \cdot] \in \text{MRV}(\alpha_2 = 2, b_2(t) = \sqrt{t}, \eta_2(\cdot), \mathbb{R}_+^3 \setminus [\text{axes}])$ and hence $P[\mathbf{Z} \in \cdot]$ has HRV.

   (c) $P[\mathbf{Z} \in \cdot]$ does not have HRV on $\mathbb{R}_+^3 \setminus 2\text{FACES}$ and does not have HRV on $(0, \infty)^3$.

   (d) $P[(Z_1, Z_2) \in \cdot] \in \text{MRV}(\alpha = 1, b(t) = t, \eta_1', \mathbb{R}_+^2 \setminus \{\mathbf{0}_2\})$ and has asymptotic independence but $P[(Z_1, Z_2) \in \cdot]$ does not have HRV on $(0, \infty)^2$ even though $P[\mathbf{Z} \in \cdot]$ has HRV on $\mathbb{R}_+^3 \setminus [\text{axes}]$.

3.3 [136] Again suppose (X_1, X_2, X_3) are iid unit Pareto random variables on $[1, \infty)$ (with index 1) and define

$$\mathbf{Z} = \left((X_1)^2 \wedge (X_2)^2, (X_2)^2 \wedge (X_3)^2, (X_1)^2 \wedge (X_3)^2\right).$$

Verify the following.

1. Each of the 3 components of $\mathbf{Z}$ has a Pareto distribution with index 1.
2. Each pair of (Z_1, Z_2, Z_3) has asymptotic independence and hence, by Problem 2.2, also $\mathbf{Z}$ has asymptotic independence.
3. Therefore,

$$P[\mathbf{Z} \in \cdot] \in \mathrm{MRV}(\alpha_1 = 1, b_1(t) = t, \eta_1(\cdot), \mathbb{R}^3_+ \setminus \{\mathbf{0}_3\}),$$

where η_1 concentrates on [axes] and on each axis, η_1 spreads mass according to $\nu_1(dx) = x^{-2}dx$, $x > 0$.

4. Also in $\mathbb{M}(\mathbb{R}^3_+ \setminus [\text{axes}])$,

$$tP\left[\mathbf{Z}/t^{2/3} \in \cdot\right] \to \eta_2(\cdot),$$

for a non-zero measure $\eta_2 \in \mathbb{M}(\mathbb{R}^3_+ \setminus [\text{axes}])$ and

$$P[\mathbf{Z} \in \cdot] \in \mathrm{MRV}(\alpha_2 = 3/2, b_2(t) = t^{2/3}, \eta_2(\cdot), \mathbb{R}^3_+ \setminus [\text{axes}])$$

5. The measure η_2 satisfies for $\mathbf{w} > \mathbf{0}_3$,

$$\lim_{t \to \infty} tP\left[\frac{\mathbf{Z}}{t^{2/3}} \in (\mathbf{w}, \infty_3)\right] = \eta_2((\mathbf{w}, \infty))$$

$$= \frac{1}{\sqrt{(w_1 \vee w_3) \cdot (w_1 \vee w_2) \cdot (w_2 \vee w_3)}}$$

$$= \lim_{t \to \infty} tP[X_1 > t^{1/3}(w_1 \vee w_3)^{1/2}, X_2 > t^{1/3}(w_1 \vee w_2)^{1/2},$$

$$X_3 > t^{1/3}(w_2 \vee w_3)^{1/2}].$$

6. So $P[\mathbf{Z} \in \cdot]$ has HRV on $\mathbb{R}^3_+ \setminus [\text{axes}]$ with limit measure η_2 that concentrates on $(\mathbf{0}_3, \infty) = R^3_+ \setminus [\text{faces}]$.
7. Summary:

    (a) $P[\mathbf{Z} \in \cdot] \in \mathrm{MRV}(\alpha_1 = 1, b_1(t) = t, \eta_1(\cdot), \mathbb{R}^3_+ \setminus \{\mathbf{0}_3\})$, and $\mathbf{Z}$ has asymptotic independence. The limit measure η_1 concentrates on [axes].

    (b) $P[\mathbf{Z} \in \cdot] \in \mathrm{MRV}(\alpha_2 = 3/2, b_2(t) = t^{2/3}, \eta_2(\cdot), \mathbb{R}^3_+ \setminus [\text{axes}])$ and η_2 concentrates on $(0, \infty)^3$. Therefore, $P[\mathbf{Z} \in \cdot]$ has HRV but we cannot seek further HRV regimes on smaller cones.

3.3 Problems

(c) HRV holds on $\mathbb{R}_+^3 \setminus$ [axes] with limit measure η_2 but $\eta_2((0, \infty)^3) > 0$ and $P[Z \in \cdot]$ is also regularly varying on $(0, \infty)^3$.

3.4 [135] Let X and Y be two iid Pareto(1) random variables. Let B be another random variable independent of (X, Y) such that $P[B = 0] = P[B = 1] = \frac{1}{2}$. Define

$$Z = (Z_1, Z_2) = B(X, X^2) + (1 - B)(Y^2, Y).$$

See Table 3.4.

Table 3.4 Values of Z

B	Prob	Z
1	$\frac{1}{2}$	(X, X^2)
0	$\frac{1}{2}$	(Y^2, Y)

Verify the following:

1. Each of the marginal random variables satisfies, as $t \to \infty$,

$$tP[Z_i > \frac{1}{4}t^2 x] \to v_{1/2}(x, \infty) = x^{-1/2}, \quad i = 1, 2; \, x > 0.$$

2. Asymptotic independence holds, namely

$$tP[Z_1 \wedge Z_2 > t^2 x] \to 0, \quad x > 0.$$

3. Therefore (see Proposition 2.2, page 49)

$$P[Z \in \cdot] \in \mathrm{MRV}(\alpha = \frac{1}{2}, b(t) = \frac{1}{4}t^2, \eta_1 = \epsilon_0 \times v_{1/2} + v_{1/2} \times \epsilon_0, \mathbb{R}_+^2 \setminus \{0_2\}).$$

4. For $x, y > 0$,

$$\lim_{t \to \infty} tP\left[\frac{Z}{t} \in (x, \infty] \times (y, \infty]\right]$$

$$= \lim_{t \to \infty} \frac{t}{2} P[X > tx, X^2 > ty] + \lim_{t \to \infty} \frac{t}{2} P\left[Y^2 > tx, Y > ty\right]$$

$$= \lim_{t \to \infty} \frac{t}{2} P[X > tx] + \lim_{t \to \infty} \frac{t}{2} P[Y > ty]$$

$$= \frac{1}{2}\left(\frac{1}{x} + \frac{1}{y}\right).$$

Note the limit does not have a two dimensional density.

5. HRV exists on the cone $\mathbb{R}_+^2 \setminus [\text{axes}]$ with limit measure η_2 satisfying

$$\eta_2(((x,\infty] \times (y,\infty]) = \frac{1}{2}\left(\frac{1}{x} + \frac{1}{y}\right), \quad x > 0, \, y > 0.$$

6. We also have

$$P[Z \in \cdot] \in MRV(\alpha = 1, b(t) = t/2, \eta_2'(\cdot), [0,\infty]^2 \setminus [\text{axes}])$$

where

$$\eta_2' = \epsilon_\infty \times \nu_1 + \nu_1 \times \epsilon_\infty$$

is the measure that concentrates on the lines through ∞, with support,

$$\{\infty\} \times [0, \infty] \cup [0, \infty] \times \{\infty\}.$$

3.5 [38] Suppose we invest in two financial instruments and for a given time horizon Z_1 and Z_2 are future losses. We construct $Z = (Z_1, Z_2)$ to have a regularly varying distribution on $\mathbb{R}_+^2 \setminus \{0_2\} = \mathbb{R}_+^2 \setminus F_1$ with limit measure $\eta_1(\cdot)$ that concentrates outside the open cone $\text{CONE} := \left\{(u, v) \in \mathbb{R}_+^2 : \frac{1}{2}u < v < 2u\right\}$, so that $F_2 := \mathbb{R}_+^2 \setminus \text{CONE}$ is closed and $\eta_1(\cdot)$ concentrates on $F_2 := \mathbb{R}_+^2 \setminus \text{CONE}$. Hence we may search for another MRV on

$$\mathbb{R}_+^2 \setminus (F_1 \cup F_2) = \overline{\text{CONE}} \setminus \{0_2\}.$$

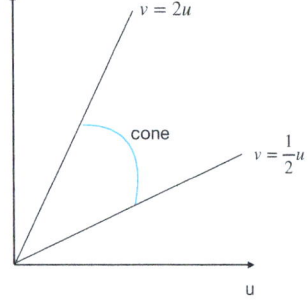

Suppose R_1, R_2, U_1, U_2, B are independent and R_1 is Pareto(1), R_2 is Pareto(2), $U_1 \sim U\big((0, 1/3) \cup (2/3, 1)\big)$, $U_2 \sim U(1/3, 2/3)$ and B is Bernoulli with values $0, 1$ with equal probability. Define

$$Z = BR_1(U_1, 1 - U_1) + (1 - B)R_2(U_2, 1 - U_2).$$

Verify the following:

1. $P[Z \in \cdot]$ is regularly varying:

$$P[Z \in \cdot] \in \text{MRV}(\alpha_1 = 1, b_1(t) = t, \eta_1(\cdot), \mathbb{R}_+^2 \setminus \{0_2\})$$

and $\eta_1(\cdot)$ concentrates on $\mathbb{R}_+^2 \setminus \text{CONE}$.

3.3 Problems 119

2. HRV is present. Since $\eta_1(\cdot)$ concentrates on $\mathbb{R}_+^2 \setminus$ CONE, seek HRV on $\mathbb{R}_+^2 \setminus (F_1 \cup F_2) = \overline{\text{CONE}} \setminus \{0\}$ and show,

$$P[\mathbf{Z} \in \cdot] \in \text{MRV}\left(\alpha_2 = 2, b_2(t) = \sqrt{t}, \eta_2(\cdot), \overline{\text{CONE}} \setminus \{\mathbf{0}\}\right)$$

and $\eta_2(\cdot)$ has uniform angular measure on $[1/3, 2/3]$.

3. Even though $\mathbf{Z}$ does not have asymptotically independent components, the vector

$$\left((2Z_1 - Z_2)_+, (2Z_2 - Z_1)_+\right)$$

does have asymptotically independent components. What are the implications for portfolio construction from this fact? (Hint: Set $T(z) = \left((2z_1 - z_2)_+, (2z_2 - z_1)_+\right)$. For any $x > 0$,

$$tP[T(\mathbf{Z}/t) > (x, x)] \to \eta_1(T^{-1}(x\mathbf{1}, \boldsymbol{\infty}_2)) = 0$$

since $T^{-1}(x\mathbf{1}, \boldsymbol{\infty}_2) \subset$ CONE.

3.6 [38] (**Asymptotic Full Dependence**) When $p = 2$, a regularly varying measure has *asymptotic full dependence* on $\mathbb{R}_+^2 \setminus \{\mathbf{0}_2\}$ if the limit measure η concentrates on [diag] (or at least a ray from the origin); this was introduced in Sect. 1.2.1. Suppose, X_1, X_2, X_3 are iid with common distribution Pareto(2). Let B be a Bernoulli random variable independent of $\{X_i : i = 1, 2, 3\}$ and $P[B = 0] = P[B = 1] = 1/2$. Construct the random vector

$$\mathbf{Z} := (Z_1, Z_2) = B((X_1)^2, (X_1)^2) + (1 - B)(X_2, X_3).$$

Verify,

1. $P[\mathbf{Z} \in \cdot]$ is regularly varying:

$$P[\mathbf{Z} \in \cdot] \in \text{MRV}\left(\alpha_1 = 1, b_1(t) = t, \eta_1(\cdot), \mathbb{R}_+^2 \setminus \{\mathbf{0}_2\}\right).$$

2. The measure η_1 concentrates on [diag] and

$$\eta_1\left([\mathbf{0}, (u, v)]^c\right) = \frac{1}{2}(u \wedge v)^{-1} \qquad (u, v) \in \mathbb{R}_+^2 \setminus \{\mathbf{0}_2\}.$$

3. Set $F_2 = \{\mathbf{0}_2\} \cup$ [diag] so that

$$\mathbb{R}_+^2 \setminus F_2 = \{(u, v) : |u - v| > 0\}.$$

Verify,

$$P[\mathbf{Z} \in \cdot] \in \text{MRV}\left(\alpha_2 = 2, b_2(t) = t^{1/2}, \eta_2(\cdot), \mathbb{R}_+^2 \setminus F_2\right)$$

where $\eta_2(\cdot)$ concentrates on [axes] and

$$\eta_2(du\, dv) = u^{-3} du\, \epsilon_0(dv) + \epsilon_0(du)\, v^{-3} dv.$$

3.7 [38] (**Steoidal Example**) Here is a poor man's version of the sexy construction in Sect. 3.2.3, page 111. This example creates a countable number of heavy tail regimes for the same distribution on $\mathbb{R}_+^2$.

Suppose $\{X_i, i \geq 1\}$ are iid random variables with common Pareto(1) distribution. Let $\{Y_1, Y_2\}$ be iid with common Pareto(2) distribution and let $\{B_i, i \geq 1\}$ be an infinite sequence of random variables with $P[B_i = 1] = 1 - P[B_i = 0] = 2^{-i}$ and $\sum_{i=1}^{\infty} B_i = 1$. For instance, let T be the index of the first success in an iid sequence of Bernoulli trials and then set $B_i = 1_{[T=i]}, i \geq 1$. Assume that $\{X_i, i \geq 1\}, \{Y_1, Y_2\}$ and $\{B_i : i \geq 1\}$ are mutually independent. Define the random vector $\mathbf{Z}$

$$\mathbf{Z} := (Z_1, Z_2) = B_1(Y_1, Y_2) + \sum_{i=1}^{\infty} B_{i+1} \left((X_i)^{\frac{1}{2-2^{-(i-1)}}}, 2^{i-1}(X_i)^{\frac{1}{2-2^{-(i-1)}}} \right).$$

Verify,

1. $P[\mathbf{Z} \in \cdot]$ is a regularly varying distribution with asymptotic full dependence:

$$P[\mathbf{Z} \in \cdot] \in \text{MRV}\big(\alpha_1 = 1, b_1(t) = t, \eta_1(\cdot), \mathbb{R}_+^2 \setminus \{\mathbf{0}_2\}\big),$$

and $\eta_1(\cdot)$ concentrates on [diag] =: Ray$_1$.
2. Remove the diagonal and let $F_2 = \{\mathbf{0}_2\} \cup \text{Ray}_1$ and find HRV on $\mathbb{R}_+^2 \setminus F_2$,

$$P[\mathbf{Z} \in \cdot] \in \text{MRV}\big(\alpha_2 = 3/2, b_2(t) = t^{2/3}, \eta_2(\cdot), \mathbb{R}_+^2 \setminus F_2\big)$$

where η_2 concentrates on the ray Ray$_2 := \{(x, 2x) : x \geq 0\}$.
3. Keep going: Remove Ray$_2$ and set $F_3 = F_2 \cup \text{Ray}_2$ and find HRV on $\mathbb{R}_+^2 \setminus F_3$.
4. In general, for $i \geq 1$, remove

$$F_i := \{\mathbf{0}_2\} \bigcup \bigcup_{j=1}^{i} \left\{ (x, 2^{j-1} x) : x \in \mathbb{R}_+ \right\}$$

and find HRV on $\mathbb{R}_+^2 \setminus F_i$.

Chapter 4
Lévy Processes with Regularly Varying Distributions: Where Do the Jumps Go?

Where indeed?

This chapter, based on [125, Section 5], provides a function space example of steroidal regular variation of the distribution of a Lévy process, whose Lévy measure is regularly varying at $+\infty$. The work was inspired by innovative and intriguing work of Hult and Lindskog [95–97] who show that when the Lévy measure of a Lévy process $X(\cdot)$ is regularly varying, the distribution $P[X \in \cdot\,]$ is regularly varying on $\mathbb{D}[0, 1] \setminus \{\mathbf{0}\}$ *with a limit measure concentrating on functions which are constant except for one jump*. Where did the other infinitely many Lévy process jumps go? Using weaker scaling and biting more out of $\mathbb{D}[0, 1]$ than just the zero-function $\mathbf{0}$ allows recovery of the other jumps and gives another example of steroidal regular variation.

This chapter requires knowledge of Lévy processes (see [1, 9, 117]). The discussion in [156, Section 5.5] is just about enough. You should also have some knowledge of weak convergence and the complete separable metric space $\mathbb{D}[0, 1]$ [13, 156], the space of right continuous functions on [0, 1) with finite left-hand limits on (0, 1]. Rapid review is in the next section.

If you were delighted by Sect. 3.2, you will love this discussion. You could also check related work and extensions to large deviations in [8, 98, 167].

© The Author(s), under exclusive license to Springer Nature Switzerland AG 2024
S. Resnick, *The Art of Finding Hidden Risks*,
https://doi.org/10.1007/978-3-031-57599-0_4

4.1 Background

In this section we consider a real-valued Lévy process $X = \{X(s), s \in [0, 1]\}$ as a random element of $\mathbb{D} := \mathbb{D}[0, 1] = \mathbb{D}([0, 1], \mathbb{R})$, the space of real-valued càdlàg functions on [0, 1]. Here are the basics on Lévy processes and also the Skorohod metric that is equivalent to a metric that makes $\mathbb{D}$ into a complete, separable metric space.

4.1.1 Skorohod Metric

We metrize $\mathbb{D}$ with the usual Skorohod metric. Let Λ be the set of all strictly increasing homeomorphisms of [0, 1] onto [0, 1] and $e(s) = s$ is the identity. Set $\|x\| = \sup_{s \in [0,1]} |x(s)|$ as the sup-norm for functions in $\mathbb{D}$. Define the Skorohod metric $d_{\text{sk}}(\cdot, \cdot)$ as

$$d_{\text{sk}}(x, y) = \inf_{\lambda \in \Lambda} \|\lambda - e\| \vee \|x \circ \lambda - y\|, \quad x, y \in \mathbb{D}.$$

The space $\mathbb{D}$ is not complete under the metric d_{sk}, but there is an equivalent metric under which $\mathbb{D}$ is complete [13, page 125]. Therefore, the space $\mathbb{D}$ fits into the framework presented in Chap. 1 and we may use the Skorohod metric to check continuity of mappings.

4.1.2 The Lévy Process X

For simplicity we suppose X has only positive jumps and its Lévy measure ν concentrates on $(0, \infty)$. The requirement that ν be a Lévy measure means

$$\int_{\mathbb{R}_+} (x^2 \wedge 1) \nu(dx) < \infty.$$

4.1.2.1 Regular Variation Assumption

As in Sect. 3.2.2, suppose $x \mapsto \nu(x, \infty)$ is regularly varying at infinity with index $-\alpha < 0$ and now due to the connection to the Lévy process, suppose also that $0 < \alpha < 2$. Let $Q(x) := \nu([x, \infty))$ and define $Q^{\leftarrow}(y) := \inf\{t > 0 : \nu([t, \infty)) < y\}$. Then the function $b(\cdot)$ given by $b(t) = Q^{\leftarrow}(1/t)$ satisfies $\lim_{t \to \infty} t\nu(b(t)x, \infty) = x^{-\alpha}$ and b is regularly varying at infinity with index $1/\alpha$.

4.1.2.2 Itô Representation

The standard Itô representation [1, 9, 117] of X, assuming $X(0) = 0$, is

$$X(s) = sa + B(s) + \int_{|x|\leq 1} x[N([0,s] \times dx) - s\nu(dx)] + \int_{|x|>1} xN([0,s] \times dx), \tag{4.1}$$

where B is standard Brownian motion independent of the Poisson random measure N on $[0, 1] \times (0, \infty)$ with mean measure Leb $\times \nu$ and Leb is Lebesgue measure on $[0, 1]$. We can and do suppose the paths of X are in $\mathbb{D}$ and that X is a random element of $\mathbb{D}$.

4.1.2.3 Representation of the Poisson Random Measure

We can give an explicit representation of the Poisson random measure N as follows. Referring to the discussion on page 103 between (3.13) and (3.14), recall that $\{Q^{\leftarrow}(\Gamma_n), n \geq 1\}$ are points of a Poisson random measure on $(0, \infty)$ written in decreasing order with mean measure ν. Let $(U_l, l \geq 1)$ be iid standard uniform random variables independent of $\{\Gamma_n\}$ and by augmentation [152, p. 317] and [156, p. 122]

$$N = \sum_{l=1}^{\infty} \epsilon_{(U_l, Q^{\leftarrow}(\Gamma_l))},$$

is a Poisson random measure on $[0, 1] \times \mathbb{R}_+$ with mean measure Leb $\times \nu$.

The Lévy-Itô decomposition allows X to be decomposed into the sum of two independent Lévy processes,

$$X = \tilde{X} + J, \tag{4.2}$$

where J is a compound Poisson process of large jumps bounded below by 1, and $\tilde{X} = X - J$ is a Lévy process of small jumps that are bounded above by 1. The compound Poisson process can be represented as the random sum

$$J(s) = \sum_{l=1}^{N_1} Q^{\leftarrow}(\Gamma_l) 1_{[U_l, 1]}(s), \quad 0 \leq s \leq 1 \tag{4.3}$$

where $N_1 = N([0, 1] \times [1, \infty))$ is the Poisson distributed random variable with mean $Q(1)$. There is also the usual compound Poisson process representation

$$J(s) = \sum_{i=1}^{P(s)} \mathcal{J}_i, \quad 0 \le s \le 1 \tag{4.4}$$

where $\{\mathcal{J}_i, i \ge 1\}$ are iid with distribution $P[\mathcal{J}_i > x] = Q(x)/Q(1)$, $x \ge 1$ and $\{P(s), s > 0\}$ is homogeneous Poisson, rate $Q(1)$ and $P(\cdot) \perp\!\!\!\perp \{\mathcal{J}_i\}$.

4.1.2.4 Notation Review

In case you do not have a photographic memory, now is a good time to recall the notation in (3.14), page 104 for $\mathbb{R}_+^{\infty \downarrow}$, $\mathbb{H}_{=j}$ and $\mathbb{H}_{\le j}$ and the result in Theorem 3.1, page 109. Recall $\nu_\alpha(\cdot)$ is the Pareto measure on $(0, \infty)$ satisfying $\nu_\alpha(x, \infty) = x^{-\alpha}$, for $x > 0$, while $\eta^{(j)}(\cdot)$ is a measure concentrating on $\mathbb{H}_{=j}$ given by

$$\eta^{(j)}(dx_1, dx_2, \ldots) = \prod_{i=1}^{j} \nu_\alpha(dx_i) 1_{[x_1 \ge x_2 \ge \cdots \ge x_j > 0]} \prod_{i=j+1}^{\infty} \epsilon_0(dx_i).$$

For $m \ge 0$, let $\mathbb{D}_{\le m} = \mathbb{D}_{\le m}[0, 1]$ be the closed subspace of the Skorohod space $\mathbb{D}$ consisting of nondecreasing step functions on $[0, 1]$ with at most m jumps and $\mathbb{D}_{=m}$ is the subspace of the Skorohod space $\mathbb{D}$ consisting of nondecreasing step functions on $[0, 1]$ with exactly m jumps. Note $\mathbb{D}_{\le 0} = \mathbb{D}_0$ are constant functions and that $P[X(0) = 0] = 1$.

The plan is to convert an $\mathbb{R}_+^\infty$ regular variation statement like (3.24), page 109, into a regular variation statement about $P[X \in \cdot]$. Based on Problem 2.10, page 87 (see also Problem 1.14, page 27) we expect the distribution of $\boldsymbol{J}$ in (4.2) to have a regular variation property and hopefully $\widetilde{X}$ has the good manners to be asymptotically negligible.

4.2 Steroidal Regular Variation of $P[X \in \cdot]$

Now we engage regular variation of measures on the function space $\mathbb{D}$. We seek results in the simplest case that

$$P[X \in \cdot] \in \mathrm{MRV}(\alpha, b(t), \eta, \mathbb{D} \setminus \mathbb{D}_{\le 0})$$

which is close to the original Hult-Lindskog result.

Theorem 4.1 ([95, 97]) *Suppose the regular variation assumptions in Sect. 4.1.2.1 hold. Then $P[X \in \cdot]$ is a regularly varying measure:*

4.2 Steroidal Regular Variation of $P[X \in \cdot]$

$$tP\left[\frac{X}{b(t)} \in \cdot\right] \to \eta(\cdot) \qquad (4.5)$$

in $\mathbb{M}(\mathbb{D} \setminus \mathbb{D}_0)$, where

- $\eta(\cdot)$ is a measure on $\mathbb{D}$ that concentrates on $\mathbb{D}_{=1}$, functions which are constant except for one single positive jump.
- The time of the single jump is uniform on $(0, 1)$ and the size of the jump is governed by the Pareto measure $\nu_\alpha(\cdot)$.

Theorem 4.1 provides easy access to half the Embrechts and Goldie result [72], [15, p. 341], and [76, Section XVII.4] that says if the Lévy measure of a Lévy process is regularly varying, then the one-dimensional marginal distributions of X are regularly varying. We will consider the converse later.

Corollary 4.1 *Suppose $Q(x) \in RV_{-\alpha}$, $0 < \alpha < 2$; that is, suppose the conditions of Sect. 4.1.2.1, page 122 hold. Then as $x \to \infty$,*

$$P[X(1) > x] \sim Q(x)$$

and

$$\lim_{t \to \infty} tP\left[\frac{X(1)}{b(t)} > x\right] = x^{-\alpha}, \quad x > 0.$$

We will verify Corollary 4.1 in Sect. 4.2.3.3 after we develop machinery for the proof of Theorem 4.1.

Of course, Theorem 4.1 raises the question of what happened to the other jumps of the Lévy process and leads to steroidal regular variation.

Theorem 4.2 ([125]) *Assume the regular variation assumptions in Sect. 4.1.2.1. For every $j \geq 1$, we have in $\mathbb{M}(\mathbb{D} \setminus \mathbb{D}_{\leq j-1})$ as $t \to \infty$,*

$$P_t^{(j)}(\cdot) := tP\left[\frac{X}{b(t^{1/j})} \in \cdot\right] \to (\eta^{(j)} \times \mathcal{L}) \circ T_j^{-1}(\cdot) =: P_\infty^{(j)}(\cdot), \qquad (4.6)$$

where $\mathcal{L}$ is Lebesgue measure on $[0, 1]^\infty$ and T_j is a (currently mysterious) mapping explained below in (4.9) on page 127. The limit measure concentrates on $\mathbb{D}_{=j}$.

The rough description of the limit in (4.6), is that the measure concentrates on non-decreasing functions with j positive jumps where the jump sizes are ordered selections chosen from the product measure ν_α^j; these jump sizes are mated with jump times that are iid uniform on $[0, 1]$. The limit in (4.6) can also be written

$$E\left[\nu_\alpha^j\left\{\mathbf{y} \in (0, \infty)^j : \sum_{i=1}^j y_i 1_{[U_i, 1]} \in \cdot\right\}\right], \qquad (4.7)$$

where $\{U_i\}$ is the sequence of iid standard uniform random variables and ν_α^j is the product measure with j-factors generated by $\nu_\alpha(\cdot)$.

4.2.1 The Plan[1]

The plan is to convert an $\mathbb{R}_+^\infty$ regular variation statement like (3.24), page 109, into a regular variation statement about $P[X \in \cdot]$. Since the Poisson points in (3.24) serve as the jumps of the Lévy process ordered in decreasing order, this is a plausible approach if we figure out how to add jump times to the jump sizes. Thus, we attempt to leverage the regular variation (3.24) into something in $\mathbb{D}$ using mappings.

So the first step is to augment the jump sizes (3.24), with a sequence of iid standard uniform random varables which serve as jump times for the Lévy process. The First Binding Lemma, Proposition 2.1, page 47 handles the augmentation.

Proposition 4.1 *Under the regular variation assumptions on ν and Q in Sect. 4.1.2.1, for any $j \geq 1$,*

$$tP\left[\left((Q^{\leftarrow}(\Gamma_l)/b(t^{1/j}), l \geq 1), (U_l, l \geq 1)\right) \in \cdot\right] \to (\eta^{(j)} \times \mathcal{L})(\cdot) \qquad (4.8)$$

in $\mathbb{M}((\mathbb{R}_+^{\infty\downarrow} \setminus \mathbb{H}_{\leq j-1}) \times [0, 1]^\infty)$ *as* $t \to \infty$, *where* $\mathcal{L}$ *is Lebesgue measure on* $[0, 1]^\infty$, $\eta^{(j)}$ *concentrates on* $\mathbb{H}_{=j}$ *and is given by (3.25), page 109.*

Think of (4.8) as regular variation on the product space $(\mathbb{R}_+^{\infty\downarrow} \setminus \mathbb{H}_{\leq j-1}) \times [0, 1]^\infty$ with scalar multiplication defined as $(\lambda, (\mathbf{x}, \mathbf{y})) \mapsto (\lambda\mathbf{x}, \mathbf{y})$.

After the augmentation in (4.8), the next step is to apply a map to (4.8) that maps jump times and jump sizes into a step function in $\mathbb{D}$. However, this procedure has difficulties:

- From (4.1) or (4.2) we see we will have to deal with infinitely many centered small jumps of $\widetilde{X}$ as well as the Brownian component.
- The Skorohod metric is hard to handle if there are infinitely many jumps.

So in (4.2), only $\mathbf{J}$ seems amenable to the program and even just for $\mathbf{J}$ there is the difficulty of a random number of jumps to overcome.[2]

---

[1] Mike Tyson: "Everyone has a plan until they get punched in the mouth."
[2] The wise Joseph Gani once said: "You know Sid, life wasn't meant to be easy."

4.2.2 The Mapping

First, we think about how to proceed when there are finitely many jumps and jump times. Define $T_m : \mathbb{R}_+^{\infty\downarrow} \times [0, 1]^\infty \mapsto \mathbb{D}$ by

$$T_m(\boldsymbol{x}, \boldsymbol{u})(s) = \sum_{i=1}^{m} x_i 1_{[u_i, 1]}(s), \quad 0 \le s \le 1. \tag{4.9}$$

Think of T_m as mapping a sequence $\boldsymbol{x}$ of jump-sizes and a sequence of distinct jump times $\boldsymbol{u}$ into a step function in $\mathbb{D}_{\le m}([0, 1]) \subset \mathbb{D}$. We use $\mathbb{D}_{\le m}$ and not $\mathbb{D}_{=m}$ in case $\{u_i\}$ are not all distinct.

We hope to apply T_m to (4.8) and then have Dr. Nefario replace m by the random N_1 defined in (4.3), page 123 to get regular variation of the distribution of $\boldsymbol{J}$. For justifying the application of T_m to (4.8), the Second Mapping Theorem 1.3, page 18 that relies on uniform continuity would be the most convenient to use (see for example Sect. 3.2.1.1) but unfortunately T_m is not uniformly continuous and, in fact, we only prove it continuous on a subset A_m of $\mathbb{R}_+^{\infty\downarrow} \times [0, 1]^\infty$,

$$A_m = \{(\boldsymbol{x}, \boldsymbol{u}) \in \mathbb{R}_+^{\infty\downarrow} \times [0, 1]^\infty : \tag{4.10}$$
$$u_i \in (0, 1) \text{ for } 1 \le i \le m; u_i \ne u_j \text{ for } i \ne j, 1 \le i, j \le m\}.$$

So we rely on the First Mapping Theorem 1.2, page 16 and its Corollary 1.1 and must verify that T_m is continuous and that it respects the forbidden zone according to Definition 1.2, page 16.

4.2.2.1 Properties of the Map T_m

The definitions of T_m and A_m are in (4.9) and (4.10) and we show T_m respects the relevant forbidden zone and that it is continuous on A_m. Aretha suggests we start with R-E-S-P-E-C-T for the forbidden zone.

Lemma 4.1 *Fix $j \ge 1$. For $m \ge j$, the map*

$$T_m : (\mathbb{R}_+^{\infty\downarrow} \times [0, 1]^\infty) \setminus (\mathbb{H}_{\le j-1} \times [0, 1]^\infty) \mapsto \mathbb{D} \setminus \mathbb{D}_{\le j-1}$$

respects the forbidden zone: Suppose $m \ge j$ and $A \subset \mathbb{D}$ is bounded away from $\mathbb{D}_{\le j-1}$. If $T_m^{-1}(A)$ is nonempty, then $T_m^{-1}(A)$ is bounded away from $\mathbb{H}_{\le j-1} \times [0, 1]^\infty$.

Proof If $A \cap \mathbb{D}_{\le m} = \emptyset$, the definitions imply $T_m^{-1}(A) = \emptyset$. Therefore, without loss of generality we take $A \subset \mathbb{D}_{\le m}$. Assume $d_{sk}(A, \mathbb{D}_{\le j-1}) > \delta > 0$ and notice that $\boldsymbol{x} \in \mathbb{D}_{\le m}$ if and only if

$$x(\cdot) = \sum_{i=1}^{m} y_i 1_{[u_i,1]}(\cdot), \quad \text{for } y_1 \geq \cdots \geq y_m \geq 0, u_i \in [0,1].$$

If $x \in A \subset \mathbb{D}_{\leq m}$, $\sum_{i=j}^{m} y_i > \delta$ as a consequence of $d_{sk}(A, \mathbb{D}_{\leq j-1}) > \delta$ since if we take the distance between x and $\tilde{x} = \sum_{i=1}^{j-1} y_i 1_{[u_i,1]}(\cdot) \in \mathbb{D}_{\leq j-1}$, we have

$$\delta < d_{sk}(A, \mathbb{D}_{\leq j-1}) \leq d_{sk}(x, \tilde{x}) \leq d_{sk}(\sum_{i=1}^{m} y_i 1_{[u_i,1]}, \sum_{i=1}^{j-1} y_i 1_{[u_i,1]}, 0) \leq \sum_{i=j}^{m} y_i$$

Since the y's are non-increasing, $y_j > \delta/(m - j + 1)$. Consequently,

$$T_m^{-1}(A) \subset \left\{ (x_i, i \geq 1) \in \mathbb{R}_+^{\infty \downarrow} : x_j > \delta/(m - j + 1) \right\} \times [0,1]^\infty,$$

and the latter set is bounded away from $\mathbb{H}_{\leq j-1} \times [0,1]^\infty$. □

Lemma 4.2 *For $m \geq 1$, $T_m : A_m \mapsto \mathbb{D}$ is continuous.*

Proof First project A_m into a finite dimensional space. Let

$$(0,1)^{m,\neq} = \{(u_1, \ldots, u_m) \in (0,1)^m : u_i \neq u_j \text{ for } i \neq j\},$$

vectors of length m with no repetitions, and define the map

$$A_m \ni (x, u) \mapsto ((x_1, \ldots, x_m), (u_1, \ldots, u_m)) \in \mathbb{R}_+^{m \downarrow} \times (0,1)^{m,\neq}.$$

This map is continuous. Since the composition of two continuous functions is continuous, it remains to check that the function

$$\mathbb{R}_+^{m \downarrow} \times (0,1)^{m,\neq} \ni ((x_1, \ldots, x_m), (u_1, \ldots, u_m)) \mapsto \sum_{i=1}^{m} x_i 1_{[u_i,1]} \in \mathbb{D} \quad (4.11)$$

is continuous at any fixed domain point $(x, u) \in \mathbb{R}_+^{m \downarrow} \times (0,1)^{m,\neq}$. Define a metric on $\mathbb{R}^{2m}$,

$$d_{2m}((x, u), (\tilde{x}, \tilde{u})) = \vee_{i=1}^{m} |x_i - \tilde{x}_i| \bigvee \vee_{i=1}^{m} |u_i - \tilde{u}_i|.$$

There exists some $\delta = \delta(u) > 0$ such that, for $(\tilde{x}, \tilde{u}) \in \mathbb{R}_+^{m \downarrow} \times (0,1)^{m,\neq}$, $d_{2m}((x, u), (\tilde{x}, \tilde{u})) < \delta$, implies that the components of $\tilde{u}$ appear in the same order as do the components of u.

4.2 Steroidal Regular Variation of $P[X \in \cdot]$

To see this, order the components of $\boldsymbol{u}$ as $0 = u_{(0)} < u_{(1)} < \cdots u_{(m)} < u_{(m+1)} = 1$, with corresponding notation for the ordered $\tilde{u}$'s. Remember $\boldsymbol{u}$ is fixed and pick[3]

$$\delta < \frac{1}{37} \bigvee_{i=0}^{m} |u_{(i+1)} - u_{(i)}|$$

so the δ-neighborhoods of each component of $\boldsymbol{u}$ contain exactly one component of $\boldsymbol{u}$. If $\tilde{\boldsymbol{u}}$ is δ-close to $\boldsymbol{u}$, these neighborhoods can only contain a single component of $\tilde{\boldsymbol{u}}$ and the orderings of the components of $\boldsymbol{u}$ and $\tilde{\boldsymbol{u}}$ cannot be different.

We now show the induced step functions are close when $d_{2m}((\boldsymbol{x}, \boldsymbol{u}), (\tilde{\boldsymbol{x}}, \tilde{\boldsymbol{u}})) < \delta$. Consider the piece-wise linear function λ_l for which $\lambda_l(0) = 0$, $\lambda_l(1) = 1$, and $\lambda_l(u_i) = \tilde{u}_i$ for each i. Notice that λ_l is strictly increasing and satisfies $\|\lambda_l - e\| < \delta$. Therefore,

$$\sup_{s \in [0,1]} \left| \sum_{i=1}^{m} x_i 1_{[\lambda_l(u_i),1]}(s) - \sum_{i=1}^{m} \tilde{x}_i 1_{[\tilde{u}_i,1]}(s) \right| < \sum_{i=1}^{m} |x_i - \tilde{x}_i| < m\delta.$$

In particular,

$$d_{\text{sk}}\left(\sum_{i=1}^{m} x_i 1_{[u_i,1]}, \sum_{i=1}^{m} \tilde{x}_i 1_{[\tilde{u}_i,1]} \right) < m\delta,$$

which shows that the mapping in (4.11) is continuous and hence also T_m is continuous on A_m. $\square$

Now apply T_m to (4.8), page 126 using the Corollary 1.1, page 17.

Corollary 4.2 *Assume the regular variation assumptions on v and Q in Sect. 4.1.2.1. For any $j \geq 1$ and $m \geq j$, as $t \to \infty$,*

$$tP\left[\sum_{l=1}^{m} \frac{Q^{\leftarrow}(\Gamma_l)}{b(t^{1/j})} 1_{[U_l,1]} \in \cdot \right] \to (\eta^{(j)} \times \mathcal{L}) \circ T_j^{-1}(\cdot) \quad (4.12)$$

in $\mathbb{M}(\mathbb{D} \setminus \mathbb{D}_{\leq j-1})$.

Proof To see how Corollary 1.1, page 17, applies rely on the convergence (4.8), page 126 in $\mathbb{M}((\mathbb{R}_+^{\infty \downarrow} \setminus \mathbb{H}_{\leq j-1}) \times [0,1]^{\infty})$. To match up with the notation in Corollary 1.1 set

$$\mu(\cdot) = \eta^{(j)} \times \mathcal{L}(\cdot)$$

---

[3] Dividing by 37 is excessive but why not?

and Lemma 4.2, page 128 insists we check that

$$\mu\{(x,u) \in \mathbb{R}_+^{\infty\downarrow} \setminus \mathbb{H}_{\leq j-1} : T_m \text{ discontinuous at } (x,u)\}$$
$$\leq \mu\{(x,u) : \mathbb{R}_+^{\infty} \times [0,1]^{\infty} \setminus A_m\}$$
$$\leq \mu\{(x,u) \in \mathbb{R}_+^{\infty} \times [0,1]^{\infty} : \exists l, l' \text{ and } u_l = u_{l'}, 1 \leq l, l' \leq m\} = 0.$$

So $\hat{T}_m$ is continuous at $\mu(\cdot)$.

The only further comment required is why the right side of (4.12) is independent of $m \geq j$. This is an immediate consequence of where $\eta^{(j)}$ concentrates. Let $f \in C(\mathbb{D} \setminus \mathbb{D}_{\leq j-1})$, so the support of f is bounded away from $\mathbb{D}_{\leq j-1}$. Then for $m \geq j$, we must show

$$\eta^{(j)} \times \mathcal{L} \circ T_m^{-1}(f) = \eta^{(j)} \times \mathcal{L} \circ T_j^{-1}(f).$$

We have merely to unpack the symbols:

$$\eta^{(j)} \times \mathcal{L} \circ T_m^{-1}(f) = \eta^{(j)} \times \mathcal{L} \circ (f \circ T_m)$$
$$= \iint f\left(\sum_{l=1}^m x_i 1_{[u_i,1]}\right) \prod_{i=1}^j \nu_\alpha(dx_i) 1_{[x_1 \geq x_2 \geq \cdots \geq x_j > 0]} \prod_{i=j+1}^{\infty} \epsilon_0(dx_i) \prod_{i=1}^{\infty} du_i$$
$$= \iint f\left(\sum_{l=1}^j x_i 1_{[u_i,1]}\right) \prod_{i=1}^j \nu_\alpha(dx_i) 1_{[x_1 \geq x_2 \geq \cdots \geq x_j > 0]} \prod_{i=j+1}^{\infty} \epsilon_0(dx_i) \prod_{i=1}^{\infty} du_i$$
$$= \eta^{(j)} \times \mathcal{L} \circ T_j^{-1}(f),$$

and this suffices. $\square$

4.2.2.2 Ta-dah! The Payoff for J: Steroidal Regular Variation

This section shows that J, the process of large jumps, possesses steroidal regular variation. For this section, fix $j \geq 1$ and assume the regular variation assumptions of Sect. 4.1.2.1. Due to Corollary 4.2, we basically have to replace m by the Poisson variable N_1. To simplify notation, we use the shorthand

$$\sum_{l=1}^m := \sum_{l=1}^m \frac{Q^{\leftarrow}(\Gamma_l)}{b(t^{1/j})} 1_{[U_l, 1]},$$

Ta-dah!

4.2 Steroidal Regular Variation of $P[X \in \cdot]$

and keep in mind the scaling by $b(t^{1/j})$ is built into this shorthand. It is tempting to use Corollary 4.2, page 129, by comparing $\sum_1^{N_1}$ with $\sum_1^m$ for large m but this does not quite suffice. Instead, pick $\beta \in (j/(j+1), 1)$ and define the integer-valued non-decreasing step function,

$$M_t := \sum_{i=1}^{N_1} 1_{[Q^{\leftarrow}(\Gamma_i) \in (b^{\beta}(t^{1/j}), \infty)]}. \tag{4.13}$$

Note

1. As $t \to \infty$, $b(t) \to \infty$ and therefore, $M_t \to 0$.
2. As a function of t, $b^{\beta}(t^{1/j}) \in RV_{\beta/(j\alpha)}$.
3. We must have $M_t \leq N_1$.
4. For $M_t < k \leq N_1$, we must have

$$Q^{\leftarrow}(\Gamma_k) \leq b^{\beta}(t^{1/j}). \tag{4.14}$$

5. As $t \to \infty$,

$$tP[M_t \geq j+1] \to 0. \tag{4.15}$$

The reason for (4.15) is

$$tP[M_t \geq j+1] \leq tP[Q^{\leftarrow}(\Gamma_{j+1}) > b^{\beta}(t^{1/j})].$$

Now $P[Q^{\leftarrow}(\Gamma_{j+1}) > x] \in RV_{-\alpha(j+1)}$ since as $x \to \infty$,

$$P[Q^{\leftarrow}(\Gamma_{j+1}) > x] = P[\Gamma_{j+1} \leq Q(x)]$$
$$= \int_0^{Q(x)} e^{-u} u^j / j! du \sim Q^{j+1}(x)/(j+1)!.$$

So $P[Q^{\leftarrow}(\Gamma_{j+1}) > b^{\beta}(t^{1/j})]$ is the composition of two regularly varying functions. Use Item 2 and the composition is thus regularly varying with index

$$-\frac{\beta}{j\alpha} \cdot \alpha(j+1) = -\beta \frac{j+1}{j}.$$

Since $\beta \frac{j+1}{j} > 1$, we can cope with multiplication by t in (4.15) and convergence to 0 follows.

Here is the formal statement of steroidal regular variation for $P[J \in \cdot]$.

Proposition 4.2 *For every $j \geq 1$, as $t \to \infty$,*

132   4 Lévy Processes with Regularly Varying Distributions: Where Do the Jumps Go?

$$tP[\boldsymbol{J} \in b(t^{1/j}) \cdot] \to (\eta^{(j)} \times \mathcal{L}) \circ T_j^{-1}(\cdot), \quad \text{in } \mathbb{M}(\mathbb{D} \setminus \mathbb{D}_{\leq j-1}).$$

Proof The proof proceeds in two steps. Step 1: Compare $\sum_1^{N_1}$ with $\sum_1^{M_t}$, and Step 2: Compare $\sum_1^{M_t}$ with $\sum_1^{j}$ relying on Corollary 4.2, page 129, to get the required behavior of $\sum_1^{j}$.

Let $f \in C(\mathbb{D} \setminus \mathbb{D}_{\leq j-1})$ be uniformly continuous on $\mathbb{D}$ with support $\operatorname{supp}(f) =: F$ bounded away from $\mathbb{D}_{\leq j-1}$, say

$$d_{\text{sk}}(F, \mathbb{D}_{\leq j-1}) = 2\delta. \tag{4.16}$$

This f is a bounded, positive, real valued test function with upper bound $\|f\|$ that allows convergence results to be concluded from Theorem 1.1 and (1.11), page 9. The modulus of continuity of f is $\omega_f(\delta)$ which is monotone in δ.

Before plunging into details, two observations:

1. If $x(\cdot) \in \mathbb{D} \setminus \mathbb{D}_{\leq j-1}$ is non-decreasing, the function $x(\cdot)$ must have at least j jumps.
2. Consider $\boldsymbol{x} \in R_+^{\infty,\downarrow}$ and $\boldsymbol{u} \in [0,1]^\infty$ and suppose $j < m$. Then since the Skorohod metric is bounded above by the uniform metric, we get a bound on the metric distance between two jump functions built from $\boldsymbol{x}$ and $\boldsymbol{u}$,

$$d_{\text{sk}}\left(\sum_{l=1}^{j} x_l 1_{[u_l,1]}, \sum_{l=1}^{m} x_l 1_{[u_l,1]}\right) \leq \sup_{0 \leq s \leq 1} \left| \sum_{l=1}^{j} x_l 1_{[u_l,1]}(s) - \sum_{l=1}^{m} x_l 1_{[u_l,1]}(s) \right|$$

$$= \sup_{0 \leq s \leq 1} \left| \sum_{l=j+1}^{m} x_l 1_{[u_l,1]}(s) \right| \leq \sum_{l=j+1}^{m} x_l. \tag{4.17}$$

Step 1. Compare $\sum_1^{N_1}$ and $\sum_1^{M_t}$. Set $\epsilon > 0$ and write,

$$t\left|Ef(\sum_1^{N_1}) - Ef(\sum_1^{M_t})\right| \leq tE\left|f(\sum_1^{N_1}) - f(\sum_1^{M_t})\right| 1_{[\sum_{l=M_t+1}^{N_1} Q^\leftarrow(\Gamma_l) \leq \epsilon b(t^{1/j})]}$$

$$+ tE\left|f(\sum_1^{N_1}) - f(\sum_1^{M_t})\right| 1_{[\sum_{l=M_t+1}^{N_1} Q^\leftarrow(\Gamma_l) > \epsilon b(t^{1/j})]} = A + B. \tag{4.18}$$

From (4.17) we see A is the case where the two normalized step functions are close and B is the case where they are not. For B we have

$$B \leq 2\|f\| tP[\sum_{l=M_t+1}^{N_1} Q^\leftarrow(\Gamma_l) > \epsilon b(t^{1/j})]$$

4.2 Steroidal Regular Variation of $P[X \in \cdot]$

and using the observation in (4.14), this has bound

$$\leq 2\|f\| t P[b^\beta(t^{1/j})(N_1 - M_t) > b(t^{1/j})\epsilon]$$
$$\leq 2\|f\| t P[N_1 > b^{1-\beta}(t^{1/j})\epsilon]$$

and now we are optimistic this goes to 0 because N_1, being Poisson, has all its moments. So pick large p, use Markov's inequality and this has the bound,

$$\leq 2\|f\| t \frac{E(N_1^p)}{\epsilon^p b^{(1-\beta)p}(t^{1/j})}.$$

The bound is of the form $t/V(t)$ where $V \in RV_{(1-\beta)p/(j\alpha)}$ so if we choose p so that $p > j\alpha/(1-\beta)$ then $(1-\beta)p/(j\alpha) > 1$ and the bound converges to 0 with $t \to \infty$. So we have disposed of B. Yay!

Now consider A, knowing that on $[\sum_{l=M_t+1}^{N_1} Q^\leftarrow(\Gamma_l) \leq \epsilon b(t^{1/j})]$, the difference of the f's can be bounded by $\omega_f(\epsilon)$. Remember $F = \mathrm{supp}(f)$ and observe that on $[\sum_{l=M_t+1}^{N_1} Q^\leftarrow(\Gamma_l) \leq \epsilon b(t^{1/j})]$, if

$$\sum_1^{N_1} \notin F \text{ and } \sum_1^{M_t} \notin F, \text{ then } A = 0,$$

$$\sum_1^{N_1} \notin F \text{ and } \sum_1^{M_t} \in F, \text{ then } \sum_1^{M_t} \in \overline{F^\epsilon}, \quad (\text{since } F \subset \overline{F^\epsilon})$$

$$\sum_1^{N_1} \in F \text{ and } \sum_1^{M_t} \notin F, \text{ then } \sum_1^{M_t} \in \overline{F^\epsilon},$$

(by (4.17), since $\sum_1^{M_t}$ is ϵ-close to $\sum_1^{N_1}$)

$$\sum_1^{N_1} \in F \text{ and } \sum_1^{M_t} \in F, \text{ then } \sum_1^{M_t} \in F \subset \overline{F^\epsilon}.$$

Thus

$$A \leq \omega_f(\epsilon) t P[\sum_1^{M_t} \in \overline{F^\epsilon}]$$

and provided $\epsilon < \delta$, we have $\overline{F^\epsilon} \cap \mathbb{D}_{\leq j-1} = \emptyset$. Now by observation 1, page 132, if $M_t < j$, for any fixed t, the non-decreasing $\mathbb{D}$-function of s, $\sum_1^{M_t} Q^\leftarrow(\Gamma_l)/b(t^{1/j}) 1_{[U_l, 1]}(s)$ is in $\mathbb{D}_{\leq j-1}$ and thus cannot be in $\overline{F^\epsilon}$. Thus, A has the upper bound

$$\leq \omega_f(\epsilon) t P[\sum_1^{M_t} \in \overline{F^\epsilon}, M_t \geq j]$$

and since $[M_t \geq j] = [M_t = j] \cup [M_t \geq j+1]$, we have the new bound,

$$\leq \omega_f(\epsilon) t \left(P[\sum_1^j \in \overline{F^\epsilon}] + P[M_t \geq j+1] \right).$$

We already know how to kill the second term in the bound using (4.15) which says $\lim_{t\to\infty} t P[M_t \geq j+1] \to 0$. For the first term, we rely on Theorem 1.1 and (1.12), page 9 as well as Corollary 4.2 to get

$$\limsup_{t\to\infty} \omega_f(\epsilon) t P[\sum_1^j \in \overline{F^\epsilon}] \leq \omega_f(\epsilon)(\eta^{(j)} \times \mathcal{L}) \circ T_j^{-1}(\overline{F^\epsilon})$$

For small $\epsilon < \delta$, $\overline{F^\epsilon}$ is bounded away from $\mathbb{D}_{\leq j-1}$ so the measure on the right is always finite. Since $\lim_{\epsilon \to 0} \omega_f(\epsilon) = 0$, the right side can be made as small as desired. This concludes Step 1.

Step 2. Compare $\sum_1^{M_t}$ and $\sum_1^j$. With the test function f as in Step 1, we write

$$tE\left|f(\sum_1^{M_t}) - f(\sum_1^j)\right| = tE|\cdot|\left(1_{[M_t<j]} + 1_{[M_t=j]} + 1_{[M_t\geq j+1]}\right) = C1 + C2 + C3.$$

Now as $t \to \infty$, we have applying (4.15),

$$C3 \leq 2\|f\| t P[M_t \geq j+1] \to 0,$$

and $C2 = 0$ which leaves $C1$ for analysis. On $[M_t < j]$, $f(\sum_1^{M_t}) = 0$ because having too few jumps puts the function in $\mathbb{D}_{\leq j-1}$ and hence away from supp(f). Further, on $[M_t < j]$, using (4.17),

$$d_{\text{sk}}(\sum_1^j, \sum_1^{M_t}) \leq \sum_{l=M_t+1}^j \frac{Q^{\leftarrow}(\Gamma_l)}{b(t^{1/j})}$$

and applying the bound in (4.14), the right side above is dominated by,

4.2 Steroidal Regular Variation of $P[X \in \cdot]$

$$\leq \sum_{l=M_t+1}^{j} \frac{b^\beta(t^{1/j})}{b(t^{1/j})} \leq \frac{j}{b^{1-\beta}(t^{1/j})} \to 0 \quad (t \to \infty).$$

So, on $[M_t < j]$, $\sum_{1}^{M_t}$ is not in the support of f because it is a function in $\mathbb{D}_{\leq j-1}$ and $\sum_{1}^{j}$ can be made arbitrarily close by choice of large enough t. Therefore for large enough t, $\sum_{1}^{j}$ is not in the support either. Hence for large t, $C1$ is 0. This completes Step 2.

The proof is complete by appealing to (4.12), page 129. □

4.2.3 Filling the Gap: Sweating the Small Stuff to Get Regular Variation of $P[X \in \cdot]$

The heavy tail properties of X are carried by J and we now fill the gap between X and J by addressing why small jumps plus Brownian component do not influence heavy tail properties. Recall $X = \widetilde{X} + J$ where

$$J(s) = \sum_{l=1}^{N_1} Q^{\leftarrow}(\Gamma_l) 1_{[U_l, 1]}(s), \ 0 \leq s \leq 1.$$

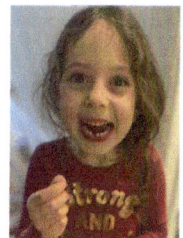

Mind the gap!

4.2.3.1 Negligibility of $\widetilde{X}$

The first step is to show a negligibility property of $\widetilde{X}$ after which we put the pieces together to get the steroidal regular variation of $P[X \in \cdot]$. The process $\widetilde{X}$ represents small jumps that should not affect asymptotics. Here is the precise result.

Lemma 4.3 *For $j \geq 1$, and any $\delta > 0$,*

$$\limsup_{t \to \infty} t P\left[\sup_{s \in [0,1]} |\widetilde{X}(s)| > b(t^{1/j})\delta\right] = 0.$$

Proof Use Skorohod's inequality for Lévy processes [18], [153, Section 7.3]:

$$P\left[\sup_{s \in [0,1]} |\widetilde{X}(s)| > 2a\right] \leq (1 - c(a))^{-1} P[|\widetilde{X}(1)| > a], \quad a > 0,$$

where $c(a) = \sup_{s \in [0,1]} P[|\widetilde{X}(s)| > a]$. Thus, since $\widetilde{X}(1)$ has all moments finite, for any $m > 1$ and $\tilde{c}(t) = c(b(t^{1/j})\delta)$,

$$tP\left[\sup_{s \in [0,1]} |\widetilde{X}(s)| > b(t^{1/j})\delta\right] \leq t(1-\tilde{c}(t))^{-1} P[|\widetilde{X}(1)| > b(t^{1/j})\delta/2]$$

$$\leq t(1-\tilde{c}(t))^{-1} \frac{E|\widetilde{X}(1)|^m}{b^m(t^{1/j})(\delta/2)^m}.$$

For large enough m, $t/b^m(t^{1/j}) \to 0$ as $t \to \infty$ and

$$\tilde{c}(t) := \sup_{s \in [0,1]} P[|\widetilde{X}(s)| > b(t^{1/j})\delta/2] \leq \sup_{s \in [0,1]} \frac{E|\widetilde{X}(s)|^m}{b^m(t^{1/j})(\delta/2)^m}$$

and from stationary, independent increments, and the fact that $b(t) \to \infty$,

$$= \sup_{s \in [0,1]} \frac{s^m E|\widetilde{X}(1)|^m}{b^m(t^{1/j})(\delta/2)^m} \leq \frac{E|\widetilde{X}(1)|^m}{b^m(t^{1/j})(\delta/2)^m} \to 0, \quad (t \to \infty).$$

□

4.2.3.2 Putting Humpty Dumpty Back Together: Comparing X and J

Now we complete the proof of Theorem 4.2, page 125. The final step in proving steroidal regular variation of $P[X \in \cdot]$ is to compare $P[X = \widetilde{X} + J \in \cdot]$ with $P[J \in \cdot]$ keeping in mind we know steroidal regular variation of $P[J \in \cdot]$ from Proposition 4.2, page 131.

As in the treatment of steroidal regular variation of $P[J \in \cdot]$, let $f \in C(\mathbb{D} \setminus \mathbb{D}_{\leq j-1})$ be bounded, uniformly continuous on $\mathbb{D}$ with support F bounded away from $\mathbb{D}_{\leq j-1}$ with $d_{\text{sk}}(F, \mathbb{D}_{\leq j-1}) = 2\delta$. Then

$$tE\left|f\left(\frac{X}{b(t^{1/j})}\right) - f\left(\frac{J}{b(t^{1/j})}\right)\right|$$

$$= tE\left|f\left(\frac{X}{b(t^{1/j})}\right) - f\left(\frac{J}{b(t^{1/j})}\right)\right| 1_{[d_{\text{sk}}(\frac{X}{b(t^{1/j})}, \frac{J}{b(t^{1/j})}) \leq \epsilon]}$$

$$+ tE\left|f\left(\frac{X}{b(t^{1/j})}\right) - f\left(\frac{J}{b(t^{1/j})}\right)\right| 1_{[d_{\text{sk}}(\frac{X}{b(t^{1/j})}, \frac{J}{b(t^{1/j})}) > \epsilon]} =: A + B.$$

To handle B, recall the Skorohod metric is bounded above by the uniform metric and

$$d_{\text{sk}}(X, J) \leq \sup_{0 \leq s \leq 1} |X(s) - J(s)| = \sup_{0 \leq s \leq 1} |\widetilde{X}(s)|$$

4.2 Steroidal Regular Variation of $P[X \in \cdot]$

so from Lemma 4.3,

$$B \leq 2\|f\|tP[\sup_{0\leq s\leq 1} \frac{|\widetilde{X}(s)|}{b(t^{1/j})} > \epsilon] \to 0, \quad (t \to \infty).$$

For A, proceed as we did in Step 1 of Proposition 4.2 for (4.18), page 132 where the distance between arguments was small. Decompose the event $[d_{sk}(X/b(t^{1/j}), J/b(t^{1/j})) \leq \epsilon]$ according as

$$\frac{X}{b(t^{1/j})} \in F, \frac{J}{b(t^{1/j})} \notin F, \text{ so that } \frac{J}{b(t^{1/j})} \in \overline{F^\epsilon},$$

$$\frac{X}{b(t^{1/j})} \notin F, \frac{J}{b(t^{1/j})} \in F, \text{ so that } \frac{J}{b(t^{1/j})} \in F \subset \overline{F^\epsilon},$$

$$\frac{X}{b(t^{1/j})} \in F, \frac{J}{b(t^{1/j})} \in F, \text{ so that } \frac{J}{b(t^{1/j})} \in F \subset \overline{F^\epsilon}.$$

The difference of the f terms in A can be bounded by $\omega_f(\epsilon)$ so for $\epsilon < \delta$,

$$\limsup_{t\to\infty} A \leq \omega_f(\epsilon) \limsup_{t\to\infty} tP[\frac{J}{b(t^{1/j})} \in \overline{F^\epsilon}]$$

$$\leq \omega_f(\epsilon)\eta^{(j)} \times \mathcal{L} \circ T_j^{-1}(f) \to 0, \quad (\epsilon \to 0).$$

Here we relied on Theorem 1.1 and (1.12) (page 9), Proposition 4.2 and the finiteness of the measure on regions bounded away from $\mathbb{D}_{\leq j-1}$.

4.2.3.3 Proof of Corollary 4.1

Consider the projection map $h : \mathbb{D} \setminus \mathbb{D}_{\leq 0} \mapsto \mathbb{R}_+ \setminus \{0\}$ defined by

$$hx(\cdot) = x(1).$$

This map is uniformly continuous on $\mathbb{D}$ since for $x_1, x_2 \in \mathbb{D}$ we have in $\mathbb{M}(\mathbb{R}_+ \setminus \{0\})$,

$$d_{sk}(x_1, x_2) = \inf_{\lambda \in \Lambda} \|\lambda - e\| \vee \|x_1 \circ \lambda - x_2\|$$

$$\geq \inf_{\lambda \in \Lambda} |\lambda(1) - e(1)| \vee |x_1(1) - x_2(1)| = |x_1(1) - x_2(1)|$$

since $\lambda(1) = e(1) = 1$. Thus from Theorem 1.3, page 18, we have

$$tP\left[\frac{h(X)}{b(t)} \in \cdot\right] = tP\left[\frac{X(1)}{b(t)} \in \cdot\right] \to \eta^{(1)} \times \mathcal{L} \circ T_1^{-1} \circ h^{-1}(\cdot).$$

Now we have to unpack the limit. For $z > 0$, the region (z, ∞) is bounded away from the forbidden zone $\{0\}$ so

$$tP\left[\frac{X(1)}{b(t)} > z\right] \to \eta^{(1)} \times \mathcal{L} \circ T_1^{-1} \circ h^{-1}(z, \infty)$$

$$= \eta^{(1)} \times \mathcal{L} \circ T_1^{-1}\{x \in \mathbb{D} : x(1) > z\}$$

$$= \eta^{(1)} \times \mathcal{L}\{(\boldsymbol{x}, \boldsymbol{u}) : x_1 1_{[u_1, 1]}(1) > z\}$$

$$= \eta^{(1)}\{\boldsymbol{x} : x_1 > z\} = \nu_\alpha(z, \infty) = z^{-\alpha}.$$

4.3 Heavy Lifting Done? Time for a Cruise[4]

Now that we did some hard work, let's take it easy and go on a cruise. What applications can we get with minimal effort from Theorem 4.2? We have seen one example already in the Embrechts & Goldie result Corollary 4.1, page 125, which said that if the Lévy measure $Q(x) := \nu(x, \infty) \in RV_{-\alpha}$, $\alpha > 0$, then $P[X(1) > x] \in RV_{-\alpha}$ and because the scaling function $b(t)$ for $Q(x)$ and $P[X(1) > x]$ is the same,

$$P[X(1) > x] \sim Q(x), \quad x \to \infty.$$

We next examine several functionals which can be applied to the steroidal regular variation.

4.3.1 First Cruise Stop: The Supremum Functional

The Lévy process X is assumed to be initialized as $X(0) = 0$, so we consider the supremum functional from $\mathbb{D}^{(0)} := \mathbb{D} \cap \{\boldsymbol{x} \in \mathbb{D} : x(0) = 0\} \mapsto \mathbb{D} : x \mapsto x^\vee$, where

$$x^\vee(s) = \sup_{0 \le u \le s} x(u).$$

This makes $x^\vee(s) \ge 0$. Recall the easiest mapping theorem is the second Theorem 1.3, page 18 which requires the map to be uniformly continuous to preserve convergence. The supremum functional is not only uniformly continuous, but also Lipschitz.

Lemma 4.4 ([191, page 80, Theorem 6.1], [192]) *For $x, y \in \mathbb{D}$,*

---

[4] More attractive before Covid-19?

4.3 Heavy Lifting Done? Time for a Cruise

$$d_{sk}(x^\vee, y^\vee) \leq d_{sk}(x, y) \tag{4.19}$$

and therefore the map $h : x \mapsto x^\vee$ is uniformly continuous on $\mathbb{D}$.

It is easier to make non-decreasing functions $x^\vee$, $y^\vee$ close than to make x, y close.

Proof We show (4.19) by showing that for any $k > 0$, if $d_{sk}(x, y) \leq k$ then $d_{sk}(x^\vee, y^\vee) \leq k$. Equivalently, it is sufficient to show that if $d_{sk}(x^\vee, y^\vee) > k$, then $d_{sk}(x, y) > k$.

First we show the analogue of (4.19) for the uniform metric $\|\cdot\|$ on $[0, 1]$. So suppose

$$\|(x^\vee - y^\vee\| = \sup_{0 \leq s \leq 1} |x^\vee(s) - y^\vee(s)| > k.$$

Sneak up on the sup. There exists $s_0 \in [0, 1]$ such that $|x^\vee(s_0) - y^\vee(s_0)|$ is close enough to the sup that it satisfies $|x^\vee(s_0) - y^\vee(s_0)| > k$. Depending on which term of this difference is bigger, either

(1) $x^\vee(s_0) - y^\vee(s_0) > k$ in which case there is an $s_1 \leq s_0$ such that $x(s_1) - y^\vee((s_0) > k$ which implies by monotonicity $x(s_1) - y(s_1) > k$ and hence

$$\|x - y\| > k; \tag{4.20}$$

or

(2) $y^\vee(s_0) - x^\vee(s_0) > k$ and there exists $s_1 \leq s_0$ such that $y((s_1) - x^\vee(s_0) > k$ so that $y(s_1) - x(s_1) > k$ and again (4.20) holds.

This verifies (4.19) for the uniform metric.

For the Skorohod metric we have

$$d_{sk}(x^\vee, y^\vee) = \inf_{\lambda \in \Lambda} \|\lambda - e\| \vee \|x^\vee - y^\vee \circ \lambda\|.$$

Now for each t,

$$y^\vee \circ \lambda(t) = \sup_{s \leq \lambda(t)} y(s) = \sup_{s \leq t} y \circ \lambda(s) = (y \circ \lambda)^\vee(t)$$

verifying that $y^\vee \circ \lambda = (y \circ \lambda)^\vee$ and therefore, using the earlier discussion on the uniform metric,

$$d_{sk}(x^\vee, y^\vee) = \inf_{\lambda \in \Lambda} \|\lambda - e\| \vee \|x^\vee - (y \circ \lambda)^\vee\|$$

$$\leq \inf_{\lambda \in \Lambda} \|\lambda - e\| \vee \|x - y \circ \lambda\| = d_{sk}(x, y).$$

This gives the Lipschitz property. □

Now from the Second Mapping Theorem 1.3 and the uniform continuity of $h(x) = x^\vee$ given in (4.19), we have under the regular variation assumptions in Sect. 4.1.2.1 that the statement

$$tP[X/b(t) \in \cdot] \to E\nu_\alpha(\{y > 0 : y1_{[U,1]} \in \cdot\}) \quad \text{in } \mathbb{M}(\mathbb{D} \setminus \{\mathbf{0}\})$$

implies also in $\mathbb{M}(\mathbb{D} \setminus \{\mathbf{0}\})$ that

$$tP[h(X)/b(t) \in \cdot] = tP[X^\vee/b(t) \in \cdot] \to E\nu_\alpha(\{y > 0 : y1_{[U,1]} \in \cdot\}).$$

In both cases, we get convergence to the same measure that concentrates on the same one-jump functions. We also get by applying the uniformly continuous map $x \mapsto x(1)$ that,

$$\lim_{t\to\infty} tP[X(1)/b(t) > x] = \lim_{t\to\infty} tP[X^\vee(1)/b(t) > x] = \nu_\alpha(x, \infty) = x^{-\alpha}, \ x > 0$$

and as $x \to \infty$,

$$P[X(1) > x] \sim P[X^\vee(1) > x].$$

We can get the full steroidal effect from Theorem 4.2, page 125, by applying the uniformly continuous map $h : x \mapsto x^\vee$ using the Second Mapping Theorem.

Corollary 4.3 *Assume the regular variation assumptions in Sect. 4.1.2.1 and set $h(x) = x^\vee$. For every $j \geq 1$. We have in $\mathbb{M}(\mathbb{D} \setminus \mathbb{D}_{\leq j-1})$, as $t \to \infty$,*

$$tP\left[\frac{X^\vee}{b(t^{1/j})} \in \cdot\right] = (\eta^{(j)} \times \mathcal{L}) \circ T_j^{-1} \circ h^{-1}(\cdot) = (\eta^{(j)} \times \mathcal{L}) \circ T_j^{-1}(\cdot), \quad (4.21)$$

where $\mathcal{L}$ is Lebesgue measure on $[0, 1]^\infty$ and T_j is the mapping in (4.9), on page 127.

Since in Theorem 4.2 the limit measure concentrates on *non-decreasing* functions with j jumps, applying the supremum functional h has no effect on the limit measure. So the limit measure concentrates on functions of the following form: Throw down j uniform random variables $U_1, \ldots, U_j$ on $[0, 1]$. Then construct a path starting at height 0 and scanning left to right until the path encounters the first, smallest, U at which time the path jumps up by a y drawn from $\nu_\alpha(\cdot)$. Then the path keeps scanning to the right until it encounters the next U, the second largest, and the path jumps upwards by an amount governed by another draw y from $\nu_\alpha(\cdot)$. Continue until the largest U after which the path is constant until time 1.

4.3.2 Second Cruise Layover: The Largest Jump Functional

In Sect. 4.1 we specified that the Lévy measure $\nu(\cdot)$ concentrates on $(0, \infty)$ so X has only positive jumps. Consequently when considering the largest jump functional, we really consider the largest *positive* jumps and thus it makes sense to cut down $\mathbb{D}$ to

$$\mathbb{D}^{\text{posJ}} := \mathbb{D} \bigcap \{x \in \mathbb{D} : x(0) = 0, x(s) - x(s-) \geq 0, 0 \leq s \leq 1\}. \tag{4.22}$$

The range of the largest jump functional is the subset of non-decreasing functions in $\mathbb{D}$,

$$\mathbb{D}^{\uparrow} := \{x \in \mathbb{D} : x(s_1) \leq x(s_2), \forall 0 \leq s_1 \leq s_2 \leq 1\}, \tag{4.23}$$

and for this section the largest jump functional $h : \mathbb{D}^{\text{posJ}} \mapsto \mathbb{D}^{\uparrow}$ is

$$h(x)(s) := \sup\{x(u) - x(u-) : u \leq s\}, \quad 0 < s \leq 1, \tag{4.24}$$

and no harm done if we specify $h(x)(0) = 0$. Applying this functional to X we get a process Y,

$$Y(s) := h(X)(s) = \sup_{u \leq s}(X(u) - X(u-)), \quad 0 \leq s \leq 1, \tag{4.25}$$

which is an example of an *extremal process*. See [67–69, 119, 139, 147–149, 157, 159, 173].

4.3.2.1 Suspense! Does Regular Variation of the Lévy Process Distribution Imply Regular Variation of the Extremal Process Distribution?

Sure does. The way forward is uniform continuity of the largest jump functional and the Second Mapping Theorem 1.3. For $x_i \in \mathbb{D}^{\text{posJ}}$, $i = 1, 2$, let $y_i = h(x_i)$. The factor 4 in the Lipschitz comparison in (4.26) is pretty relaxed. But we are on a cruise where behavioral standards are usually, ahem, relaxed.

Oy, the suspense!

Lemma 4.5 *Apply the largest jump functional to $x_1, x_2 \in \mathbb{D}^{posJ}$ to get y_1, y_2. Then*

$$d_{sk}(y_1, y_2) \leq 4 d_{sk}(x_1, x_2), \tag{4.26}$$

and therefore the largest jump functional $h : \mathbb{D}^{posJ} \mapsto \mathbb{D}^{\uparrow}$ is uniformly continuous.

Proof Suppose $0 < d_{sk}(x_1, x_2) \leq \epsilon$. Then there is $\lambda \in \Lambda$ such that

142   4 Lévy Processes with Regularly Varying Distributions: Where Do the Jumps Go?

$$\|\lambda - e\| \vee \|x_1 \circ \lambda - x_2\| \leq 2\epsilon.$$

So for all $s \in [0, 1]$,

$$|x_1(\lambda(s)) - x_2(s)| \leq 2\epsilon, \qquad (4.27)$$

and for all $s \in (0, 1]$, and all n large enough to make $s - 1/n > 0$,

$$|x_1(\lambda(s - \frac{1}{n})) - x_2(s - \frac{1}{n})| \leq 2\epsilon,$$

and therefore, letting $n \to \infty$,

$$|x_1(\lambda(s-)) - x_2(s-)| \leq 2\epsilon, \quad (0 < s \leq 1. \qquad (4.28)$$

Thus, remembering definition (4.24), with some bookkeeping, for all $s > 0$,

$$|y_1(\lambda(s)) - y_2(s)| = \left|(x_1(\lambda(s)) - x_1(\lambda(s-))) - (x_2(s) - x_2(s-))\right|$$

$$\leq \left|x_1(\lambda(s)) - x_2(s)\right| + \left|x_1(\lambda(s-)) - x_2(s-)\right| \leq 4\epsilon.$$

This means

$$d_{\mathrm{sk}}(\mathbf{y}_1, \mathbf{y}_2) = \inf_{\lambda \in \Lambda} \|\lambda - e\| \vee \|\mathbf{y}_1 \circ \lambda - \mathbf{y}_2\| \leq 4\epsilon.$$

Let $\epsilon \downarrow d_{\mathrm{sk}}(\mathbf{x}_1, \mathbf{x}_2)$ to finish. □

Now apply the Second Mapping Theorem 1.3 to Theorem 4.1.

Corollary 4.4 *Suppose the regular variation assumptions in Sect. 4.1.2.1 hold meaning $Q(x) = \nu(x, \infty) \in RV_{-\alpha}$. Then $P[Y \in \cdot]$ is a regularly varying measure: As $t \to \infty$,*

$$tP\left[\frac{Y}{b(t)} \in \cdot\right] \to \eta(\cdot), \qquad (4.29)$$

in $\mathbb{D}^\uparrow \setminus \{0\}$.

The limit measure of regular variation for $P[X \in \cdot]$, $P[X^\vee \in \cdot]$, and for $P[Y \in \cdot]$ is the same and has representation

$$\eta(\cdot) := E\nu_\alpha\{y > 0 : y 1_{[U,1]} \in \cdot\},$$

for $U \sim U[0, 1]$, meaning $\eta(\cdot)$ concentrates on single jump functions that are 0 until a uniform jump time and then jump to a height drawn from $\nu_\alpha(\cdot)$. The simplicity

4.3 Heavy Lifting Done? Time for a Cruise

of the limit measure shows the strengths and dangers of doing inference using an asymptotic model which tends to compress some fine structure.

4.3.2.2 Bonus: Regular Variation of the Lévy Measure, the Lévy Process Distribution and the Extremal Process Distribution

Bonus!

A charming benefit of this discussion is that three regular variation statements imply each other.

Corollary 4.5 *The following statements are equivalent.*

1. *The function $Q(x) := \nu(x, \infty) \in RV_{-\alpha}$.*
2. *The measure $P[X \in \cdot] \in MRV(\alpha, b(t), \eta, \mathbb{D} \setminus \mathbb{D}_{\{0\}})$ as in (4.5), page 125.*
3. *The measure $P[Y \in \cdot] \in MRV(\alpha, b(t), \eta, \mathbb{D}^\uparrow \setminus \mathbb{D}_{\{0\}})$ as in (4.29), page 142.*

Proof The implication (1)⇒(2) is the content of Theorem 4.1, page 124. The reason for (2)⇒ (3) is that the conclusion in (3) results from applying the Second Mapping Theorem to regular variation of $P[X \in \cdot]$ and does not explicitly use an assumption on Q. Finally assume (3). Apply the uniformly continuous map $x \mapsto x(1)$ to (4.29) and we get for $x > 0$, as $t \to \infty$,

$$tP[Y(1)/b(t) > x] \to E\nu_\alpha\{y > 0 : y1_{[U,1]}(1) > x\}$$
$$= E\nu_\alpha\{y > 0 : y \cdot 1 > x\} = x^{-\alpha}. \qquad (4.30)$$

However, from the representation in Sect. 4.1.2.3, we have

$$Y(1) \stackrel{d}{=} Q^\leftarrow(\Gamma_1)$$

so

$$tP[Y(1)/b(t) > x] = tP[Q^\leftarrow(\Gamma_1)/b(t) > x] = tP[\Gamma_1 \leq Q(b(t)x)]$$
$$= t(1 - e^{-Q(b(t)x)}) \sim tQ(b(t)x),$$

and the limit being $x^{-\alpha}$ from (4.30), implies that $Q(x) \in RV_{-\alpha}$. □

4.3.2.3 Steroidal Regular Variation of the Extremal Process Distribution

With the definition of the largest jump functional h in (4.24), page 141, the Second Mapping Theorem 1.3, Theorem 4.2, page 125, gives using (4.7) that in $\mathbb{M}(\mathbb{D}^\uparrow \setminus \mathbb{D}_{\leq j-1})$, for any $j \geq 1$,

144   4 Lévy Processes with Regularly Varying Distributions: Where Do the Jumps Go?

$$tP[\frac{Y}{b(t^{1/j})} \in \cdot] = tP[\frac{h(X)}{b(t^{1/j})} \in \cdot]$$

$$\to E v_\alpha^j \{(y_1, \ldots, y_j) \in (0, \infty)^j : h\Big(\sum_{i=1}^{j} y_i 1_{[U_i, 1]}\Big) \in \cdot\}. \quad (4.31)$$

Now unpack the notation for the limit measure. For $0 \le s \le 1$, the function inside h has jumps y_i at times $U_1, \ldots, U_j$ so

$$h\Big(\sum_{i=1}^{j} y_i 1_{[U_i, 1]}\Big)(s) = \bigvee_{i=1}^{j} y_i 1_{[U_i, 1]}(s) = \bigvee_{U_i \le s} y_i,$$

with the understanding that the max of an empty set gives 0. Use $e(s) = s$ for the identity function on $[0, 1]$ and the limit measure in (4.31) is

$$E v_\alpha^j \{(y_1, \ldots, y_j) \in (0, \infty)^j : \bigvee_{U_i \le e} y_i \in \cdot\},$$

So the limit measure concentrates on functions which can *possibly* jump at j uniformly chosen points $U_1, \ldots, U_j$ and as the function argument scans from left to right, each time the argument encounters a U, there is a jump if the corresponding y drawn from ν_α makes the running max increase. However, the limit measure cannot put mass on the forbidden zone and must avoid $\mathbb{D}_{\le j-1}$ and thus cannot afford to concentrate on functions that lose jumps to y's that are out of monotone increasing order as the function argument scans left to right. So the limit measure must concentrate on functions of the form $\bigvee_{u_i \le e} y_i$ where $0 < u_1 < u_2 < \cdots < u_j < 1$ and $0 < y_1 < y_2 < \cdots < y_j$. Thus, another way to express (4.31) is that for $j \ge 1$, in $\mathbb{M}(\mathbb{D}^\uparrow \setminus \mathbb{D}_{\le j-1})$,

$$tP[\frac{Y}{b(t^{1/j})} \in \cdot] \to E v_\alpha^j \{(y_1, \ldots, y_j), 0 < y_1 < y_2 < \cdots < y_j, \bigvee_{U_{(i)} \le e} y_i \in \cdot\}, \quad (4.32)$$

where $U_{(1)} < \cdots < U_{(j)}$ are increasing order statistics of a uniform sample size j on $[0, 1]$.

Remark 4.1 Note for $j \ge 2$, the event $\{x \in \mathbb{D}^\uparrow : h(x)(1) > M\}$ is not bounded away from $\mathbb{D}_{\le j-1}$ so there is no hope of getting $tP[\frac{Y(1)}{b(t^{1/j})} > M]$ to converge the way we did for $j = 1$ in (4.30).

4.3.3 Next Cruise Stop: Smoothing

Of course, there are many ways to average or smooth a function; see [95, 97, 124]. Here we merely concentrate on the map $I : \mathbb{D} = \mathbb{D}[0, 1] \mapsto \mathbb{C}[0, 1]$, where $\mathbb{C}[0, 1]$ is the continuous functions on $[0, 1]$, and I is defined for $x \in \mathbb{D}$ as

$$Ix(t) = \int_0^t x(u)du, \quad 0 \le t \le 1.$$

The image of $\mathbb{D}$ under I consists of absolutely continuous functions on $[0, 1]$ whose densities are in $\mathbb{D}$.

If we smooth the Lévy process $X(\cdot)$ yielding IX, how do regular variation properties transfer from X to IX? We limit the story to $j = 1$. For one thing, all cruises, however idyllic, must end with a subsequent return to life and, also, we are nervous about the nasty Remark 4.1, page 144. Considering only $j = 1$ smooths the waves.

Reassuringly, the map I is continuous [192, Theorem 11.5.1, p. 383]. To see this, suppose $x_n, x \in \mathbb{D}$ and $d_{\text{sk}}(x_n, x) \to 0$. Then with $\|x(\cdot)\| = \sup_{\{0 \le s \le 1\}} |x(s)|$,

1. Skorohod convergence implies $\|x_n\| \to \|x\| < \infty$ and therefore $\sup_n \|x_n\| \vee \|x\| = K < \infty$ is bounded;
2. From $d_{\text{sk}}(x_n, x) \to 0$, we get $x_n(t) \to x(t)$ at all points of continuity of x and therefore for almost all $s \in [0, 1]$, $|x_n(s) - x(s)| \to 0$. Also

$$\|x_n - x\| = \sup_{0 \le s \le 1} |x_n(s) - x(s)| \le \sup_n \|x_n\| \vee \|x\| = K.$$

3. Hence by dominated convergence,

$$\int_0^1 |x_n(s) - x(s)|ds \to 0,$$

and for any $t \in [0, 1]$,

$$\int_0^t |x_n(u) - x(u)|du \to 0.$$

4. The payoff:

$$\sup_{0 \le t \le 1} |Ix_n(t) - Ix(t)| \le \sup_{0 \le t \le 1} \left| \int_0^t x_n(u)du - \int_0^t x(u)du \right|$$

$$\le \sup_{0 \le t \le 1} \int_0^t |x_n(u) - x(u)|du$$

$$\leq \int_0^1 |x_n(u) - x(u)| du \to 0$$

and $Ix_n \to Ix$ in the uniform metric on $\mathbb{C}[0, 1]$.

4.3.3.1 Dressing for the Cruise Interlude

Fix $j = 1$ and suppose (4.6) in Theorem 4.2, page 125 takes place in $\mathbb{M}(\mathbb{D}^{posJ} \setminus \mathbb{D}_0)$, as appropriate for the distribution of X. Define,

$$\text{LIN} = \{x(\cdot) \in \mathbb{C}[0, 1] : x(t) = ct, \ 0 \leq t \leq 1; \ c \geq 0\},$$

and consider

$$I : \mathbb{M}(\mathbb{D}^{posJ} \setminus \mathbb{D}_0) \mapsto \mathbb{C}[0, 1] \setminus \text{LIN}.$$

We know I is continuous so if it respects the forbidden zone, there is smooth sailing ahead using (4.6) for $j = 1$ (page 125) and old reliable Theorem 1.2, page 16.

Set $d_u(\boldsymbol{x}, \boldsymbol{y}) = \|\boldsymbol{x} - \boldsymbol{y}\| = \sup_{0 \leq t \leq 1} |\boldsymbol{x}(t) - \boldsymbol{y}(t)|$ and suppose $B \subset \mathbb{C}[0, 1]$ satisfies $B \in \mathbb{BA}(\mathbb{C}[0, 1] \setminus \text{LIN})$. We check $I^{-1}(B) \in \mathbb{BA}(\mathbb{D}^{posJ} \setminus \mathbb{D}_0)$. If this fails and $d_{sk}(\mathbb{D}_0, I^{-1}(B)) = 0$, some Titanic sized disaster should occur. Zero distance requires there exist $\boldsymbol{x}_n \in \mathbb{D}_0$ (meaning a constant function with value $c_n \geq 0$), and $\boldsymbol{y}_n(\cdot) \in I^{-1}(B)$ such that $d_{sk}(\boldsymbol{x}_n(\cdot), \boldsymbol{y}_n(\cdot)) \to 0$. From the definition of Skorohod distance, there exist homeomorphisms $\lambda_n \in \Lambda$ and $d_u(\boldsymbol{y}_n \circ \lambda_n, c_n) \vee d_u(\lambda, e) \to 0$. Since c_n is a constant function–this is key–we have $d_u(\boldsymbol{y}_n(\cdot), c_n) \to 0$. This entails (see discussion page 145) $d_u(I\boldsymbol{y}_n(\cdot), c_n(\cdot)) \to 0$ or $d_u(B, \text{LIN}) = 0$ as well. However, this sinking of the Titanic constradicts the assumption $B \in \mathbb{BA}(\mathbb{C}[0, 1] \setminus \text{LIN})$.

Assembling the pieces (4.6) and Theorem 4.2 gives the $j = 1$ result.

$$tP\left[\frac{IX}{b(t)} \in \cdot\right] \to Ev_\alpha\{y \in (0, \infty) : y(e - U)^+ \in \cdot\}, \tag{4.33}$$

in $\mathbb{M}(\mathbb{C} \setminus \text{LIN})$ where $U \sim U(0, 1)$.

4.4 Problems

4.1 Let $Z \geq 0$ have a regularly varying distribution tail and suppose $Z \perp\!\!\!\perp U \sim U(0, 1)$. Define the random one-jump function $X(t) = Z 1_{[U,1]}(t), 0 \leq t \leq 1$ and show the distribution of X is regularly varying on $\mathbb{D} \setminus \mathbb{D}_{\leq 0}$. Where does the limit measure concentrate?

4.4 Problems

4.2 [95] Modify the setup in Problem 4.1 by supposing Z is Pareto distributed with index $\alpha > 0$ on $[1, \infty)$ and defining $X(t)$ for $t \geq 0$ by

$$X(t) = \begin{cases} 0, & \text{if } t \in [0, U), \\ Z, & \text{if } t \in [U, U + 1/(2Z)), \\ 0, & \text{if } t \in [U + 1/(2Z), U + 1/Z), \\ Z, & \text{if } t \in [U + 1/Z, \infty). \end{cases}$$

Then $\{X(t), 0 \leq t \leq 1\}$ is not a random element of $\mathbb{D}([0,1]) \setminus \{\mathbf{0}\}$ with a regularly varying distribution.

4.3 Is the map T_m defined in (4.9), page 127, uniformly continuous?

4.4 Suppose $0 < t_1 < t_2 < \cdots < t_p \leq 1$ for $p \geq 1$ and then generalize (2.43), page 55 and Corollary 4.1, page 125 to show that when $Q(x) \in \text{RV}_{-\alpha}$, the vector $(X(t_1), X(t_2), \ldots, X(t_p))$ has a multivariate regularly varying distribution on $\mathbb{R}_+^p \setminus \{\mathbf{0}_p\}$; that is,

$$P[(X(t_1), X(t_2), \ldots, X(t_p)) \in \cdot] \in \text{MRV}\big(\alpha, b(t), \eta(\cdot), \mathbb{R}_+^p \setminus \{\mathbf{0}_p\}\big)$$

and you have to specify η. You can either do this by generalizing Corollary 4.1 or by analyzing $(X(t_1), X(t_2) - X(t_1), \ldots, X(t_p) - X(t_{p-1}))$ and then applying the p-dimensional cousin of CUMSUM.

4.5 Let $X = \{X(s), 0 \leq s \leq 1\}$ be a Lévy process as described in Sect. 4.1. Define the extremal process

$$Y(s) := \sup_{0 \leq u \leq s} \big(X(u) - X(u-)\big), \quad 0 < s \leq 1.$$

Compute the finite dimensional distributions of Y; that is for $0 < t_1 < t_2 < \cdots < t_k \leq 1$, and $x_1, \ldots, x_k \in \mathbb{R}_+$ compute

$$P[Y(t_1) \leq x_1, \ldots, Y(t_k) \leq x_k],$$

in terms of the Lévy measure $\nu(\cdot)$.

4.6 Verify for $M > 0$, that $\{x \in \mathbb{D}^\uparrow : h(x) > M\} \notin \text{BA}(\mathbb{D}_{\leq 1})$.

4.7 The set $\mathbb{D}_{\leq m}$ is closed. Is it compact for $m > 1$? Hint: Either consult [13, Theorem 12.3, page 130] or imagine the map $\big((u_1, \ldots, u_m), (x_1, \ldots, x_m)\big) \in [0,1]^m \times \mathbb{R}_+^m \mapsto \sum_{l=1}^m x_l 1_{[u_l, 1]} \in \mathbb{D}_{\leq m}$. If $\mathbb{D}_{\leq m}$ were compact, Proposition 1.4, page 21, would be applicable for a variety of functionals applied to Theorem 4.2.[5]

---

[5] Alas, ...

4.8 Imagine a map $\mathbb{D} \mapsto \mathbb{C} = \mathbb{C}[0, 1]$ defined by $x \mapsto Ix = \int_0^{(\cdot)} x(u)du$. Is this map *uniformly* continuous on $\mathbb{D}$?

4.9 For $M > 0$, is

$$\{x(\cdot) \in \mathbb{D} : \|x\| \leq M\} \subset \mathbb{D}$$

a compact subset of $\mathbb{D}$? (Consult if necessary [13, p. 130].)

4.10 For $m \geq 1$, is $\mathbb{D}_{=m}$ bounded away from $\mathbb{D}_{\leq m-1}$?

4.11 For $M > 0$ set

$$B_M := \{x \in \mathbb{D}^{\text{posJ}} : 0 < hx(1) = \sup_{0 < s \leq 1} (x(s) - x(s-)) \leq M\}, \quad (4.34)$$

where h is the largest jump functional. These are the functions in $\mathbb{D}^{\text{posJ}}$ with jumps that are positive and bounded by M.

1. Show $\mathbb{D}_{\leq j-1} \cap B_M$ is compact. (Hint: Perhaps show this is the continuous image of a compact set.)
2. Show the limit measure $P_\infty^{(j)}(\cdot)$ in (4.6), page 125, applies 0 mass to ∂B_M.

4.12 Show LIN is closed in $\mathbb{C}$.

Chapter 5
Exploring Data Cautiously

This chapter describes adventures when bravely statistically analyzing multivariate data that can plausibly be generated by a regularly varying distribution. The focus is on multivariate analysis and we assume the reader is familiar with one-dimensional techniques for heavy tailed data such as QQ-plots, Hill plots, altHill plots, standardization of multivariate data using the rank transformation, angular density plots. Some references: [7, 27, 51, 110, 156].

We also assume familiarity with the fact that proper choice of threshold in this subject is justifiably considered a black art. However, see Sect. 5.2 for a data driven method common in computer science and network science. The method has problems but what doesn't?

Apologies in advance: Any description of mutivariate inference techniques in the age of big data is akin to hitting a moving target and is subject to personal taste and limitation.

Section 5.1 reminds us to be humble when doing inference based on an *asymptotic* model and outlines a general inferential approach for multivariate heavy tailed data. Generally, one separates estimating marginal tail indices from the problem of estimating the dependence structure. In polar coordinate language, where the data comes from the distribution of (R, Θ) and $P[R > r] \in RV_{-\alpha}$, one first determines $\alpha > 0$ and then tries to say something intelligent about the angular measure $S(\cdot)$ of the MRV limit measure $\eta(\cdot)$. Such efforts can be frustrating and the beer alternative is not excluded. (See page 152.) Section 5.3 discusses graphical exploratory methods for two-dimensional data to assess how much extremal dependence is present. Techniques include diamond plots and Hillish plots. Later in Sect. 5.7, we give an imperfect numerical summary of extremal dependence between two variables and this technique is extended in Sect. 5.7.1 to find subgroups of components of higher dimensional random vectors that have low probability of being simultaneously extreme; this is just the thing portfolio managers need to know to minimize risk. *Asymptotic independence* returns in Sects. 5.6 and 5.7 and gets thorough discussion.

A theme for the analysis of two dimensional non-negative MRV data is to distinguish between the following cases:

- Weak dependence where the limit measure $\eta(\cdot)$ concentrates on all of $\mathbb{R}_+^2$;
- Strong dependence where the limit measure $\eta(\cdot)$ concentrates on a subcone $C \subsetneq \mathbb{R}_+^2$;
- Full dependence where the limit measure $\eta(\cdot)$ concentrates on a ray from the origin;
- Asymptotic independence where $\eta(\cdot)$ concentrates on the axes and puts no mass on $(0, \infty)^2$.

The chapter contains both real and simulated data examples that illustrate techniques and pitfalls. The last Sect. 5.7.5 discusses preferential attachment as a model of social network growth and since this mechanism produces heavy tailed data from either the weak or full dependence case depending on parameter choice and we try previously discussed techniques on data simulated with this model.

5.1 Reasons for Caution: Zombies and Fairy Tales

All subjects have persistent zombie[1] ideas that never die. The idea of a *true model* makes sense if one is simulating data but, otherwise, is only a convenient organizing principle. Heavy tail analysis offers a weakening of the idea of the true model by suggesting there is a true *asymptotic model*. So when X is a random element of some TABOF space $\mathbb{S} \setminus F$ with a distribution that is regularly varying, there is an assumption about an asymptotic property of $P[X \in \cdot]$, namely that for some $b(t)$ and limit measure $\eta \in \mathbb{M}(\mathbb{S} \setminus F)$,

$$t P[X/b(t) \in \cdot] \to \eta(\cdot). \tag{5.1}$$

Scary zombie

Assuming an asymptotic property should be a more robust assumption than assuming a *true* model for $P[X \in \cdot]$ and hopefully, we can still make inferences about $P[X \in \cdot]$, at least in certain remote regions by discovering $b(t)$ and $\eta(\cdot)$. We have to drink the Kool-Aid that there are advantages to discerning $b(t), \eta(\cdot)$ as opposed to directly estimating $P[X \in \cdot]$.

---

[1] Thank you Paul Krugman. It's not just Economics.

5.1 Reasons for Caution: Zombies and Fairy Tales

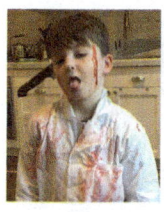

Kool-Aid?

If $\mathbb{S} = \mathbb{R}_+$, η is a Pareto measure and we are in a reasonably comfortable situation. We have to infer something about the index of the power law and this may lead to use of Hill, altHill and QQ plots. However, beyond $\mathbb{R}_+$, if $\mathbb{S}$ is more complicated, we have a semi-parametric inference problem and the set of limit measures in $\mathbb{M}(\mathbb{S} \setminus F)$ is in 1–1 correspondence with probability measures on $\aleph_{\mathbb{S}_F} = \{x \in \mathbb{S}_F : d(x, F) = 1\}$. This follows by switching to generalized polar coordinates $X \mapsto \left(d(X, F), \frac{X}{d(X,F)} \right)$.

The take away is that MRV estimation means picking a measure out of a very large class, namely the class of probability measures on $\aleph_{\mathbb{S}_F}$ as well as estimating the index of the tail probability $P[d(X, F) > x]$.

So we proceed with a cautious, exploratory approach. Employ something like the following strategy to decide if multivariate data can be plausibly generated by a MRV model:

1. Do the one-dimensional marginal data appear to come from a heavy tailed distribution? Analyzing QQ and Hill plots helps with this assessment; see [156].

   If the one-dimensional marginal data appear to come from a heavy tailed distribution, is it plausible that the marginal distributions can be modelled with the same tail index?

2. Take advantage of dimension reductions of data in $\mathbb{R}^p$ for special cases where applicable. For instance,

   (a) In Sect. 3.1.1.1, page 95, we diagnosed $P[X \in \cdot] \in \text{MRV}(\alpha, b(t), \eta, \mathbb{R}_+^p \setminus \{\mathbf{0}\})$ by using the one-dimensional reduction that tests if $\vee_{i=1}^p s_i X_i$ has a regularly varying tail with index α for $s \in \mathbb{R}_+^p \setminus \{\mathbf{0}\}$. See also [156, p. 326] and [50].

   (b) In Sect. 3.1.1.2, page 96, we discussed the one-dimensional reduction diagnostic for $P[X \in \cdot] \in \text{MRV}(\alpha_2, b_2(t), \eta_2, \mathbb{R}_+^p \setminus [\text{axes}])$ by testing if $\wedge_{i=1}^p a_i X_1$ has a regularly varying distribution tail with index α_2 for $\mathbf{a} > \mathbf{0}$. See also [156, p. 326].

   (c) Test if $X = (X_1, \ldots, X_p)$ has asymptotic independence by seeing if this is true for any pair (X_i, X_j) for $1 \le i, j \le p; i \ne j$. See the remark at the beginning of Sect. 3.1.1.1 and also [157, p. 296].

3. If you are convinced p-dimensional data can plausibly have an MRV distribution, but asymptotic independence does not hold, how dependent are the components? As an aid, try to estimate the support of the limit measure. A diffuse support or a support that is a proper subset tell different dependence stories. See [122] and Sect. 5.7.1.

4. Try to assess if different disjoint subgroups of components are asymptotically independent. Risk managers would use this information to assemble portfolios. Suggestions for how to do this are in Sects. 5.6 and 5.7.

5. Construct numerical measures of dependence between components to assess for further testing which components are the most dependent. Numerical measures are crude summaries of properties of measures but you are getting paid to do *something*! See Sect. 5.7.
6. If all else fails, cope with frustration by following the example of the young men on the right.

How to cope

5.2 Threshold Selection

🖼️KONECT ‣ Networks ‣

Flickr

This is the undirected network of Flickr images sharing common metadata such as tags, groups, locations etc. A node represents an image, and an edge indicates that two images share the same metadata.

Metadata

Code	FI
Internal name	flickrEdges
Name	Flickr
Data source	http://snap.stanford.edu/data/web-flickr.html
Availability	✅ Dataset is available for download

Statistics

Size	n = 105,938
Gini coefficient	G = 0.780 096
Balanced inequality ratio	P = 0.153 442
Relative edge distribution entropy	H_{er} = 0.876 354
Power law exponent	γ = 1.407 23
Tail power law exponent	$γ_t$ = 1.731 00 (dmin=7)

Fig. 5.1 A snapshot from KONECT of summary statistics for the Flickr friendship dataset

The peaks over threshold (POT) method rests on an assumption that makes life bearable and allows us to move forward. It is patiently explained and effectively used in [27]. The method suggests that the limit relation defining the asymptotic property

$$tP\left[\frac{X}{b(t)} \in \cdot\right] \to \eta(\cdot),$$

be replaced by equality *for all large t*. So for a nice set A,

$$P[X \in A] \approx t^{-1}\eta(A/b(t)),$$

and then we must conjure estimates of $b(t)$ and η. No matter how simple the context, this is a crucial hurdle to applying any extreme value method: How do we select thresholded data that can give accurate information about the quantities smiling at us from asymptopia. There is copious literature grappling with this difficult issue [7, 55, 59, 64, 77, 82, 140, 171, 182].

A practical, but not optimal, method that is widely used in Computer Science and Network Science is advocated by Aaron Clauset (see [24, 179] and http://tuvalu.santafe.edu/aaronc/powerlaws/). The method is based on choosing the threshold to minimize a distance measure. Datasets included in the public repository KONECT

5.2 Threshold Selection

(konect.cc) (see [115, 116]) contain estimates of degree distribution power law indices γ_t computed by this method. In Fig. 5.1, part of a screenshot for a particular KONECT dataset is given where the power law estimate is highlighted. Widely used implementations such as the R-package *poweRlaw* [83] (which also exists in Python) as well as the *igraph* [29] function `fit_power_law` have made the method easy to apply. Here is a description of the method for one-dimensional data.

5.2.1 The Minimum Distance Threshold Method

In one-dimension, threshold selection amounts to deciding which portion of the data contains accurate tail information or equivalently, how many upper order statistics to use. The POT philosophy hopes that above this threshold, data behaves like it came from a Pareto distribution. More precisely, suppose $X_1, \ldots, X_n$ are a random sample from a distribution F with

$$P[X_1 > x] = \bar{F}(x) = x^{-\alpha} L(x), \quad (x \to \infty), \tag{5.2}$$

with $L(x)$ slowly varying at ∞ and denote the order statistics in decreasing order by $X_{(1)} \geq X_{(2)} \geq \cdots \geq X_{(n)}$. If one uses the kth-largest order statistic as a threshold, the Hill estimator of α^{-1} [51, 86, 92, 126, 141, 156] based on k upper order statistics is

$$\hat{H}_{k,n} := \left(\frac{1}{k-1} \sum_{i=1}^{k-1} \log \frac{X_{(i)}}{X_{(k)}} \right). \tag{5.3}$$

The estimator is consistent, meaning if $n \to \infty$ and $k \to \infty$ but $n/k \to \infty$ then $H_{k,n} \xrightarrow{P} \alpha^{-1}$. It is also frequently asymptotically normal under conditions that control asymptotic bias (but these conditions may be difficult to check in practice). The consistency has also been extended to dependent contexts in an expanding literature which attempts to address practical situations with varying degees of success [36, 55, 60, 64, 65, 93, 114, 161, 163, 184, 185].

This is hunky-dory but how do we choose k when we have data and no prescribed model? There are various graphical plotting methods such as Hill, altHill, smooHill, QQ [61, 156, 162, 164] which are helpful but not definitive and sometimes subjective. A sensible idea [24, 83] is to minimize the distance between the best fitting Pareto tail and the observed empirical tail measure. So define,

$$D_k := \sup_{y \geq 1} \left| \frac{1}{k-1} \sum_{i=1}^{n} 1_{(y,\infty)}\left(\frac{X_{(i)}}{X_{(k)}}\right) - y^{-\hat{H}_{k,n}^{-1}} \right| \tag{5.4}$$

and use as threshold estimate $X_{(k_n^*)}$ with

$$k_n^* := \operatorname*{argmin}_{k \in \{2,\ldots,n\}} D_k. \tag{5.5}$$

(If the point of minimum is not unique, choose the smallest one.) We call this procedure the minimum distance selection procedure (MDSP).

5.2.1.1 Beware of Mischief

If you (or nature) try hard enough, you can fool most statistical procedures, even maximum likelihood. So what about the MDSP? Well .... The procedure is consistent [10] but a serious objection is it does not produce estimates that are asymptotically normal or that achieve minimal mean square error [63]. Theoretical analysis of the MDSP is limited to situations where data is assumed to come from an iid model of repeated sampling. Weirdly, with dependent data such as network node degrees, simulation evidence indicates the method can do better than in the iid case [63], though simulation evidence points to the fact that the method's accuracy is sensitive to model parameter choice.

Beware mischief

Here are some other caveats and observations.

- The MDSP often leads to choosing a k_n^* that is too small, resulting in increased variance and root mean squared error (RMSE) for the Hill estimator relative to a choice which minimizes the asymptotic mean squared error. In general the MDSP estimator does not achieve minimum asymptotic variance. This is certainly true for data actually drawn from a Pareto distribution where the Hill estimator $\tilde{H}_{n,n}$ is the maximum likelihood estimator of α and therefore yields minimum asymptotic variance. However it is also demonstrably true for other distributions as well.
- The MDSP tail index estimator has a difficult to calculate non-normal limit distribution. So we cannot compute reliable confidence intervals.
- The bad news in the previous two bullets is balanced by the fact that the MDSP requires no assumptions about difficult to check second order regular variation behavior or control of asymptotic bias [36, 51, 59, 141].
- In Computer Science and Network Science, estimation of tail indices and threshold selection is often considered a settled issue. In these communities, mathematical kvetching about properties of the MDSP procedure may have little impact.
- For multivariate data, thresholds chosen to estimate the tail index of the distribution of the distance to the forbidden zone may not be ideal for estimating the measure on $\aleph_{\mathbb{S}_F}$. There is only empirical evidence about this and no theory.

We simulated 100,000 samples of size 1000 from a standard Pareto distribution and for each sample we compute the MDSP estimate of α and the MLE. The MLE, mathematical statistics theory tells us, is asymptotically normal with minimum asymptotic variance. Figure 5.2 from [63] displays the normal QQ plot for the 100,000 estimates $n^{1/2}(\hat{H}^{-1}_{k_n^*,n} - \alpha)$ (red) and also of $n^{1/2}(\hat{H}^{-1}_{n,n} - \alpha)$ (blue) where $n = 1,000$. The black dashed line is the main diagonal. The blue curve behaves nicely but observe the Hill estimator based on the top k_n^* order statistics has much heavier tails which is important when constructing confidence intervals.

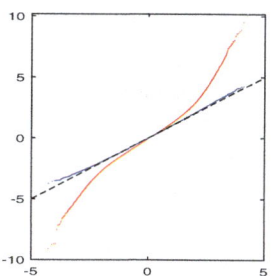

Fig. 5.2 QQ plots: MDSP vs Hill

At a more granular level of one simulation, we simulated 1000 observations from standard Pareto. Applying *poweRlaw*, we get the thresholded observations are those bigger than 1.204; there are 809 of these. Thus, $1000 - 809 = 191$ observations are needlessly discarded.

The MDSP estimate is reasonable and defensible but like many things in statistics must be used with caution and common sense especially when making confidence statements based on asymptotic normality. The package *poweRlaw* [83] analyzes parameter uncertainty using the bootstrap and avoids the issue of non-asymptotic normality of the estimator.

Cautions about using the MDSP are only strengthened when departing from the iid setting where little theory supports the method.

5.2.1.2 MDSP Test Drives with Data

Evel Knievel rides again

Time for illustrative experiments to see how the method performs on real data. Is the method really reasonable? Or worriesome? Or both?

We consider the Facebook wall post data accessed at konect.cc/networks/facebook-wosn-links/ [115, 116]. This is described more fully in the next section. The given data is edge data accompanied by timestamps and this has been converted to (out-degree,in-degree) data indexed by node via *igraph* [29]. The length of the node indexed degree data is 46,953 and for now just consider this as bivariate heavy-tailed data. For the impatient, the two-dimensional scatter plot of the pairs is in Fig. 5.8 below on page 160 but for the most part, this section sticks to one-dimensional analysis.

Table 5.1 lists various subsets of the (out, in) data and distances of points to forbidden zones. For each, the poweRlaw tail index estimate $\hat{\alpha}$ is given as well as the choice of threshold

Table 5.1 Various poweRlaw tail estimates along with threshold choices

Data	Length (data)	ntail= k	$\hat{\alpha}$
facebk (sum comp)	46,953	670	2.52
facebk[,1] (> 0)	42,390	275	2.98
facebk[,2] (> 0)	39,986	304	2.75
facebkpos[,1] (both pos)	35424	275	2.98
facebkpos[,2] (both pos)	35424	263	2.66
[wedge] (sum comp)	11231	320	2.33
[< wedge] (sum comp)	23,687	244	2.93
[< wedge][, 1]	23,687	95	3.7
[< wedge][, 2]	23,687	1064	1.84
R[< wedge](> 0)	23,687	214	2.92
[> wedge] (sum comp)	12,104	334	2.11
[> wedge][, 1]	12,104	382	1.83
[> wedge][, 2]	12,104	247	2.682
R[> wedge](> 0)	12089	48	3.435

in the form of k or in *poweRlaw* lingo "ntail". The package gives estimates for the index of the density so our numbers are adjusted by 1 to give the tail index of the distribution which is consistent with the rest of the book. The designation *sum comp* means components have been summed to give the L_1 distance from the origin. The subset facebkpos means pairs for which both components are positive. The designation [,1] means first components of pairs and similarly [,2] means second component. The subset [wedge] consists of pairs (x, y) such that $x/(x+y) \in [a, b]$ for certain choices of a, b with $0 < a < b < 1$; see Sect. 5.3.1.1, page 163. Finally [< wedge] is the subset of points (x, y) such that $x/(x+y) > b$, [> wedge] is the subset (x, y) such that $x/(x+y) < a$ and R[> wedge](> 0) are L_2-distances of points in [> wedge] to [wedge]; these result from a distance to the forbidden zone [wedge] rather than conventional polar coordinate transformation using the origin as forbidden zone.

For the most part, Table 5.1 estimates of α look reasonable but as expected (and feared) the *poweRlaw* optimal choices of k ("ntail") are rather small indicating a tendency to discard large portions of the data and use a relatively few upper order statistics in the estimation. Sometimes the choices of k are glaringly small as in $k = 48$ and $k = 95$. For network data, we are conditioned to expect estimates between 2 and 3 so the majority of the α-estimates seem sensible. Comparing *poweRlaw* estimates with those obtained by other means usually ($\neq$ always) brings some concordance. For example, for the line R[< wedge], we use *parfit* [156, p. 366] and *altHillalpha* [156, p. 364] to estimate the α for the second component to get the plots in Fig. 5.3.

On the right of Fig. 5.3, the vertical line corresponds to the *poweRlaw* choice of threshold and the horizonal line to the *poweRlaw* estimate of α. The concordance between *parfit*, *altHill* and *poweRlaw* is satisfying.

5.2 Threshold Selection

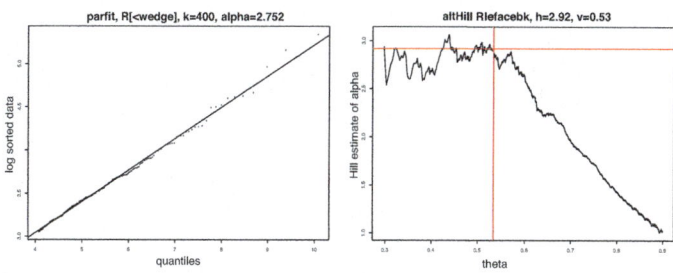

Fig. 5.3 $R[< \text{wedge}]$. Left: parfit, $k = 400$, $\alpha = 2.75$. Right: altHillalpha with horizontal line at height 2.9 and the vertical line at θ satisfying $n^\theta = 214$

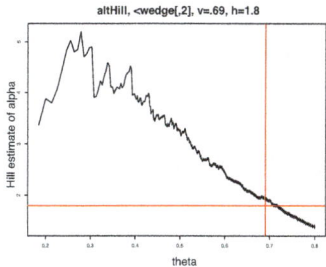

Fig. 5.4 Poor plot

However, estimates less than 2 indicate suspiciously heavy tails that indicate more exploration is warranted. Figure 5.4 showing the altHill plot for second components of a subset of points below the wedge requires paranormal abilities to find the α (horizontal line) and threshold (vertical line) given by *poweRlaw*. While this could be construed as a strength of *poweRlaw* over plotting methods requiring viewer interpretation, it may be the MDSP identifies something not there.

Other issues which may stretch the MDSP beyond its intended design:

1. The presented theory is for *standard* regular variation of measures and this implies the same normalizing function applies to all components and this in turn requires the same α for both components. At a minimum we need data components to have similar αs if not identical αs and then test for equality. Since MDSP estimates are not asymptotically normal, testing for equality of tail indices must be done by bootstrap but for node indexed data, is the bootstrap justified? (Probably.)

2. If you are reasonably confident the components have the same αs, is the α for the sum of the components (polar distance in the L_1 metric to the origin as forbidden zone) the same? Looking again at Table 5.1 might keep you up at night. For instance the first 3 lines give α-values 2.52 (sum comp), 2.98 (first component), 2.75 (second component) and while these are not terrifyingly different, can they be construed as equal without reliable confidence statements?

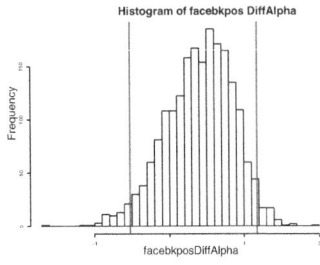

Fig. 5.5 Histogram of difference of α-estimates

Concerning the second item about equality of αs for the two components: Observe in Table 5.1 the α-estimates 2.98 and 2.66 for the first and second components. The data set consists of 35,424 pairs. For each component we created 2000 bootstrap samples, applied MDSP to get for each component 2000 estimates of component αs. These were then differenced and a histogram plotted in Fig. 5.5. The blue vertical lines are the 0.025 and 0.975 empirical quantiles of the differenced estimates of α and since 0 is comfortably between the blue lines, we cannot reject equality of the αs.

The MDSP is a useful and sensible selection criterion but it pays to keep in mind that this criterion did not come down on tablets from Mount Sinai.

5.3 Two-Dimensional Visualization of Multivariate Dependence

Visualization as an exploratory tool works best in two dimensions. If you are clever, you can extend this to three dimensions but beyond three, one has to rely on projection methods, techniques that consider small subsets of components and algorithms that take the place of eyeball exploration.

Even at the level of raw two dimensional scatterplots, stark differences in dependence structures can be glaringly obvious. Consider the three panels in Fig. 5.6. On the right panel are realized points from independent standard Pareto pairs hugging the axes so large values in both components do not occur simultaneously. Such a plot suggests an MRV model should have a limit measure concentrating on the axes. In the left and middle panels are scatterplots of old exchange rate return data from before the introduction of the Euro in 1999. The exchange rates are against the US dollar over a period of 6041 days from January 1971 to February 1994. On the left are absolute log-returns of the French franc against the German deutschmark and the funneling effect in the picture suggests that a large change in one component is often matched by a large change in the other component. This presumably reflects the high dependence of the economies of the two countries. An MRV model would perhaps have a limit measure that concentrates on a ray emanating from the origin. In the middle panel are absolute returns of the Japanese yen against the German deutschmark and in this picture we see a large value in one component matched by

5.3 Two-Dimensional Visualization of Multivariate Dependence

a variety of values in the other. An MRV model for this data should have a limit measure with support equal to whole first quadrant.

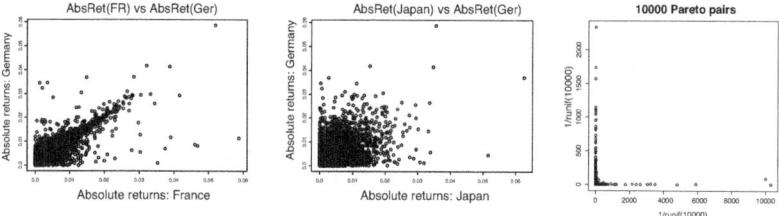

Fig. 5.6 Left: (France, Germany). Middle: (Japan, Germany). Right: independent Pareto pairs

In practice, it is not easy to find scatterplot data yielding suggestive pictures similar to those in Fig. 5.6. In particular, it is difficult to find plots clearly suggesting either asymptotic independence, where the limit measure concentrates on the axes, or asymptotic full dependence, where the limit measure concentrates on a line. When modeling growth of preferential attachment networks, the common model has the empirical frequency of nodes with in- and out-degree (k, l) converging to a discrete MRV distribution with mass function p_{kl}. The limit measure of this MRV distribution has support equal to the whole first quadrant [170]. However, if we add reciprocity to the model, requiring when an edge (k, l) is added, it be matched with probability p with an additional edge (l, k), then the limit distribution looks like *asymptotic full dependence* as in the left panel of Fig. 5.6 [186].

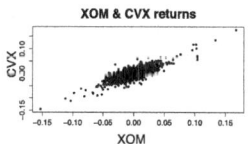

Fig. 5.7 XOM vs CVX

More common than *asymptotic full dependence* is *strong dependence* (Example 2.4, page 45) where the data seems to occupy a narrow ellipsoid, funnel or band shape. The cloud in Fig. 5.7 is more typical than full dependence and represents scatterplot data in $\mathbb{R}^2$ of returns from closing daily prices of Exxon (XOM) and Chevron (CVX) from January 2, 1998 to August 9, 2013. The length of the return vector series is 3925 and one expects strong dependence from two dominant companies engaged in similar economic activity.

Another example from a different discipline is the Facebook wall post data accessed at konect.cc/networks/facebook-wosn-links/ [115, 116]. This was analyzed in Sect. 5.2.1.2; see also [42, 180, 188]. Conversion of edge data to node-indexed in- and out-degree counts was done using the R-package *igraph* [29]. The data is from 2009 and represents posts by a subset of Facebook users in a limited geographic area to other users' walls in the same area. Nodes are users and a directed edge represents a post from the user to the user whose wall is receiving the post. There are 46,952 users and 876,993 edges. We focus on out- and

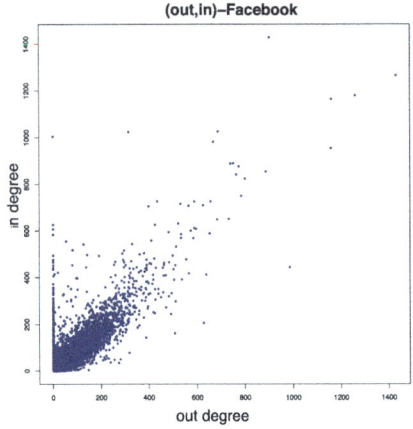

Fig. 5.8 Scatter plot of out- and in-degree indexed by node for Facebook wallpost data

in-degree indexed by the nodes as $\{(Z_{1,i}, Z_{2,i}) : 1 \leq i \leq 46952\}$. Of course this data is not the result of iid replication but conventional tools of heavy tail analysis seem effective. The scatter-plot of (out,in)-degrees in Fig. 5.8 shows the expected strong asymptotic dependence between out- and in-degrees. As is typical of social networks, there are many points of the form $(0, m)$ climbing the vertical axis which presumably correspond to users who join the social network but are not active. These points could be cleaned before further analysis but most points seem to cluster in a narrow band and strong dependence thus seems likely.

The previous plots in Figs. 5.6, 5.7, and 5.8 illustrate distinguishable cases for two-dimensional data. Dependence structures can be expressed either in Cartesian or polar coordinates and for polar coordinates we recall the mapping

$$x \mapsto \left(\|x\|, \frac{x}{\|x\|}\right) =: (r, (\theta, 1-\theta)) \in (0, \infty) \times \{(x, y) \in [0, 1]^2 : x + y = 1\},$$

where a convenient choice of norm is L_1 so $\|x\| = |x_1| + |x_2|$. The polar coordinate transformation is most conveniently written as

$$x \mapsto (r, \theta), \quad \theta \in [0, 1]. \tag{5.6}$$

When (5.1) holds in $\mathbb{R}_+^2 \setminus \{\mathbf{0}\}$, under this mapping, the limit measure transforms

$$\eta(\cdot) \mapsto \nu_\alpha(\cdot) \times S(\cdot) \tag{5.7}$$

where $S(\cdot)$ concentrates on $[0, 1]$, and $\alpha > 0$. Then we [190] summarize striking dependence cases as:

5.3 Two-Dimensional Visualization of Multivariate Dependence

1. Asymptotic full dependence: The limit measure $\eta(\cdot)$ concentrates on a ray emanating from the origin. The angular measure $S(\cdot)$ concentrates at a point.
2. Asymptotic strong dependence: The limit measure $\eta(\cdot)$ concentrates on a narrow cone whose point initiates from the origin. The angular measure $S(\cdot)$ concentrates on a proper subset $[a, b]$ of $[0, 1]$.
3. Asymptotic independence: The limit measure $\eta(\cdot)$ concentrates on the axes and the angular measure $S(\cdot)$ is a two-point distribution concentrating on $\{\{0\}, \{1\}\}$.
4. Asymptotic weak dependence: The limit measure $\eta(\cdot)$ concentrates on all of the state space $\mathbb{R}_+^2$ and the angular measure concentrates on $[0, 1]$.

Other possibilities, of course, are possible. The angular measure S could concentrate on a union of intervals or on a finite and discrete subset of $[0, 1]$.

Keep in mind we are being cautious and exploring, feeling our way among different dependence scenarios. Why are we cautious? One straightforward nonparametric estimator [156, p. 309ff] for $S(\cdot)$ based on observations $X_1, \ldots, X_n$ with polar coordinate versions $\{(R_i, \Theta_i), 1 \leq i \leq n\}$, where $\Theta_i = (\Theta_i, 1 - \Theta_i)$, is

$$\hat{S}_n(\cdot) := \frac{1}{k} \sum_{i=1}^{n} I_{[R_i \geq R_{(k)}]} \epsilon_{\Theta_i}(\cdot) = \frac{1}{k} \sum_{i=1}^{n} \epsilon_{(\Theta_i, R_i/R_{(k)})} (\cdot \times [1, \infty)) \tag{5.8}$$

with $R_{(k)}$ the kth largest R and $\epsilon_\Sigma(\cdot)$ being the Dirac measure concentrating at Σ. Considering $\hat{S}_n(\cdot)$ as a random element or random measure in $\mathbb{M}([0, 1])$, we have the consistency result $\hat{S}_n(\cdot) \Rightarrow S(\cdot)$ in $\mathbb{M}([0, 1])$ as $n \to \infty$, $k = k(n) \to \infty$ and $k/n \to 0$. So while $\hat{S}_n(\cdot)$, being the empirical distribution of Θ's corresponding to big R's, seems perfectly natural, we are nervous because estimating a *measure* $S(\cdot)$ requires a certain amount of hubris as several estimates could be consistent with the data. For instance the measure in (5.8) is always discrete but there is no assurance this is the case for $S(\cdot)$. At least both $\hat{S}_n(\cdot)$ and $S(\cdot)$ are probability measures. Be thankful for small favors.

5.3.1 The Diamond Plot

When doing empirical detection for the cases above using data in either $\mathbb{R}_+^2$ or $\mathbb{R}^2$, it is convenient and informative to map points $\mathbb{R}^2 \mapsto \aleph_{0_2}$

$$x \mapsto \left(\frac{x_1}{|x_1| + |x_2|}, \frac{x_2}{|x_1| + |x_2|} \right) = (\theta_1, \theta_2)$$

onto the L_1 unit sphere or diamond, perhaps after thresholding data according to the L_1-norm; that is, we use the subset of the data far from the origin. We call the resulting two-dimensional plot the *diamond plot*.

Observing how points cluster on the L_1 unit sphere provides a visualization of dependence. This can be done for two-sided data in $\mathbb{R}^2$ as for (XOM,CVX), whose scatter plot is Fig. 5.7, or for non-negative data in $\mathbb{R}_+^2$ such as (out,in) degree for Facebook wallpost data whose scatterplot is Fig. 5.8.

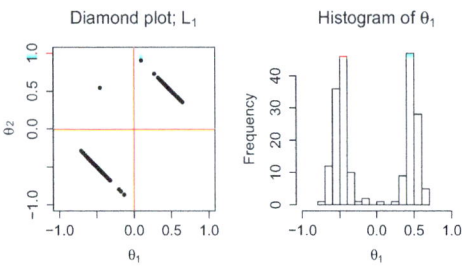

Fig. 5.9 Four quadrant plots for (XOM,CVX)

Figure 5.9 from [42] gives the diamond plot of the oil returns discussed in Fig. 5.7 where points with the 200 largest L_1-norm are used for the plotting. The choice of 200 was arrived at after some experimentation and alternative methods could use the MDSP applied to $\{R_i, 1 \leq i \leq n\}$ to decide on a threshold or one could construct the diamond plot with sliders allowing one to visually slide with the variable k to explore its effect on the plot. Of course, the diamond plot can only be carried out if the four tails ($\pm$XOM,$\pm$CVX) are the same; this was checked in [42] and a value of $\alpha = 2.7$ was fixed.

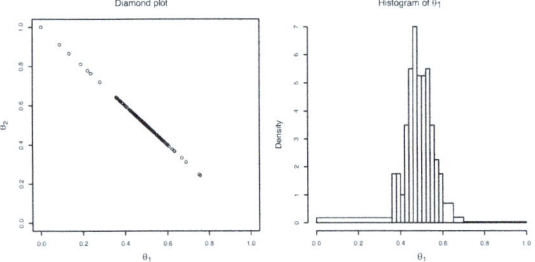

Fig. 5.10 Positive quadrant diamond plot for (out,in) degree of Facebook wallpost data

The diamond plot is $\{(\theta_1, \theta_2) : |\theta_1| + |\theta_2| = 1, r \geq R_{(200)}\}$. Figure 5.9 also has a histogram for θ_1. (An alternate to the histogram is a density plot which estimates the perhaps fictitious or non-existent density of S; an example with different data is in [156, p. 317]. However, the smoothing required for an angular density plot sometimes obscures the truth, especially when little data is available.)

Based on plots in Fig. 5.9 which have almost no points in the second and fourth quadrants, there is little evidence that a large positive change in one variable is accompanied by a large negative change in the other. There is also no strong evidence for asymptotic full dependence but visual evidence is good for strong dependence and it appears the angular measure concentrates in two intervals to the left and right of the origin corresponding to the first and third quadrants.

For the Facebook wallpost data, both out- and in-degree have tail indices approximately 2.8. (As emphasized previously, this is for the tail of the cumulative distribution function and not, as common in Network Science, the tail index of the density.) The diamond plot in Fig. 5.10 is constructed using the 200 largest L_1-norms of (out,in). This diamond plot for (θ_1, θ_2) is not quite as pristine as for the

5.3 Two-Dimensional Visualization of Multivariate Dependence

oil data, perhaps because of inactive nodes with out-degree = 0, but again, the assumption that S concentrates on a proper subinterval of $[0, 1]$ is quite plausible. The histogram in Fig. 5.10 of θ_1 confirms this impression since the majority of θ's cluster about a central value.

5.3.1.1 Consistency Suggested by the Diamond Plot: Beware of More Mischief

The support of $S(\cdot)$ contains information about the dependence structure and whether a large value in one variable is likely to be matched by a large value in the other. How do we find the support? The next result uses the parameterization of S resulting from using the transformation in (5.6); the form of the estimators is expected but what is moderately surprising is that consistency of these intuitive estimators required an additional condition beyond just the regular variation assumption.

Mischief × 2

Theorem 5.1 (Strong Dependence) *Assume* $\{X_1, \ldots, X_n\}$ *is an iid sample of vectors in* $\mathbb{R}_+^2$ *and in* L_1*-polar coordinates the sample is*

$$\{(R_1, (\Theta_1, 1 - \Theta_1)), \ldots, (R_n, (\Theta_n, 1 - \Theta_n))\}$$

with $\Theta_i \in [0, 1]$. *Suppose the common distribution of the sample* $P[X_1 \in \cdot] \in MRV(\alpha, b(t), \eta(\cdot), \mathbb{R}_+^2 \setminus \{\mathbf{0}\})$ *is regularly varying in* $\mathbb{R}_+^2 \setminus \{\mathbf{0}\}$ *with limit measure* $\eta(\cdot)$ *in Cartesian coordinates and after the mapping to* L_1*-polar coordinates the limit measure on* $(\mathbb{R}_+ \setminus \{0\}) \times [0, 1]$ *is* $\nu_\alpha \times S(\cdot)$ *as in (5.7). Assume the support of* $S(\cdot)$ *is* $[a, b] \subset [0, 1]$ *with* $0 < a < b < 1$. *Then*

1. *For* $\delta > 0$ *small, as* $n \to \infty$, $k = k(n) \to \infty$ *and* $k/n \to 0$,

$$P[\bigwedge_{i=1}^n \{\Theta_i : R_i \geq R_{(k)}\} \leq a + \delta] \to 1, \quad P[\bigvee_{i=1}^n \{\Theta_i : R_i \geq R_{(k)}\} \geq b - \delta] \to 1,$$
(5.9)

If additionally $k = k(n)$ *satisfies*

$$k\frac{n}{k} P[\Theta_1 \leq a - \delta, R_1 > b\left(\frac{n}{k}\right)] \to 0, \quad k\frac{n}{k} P[\Theta_1 \geq b + \delta, R_1 > b\left(\frac{n}{k}\right)] \to 0,$$
(5.10)

then we consistently estimate the endpoints of the support of $S(\cdot)$ *with*

$$\hat{a} := \bigwedge \{\Theta_i : R_i \geq R_{(k)}\} \xrightarrow{P} a, \quad \hat{b} := \bigvee \{\Theta_i : R_i \geq R_{(k)}\} \xrightarrow{P} b, \quad (5.11)$$

as $n \to \infty$.

2. *If (5.11) holds, we consistently estimate the support of $S(\cdot)$ as a compact subset of $[0, 1]$ using*

$$\widehat{[a, b]} := \{\Theta_i : R_i \geq R_{(k)}\} \xrightarrow{P} [a, b] \qquad (5.12)$$

as $n \to \infty$, $k \to \infty$ and $k/n \to 0$ where convergence in (5.12) is in the Hausdorff metric on $\mathcal{K}([0, 1])$.

Without multiplication by k in (5.10) the terms converge to $0 = S[0, a - \delta] = S[b+\delta, 1]$ so (5.10) is a condition on how fast k diverges. If Θ_1, R_1 are independent, the condition is certainly satisfied but, in general, without knowing $P[X \in \cdot]$, the condition is uncheckable. Therefore, you might as well just pray and assume it is true. So the estimation procedure is: Threshold the data and for this thresholded subset take the minimum and the maximum of the one-dimensional Θ's as the estimated endpoints. Sometimes robustification increases confidence that you are not being naive; for instance you can use the 10 and 90% quantiles of the Θs as estimates after thresholding. Drawing the diamond plot with sliders varying the threshold also helps.

Of course, in part 2 of Theorem 5.1 we could also estimate the support by $[\hat{a}, \hat{b}]$ using part 1 so the real interest in (5.12) is that the Θ's fill in the interval $[a, b]$ and do not leave holes.

If this is your first acquaintance with convergence of compact sets, a quick discussion follows. Standard references are [129, 138, 178]. The immediate need is for compact subsets of $[0, 1]$ so that is how the discussion is framed but generalizing to more general metric spaces is straightforward. The Hausdorff metric is $D : \mathcal{K}([0, 1]) \times \mathcal{K}([0, 1]) \mapsto \mathbb{R}_+$ defined for two compact subsets K_1, K_2 of $[0, 1]$ as

$$D(K_1, K_2) = \inf\{\delta > 0 : K_1 \subset K_2^\delta \text{ and } K_2 \subset K_1^\delta\},$$

and recall K^δ is the δ-neighborhood of K. So if $D(K_1, K_2)$ is small, K_1 and K_2 are near each other and have similar shape.

The following criterion [129, page 6] for convergence is somtimes convenient: We have convergence of compact sets $K_n \to K$ in the Hausdorff metric iff

1. (CONDITION 1.) $z \in K$ implies $\exists z_n \in K_n$ and $z_n \to z$.
2. (CONDITION 2.) For a subsequence $\{n_j\}$, if $z_{n_j} \in K_{n_j}$ and $\{z_{n_j}\}$ converges, then $\lim_{n_j \to \infty} z_{n_j} \in K$.

For the proof of Theorem 5.1, the following Lemma is helpful.

Lemma 5.1 ([122])

(i) *For $n \geq 0$, suppose $m_n \in \mathbb{M}([0, 1])$ with support $K_n = supp(m_n)$ and $m_0(\partial(K_0)) = 0$. Then for any $\delta > 0$, $m_n \to m_0$ implies there exists a finite integer $n_0 = n_0(\delta)$ such that $K_0 \subset K_n^\delta$, for $n \geq n_0$.*

5.3 Two-Dimensional Visualization of Multivariate Dependence

(ii) For $n \geq 0$, suppose M_n are random measures in $\mathbb{M}([0, 1])$ and the support of M_n is the random compact set K_n. Assume M_0 is non-random and $M_0(\partial K_0) = 0$. Then for $\delta > 0$,

$$M_n \xrightarrow{P} M_0 \text{ implies } \lim_{n \to \infty} P[K_0 \subset K_n^\delta] \to 1. \quad (5.13)$$

Proof of Lemma 5.1

(i) For any $x \in K_0$, there exists a δ-neighborhood $B(x, \delta)$ such that

(a) $m_0(\partial B(x, \delta)) = 0$;
(b) $m_0(B(x, \delta)) > 0$, since x is in the support and
(c) $m_n(B(x, \delta)) > 0$ for $n \geq n_0(x)$ since $m_n(B(x, \delta)) \to m_0(B(x, \delta))$.

Thus

$$K_0 \subset \bigcup_{x \in K_0} B(x, \delta) \quad (5.14)$$

and because K_0 is compact and $\{B(x, \delta), x \in K_0\}$ is a cover by open sets, there is a finite subcover $\{B(x_i, \delta), i = 1, \ldots, m\}$ and for $n \geq \vee_{i=1}^m n_0(x_i) =: n_0$ we have $m_n(B(x_i, \delta)) > 0$, $i = 1, \ldots, m$. Therefore for $n \geq n_0$,

$$B(x_i, \delta) \cap K_n \neq \emptyset, \quad i = 1, \ldots, m.$$

Pick any $y \in K_0$, refer to (5.14), and suppose $y \in B(x_{i*}, \delta)$; thus there exists $y_n \in B(x_{i*}, \delta) \cap K_n$ and $d(y_n, y) \leq d(y_n, x_{i*}) + d(x_{i*}, y) \leq 2\delta$. Therefore $y \in K_0$ implies $y \in K_n^{2\delta}$ if $n \geq n_0$. Among friends 2δ is as good as δ.

(ii) If the limit in (5.13) is not 1, then along some subsequence $\{n'\} \subset \{n\}$ we have $\lim_{n' \to \infty} P[K_0 \subset K_{n'}^\delta] \to p < 1$. Then along some further subsequence $\{n''\} \subset \{n'\}$ we have both

(a) $\lim_{n'' \to \infty} P[K_0 \subset K_{n''}^\delta] \to p < 1$; and
(b) $M_{n''} \to M_0$ almost surely in $\mathbb{M}([0, 1])$.

Let $\Lambda := \{\omega : M_{n''}(\omega) \to M_0\}$ so $P(\Lambda) = 1$. From (i), on Λ there exists an integer valued function $n_0''(\omega) = n_0''(\delta, \omega)$ such that $K_0 \subset K_{j''}^\delta$ if $j'' \geq n_0''(\omega)$; remember the K_n are random. Write $\Lambda_{m''} = [n_0'' \leq m'']$ so $P(\Lambda_{m''}) \uparrow P[n_0'' < \infty] = P(\Lambda) = 1$. Part (i) of the lemma asserts

$$[M_{n''} \to M_0] \subset [K_0 \subset K_{j''}^\delta; \forall j'' \geq n_0''(\omega)]$$

so

$$1 = P[M_{n''} \to M_0] \leq P[K_0 \subset K_{j''}^\delta; \forall j'' \geq n_0''(\omega)]$$

and splitting the event on the right according to $\Lambda_{m''}$ or its complement gives

$$\leq P[K_0 \subset K_{j''}^{\delta}; \forall j'' \geq m''; \Lambda_{m''}] + P(\Lambda_{m''}^c)$$

$$\leq P[K_0 \subset K_{m''}^{\delta}] + P(\Lambda_{m''}^c).$$

The right side is made < 1 by choosing m'' sufficiently large. This gives a contradiction caused by supposing the limit on the right side of (5.13) is $p < 1$. □

Proof of Theorem 5.1

Part 1 One way to obtain a proof is to use the découpage de Lévy (eg. [157]) and do the following: For now, think like a mathematician and imagine we have an infinite family $\{(\Theta_i, R_i), i \geq 1\}$. Fix n and $z > 0$ and let $\{\tau_i\}$ be indices where $R_{\tau_i} > zb(n/k)$, that is, $\tau_0 = 0$ and

$$\tau_{i+1} = \inf\{m > \tau_i : R_m > zb(n/k)\}, \ i \geq 1.$$

Then $\{(\Theta_{\tau_i}, R_{\tau_i}), i \geq 1\}$ are iid with common distribution

$$P[(\Theta_{\tau_i}, R_{\tau_i}) \in \cdot] = P[(\Theta_1, R_1) \in \cdot \mid R_1 > zb(n/k)],$$

and also independent of

Think!

$$N_n = \sum_{i=1}^{n} 1_{[R_i > zb(n/k)]} = \sup\{i : \tau_i \leq n\},$$

which is a binomial random variable with success probability $P[R_1 > zb(n/k)]$ and n trials. Then

$$\bigwedge_{j=1}^{n}\{\Theta_j : R_j \geq zb(n/k)\} = \bigwedge_{i=1}^{N_n} \Theta_{\tau_i} \text{ and } \bigvee_{j=1}^{n}\{\Theta_j : R_j \geq zb(n/k)\} = \bigvee_{i=1}^{N_n} \Theta_{\tau_i},$$

so working toward (5.9) we have

$$P[\bigwedge_{i=1}^{n}\{\Theta_i : R_i \leq zb(n/k)\} \leq a + \delta] = 1 - P[\bigwedge_{i=1}^{N_n} \Theta_{\tau_i} > a + \delta],$$

and we show that the probability after the minus sign converges to 0. This probability is

5.3 Two-Dimensional Visualization of Multivariate Dependence

$$P[\bigwedge_{i=1}^{N_n} \Theta_{\tau_i} > a + \delta] = E\left((P[\Theta_{\tau_1} > a + \delta])^{N_n}\right)$$

and from the form of the generating function of the binomial random variable N_n the previous expression is

$$\left(1 - P[\Theta_{\tau_1} \leq a + \delta] P[R_1 > zb(n/k)]\right)^n$$

$$= \left(1 - \frac{k \frac{n}{k} P[\Theta_1 \leq a + \delta, R_1 > zb(n/k)]}{n}\right)^n.$$

Now

$$\frac{n}{k} P[\Theta_1 \leq a + \delta, R_1 > zb(n/k)] \to S[0, a + \delta] > 0,$$

as a consequence of a being in the support of S and this proves $P[\bigwedge_{i=1}^{N_n} \Theta_{\tau_i} > a + \delta] \to 0$. The analogous statement involving max instead of min and $b - \delta$ instead of $a + \delta$ is proven similarly.

It remains to show we can replace $zb(n/k)$ with $R_{(k)}$. For this we use the fact that $R_{(k)}/b(n/k) \xrightarrow{P} 1$ as $n \to \infty$, $k \to \infty$, $k/n \to 0$ [156, p. 81ff]. We have for any small $\epsilon > 0$, using the definition of convergence in probability,

$$P[\bigwedge_{i=1}^{n} \{\Theta_i : R_i \geq R_{(k)}\} \leq a + \delta]$$

$$= P[\bigwedge_{i=1}^{n} \{\Theta_i : R_i \geq R_{(k)}\} \leq a + \delta, \left|\frac{R_{(k)}}{b(n/k)} - 1\right| \leq \epsilon] + o(1)$$

$$\geq P[\bigwedge_{i=1}^{n} \{\Theta_i : R_i \geq (1+\epsilon)b(n/k)\}\} \leq a + \delta] + o(1) + o(1) \to 1,$$

where we imagine $z = (1 + \epsilon)$. The rest of the verification of (5.9) is similar.

The proof of (5.11) is similar. We focus on showing $\hat{a} \xrightarrow{P} a$ using (5.10). We need $P[\bigwedge\{\Theta_i : R_i \geq R_{(k)}\} > a - \delta] \to 1$. The version of the probability with $zb(n/k)$ replacing $R_{(k)}$ is

$$P\left[\bigcap_{i=1}^{N_n} [\Theta_{\tau_i} > a - \delta]\right] = E\left(P[\Theta_{\tau_1} > a - \delta]\right)^{N_n}$$

$$= \left(1 - P[R_1 > zb(n/k)] P[\Theta_{\tau_1} \leq a - \delta]\right)^n$$

$$=\left(1 - P[\Theta_1 \leq a - \delta, R_1 > zb(n/k)]\right)^n$$
$$=\left(1 - \frac{k\frac{n}{k}P[\Theta_1 \leq a - \delta, R_1 > zb(n/k)]}{n}\right)^n$$

and this converges to 1 iff the first relation in (5.10) holds. Note the proof with $a - \delta$ requires the extra condition because $S[0, a - \delta] = 0$ but $S[0, a + \delta] > 0$.

Finish by setting $z = (1 - \epsilon)$ and swapping $R_{(k)}$ with $zb(n/k)$.

Part 2 Now on to business: Since (5.9) is always true assuming regular variation, we have as $n \to \infty$ [122, p. 267],

$$P[\bigwedge_{i=1}^n \{\Theta_i - \delta : R_i \geq R_{(k)}\} \leq a, \bigvee_{i=1}^n \{\Theta_i + \delta : R_i \geq R_{(k)}\} \geq b] \to 1$$

so

$$P\left[[a, b] \subset [\hat{a}, \hat{b}]^\delta\right] \to 1. \tag{5.15}$$

If also (5.11) holds (implied by (5.10)), then

$$P\left[[\hat{a}, \hat{b}] \subset [a, b]^\delta\right] \to 1,$$

implying

$$P[\{\Theta_i : R_i \geq R_{(k)}\} \subset [a, b]^\delta] \to 1.$$

This is encouraging but not the same thing as asserted in (5.12) so we must make sure $\{\Theta_i : R_i \geq R_{(k)}\}$ does not leave gaps in $[a, b]$ and (5.15) must be strengthened to

$$P[[a, b] \subset \{\Theta_i : R_i \geq R_{(k)}\}^\delta] \to 1. \tag{5.16}$$

This follows by applying Lemma 5.1 (ii) with the definition of $\hat{S}_n$ in (5.8), page 161, and the fact that $\hat{S}_n \xrightarrow{P} S$ in $\mathbb{M}([0, 1])$ and identifying in Lemma 5.1

$$M_n = \hat{S}_n, \qquad K_n = \{\Theta_i : R_i \geq R_{(k)}\},$$
$$M_0 = S, \qquad K_0 = [a, b].$$

This completes the proof. □

Late Breaking News In case the angular measure $S(\cdot)$ concentrates on $[a, b] \subsetneq [0, 1]$ and strong dependence is present, estimating the support region with the max

5.3 Two-Dimensional Visualization of Multivariate Dependence

and min as suggested in (5.9) leads to estimates that are highly sensitive to changes in the data that might occur when experimenting with the threshold. This objection is present even before factoring in the necessity of condition (5.10) to get consistency. An alternate optimization procedure for estimating a, b is in [190]; this procedure is consistent and seems less sensitive to data perturbation but requires the extra assumption beyond just MRV of second order mutivariate regular variation.

The estimation procedure producing consistent estimates of a, b is as follows: For $[a, b] \subsetneq [0, 1]$ define (more in Sect. 5.5.2, page 179)

$$\mathbb{C}_{a,b} = \{(x, y) \in \mathbb{R}_+^2 : \frac{x}{x+y} \in [a, b]\} \tag{5.17}$$

which is expressed in terms of L_1 θs and in terms of slopes we have

$$= \{(x, y) \in \mathbb{R}_+^2 : y/x \in [b^{-1} - 1, a^{-1} - 1]\} \tag{5.18}$$

as the Cartesian wedge corresponding to $\theta \in [a, b]$. A scaled distance $(x, y) \in \mathbb{R}_+^2$ to $\mathbb{C}_{a,b}$ is

$$d^*((x, y), \mathbb{C}_{a,b}) := \max\left\{(b^{-1} - 1)x - y, y - (a^{-1} - 1)x, 0\right\}. \tag{5.19}$$

Note if $(x, y) \in \mathbb{C}_{a,b}$, this distance is 0.

Assume $\{X_1, \ldots, X_n\}$ is an iid sample of vectors in $\mathbb{R}_+^2$ and in L_1-polar coordinates the sample is

$$\left\{(R_1, (\Theta_1, 1 - \Theta_1)), \ldots, (R_n, (\Theta_n, 1 - \Theta_n))\right\}$$

with $\Theta_i \in [0, 1]$. Suppose the common distribution of the sample $P[X_1 \in \cdot] \in$ MRV$(\alpha, b(t), \eta(\cdot), \mathbb{R}_+^2 \setminus \{\mathbf{0}\})$ is regularly varying in $\mathbb{R}_+^2 \setminus \{\mathbf{0}\}$ with limit measure $\eta(\cdot)$ concentrating on $\mathbb{C}_{a,b}$. Let $R_{(1)} \geq R_{(2)} \geq \cdots \geq R_{(n)}$ be the decreasing order-statistics of $\{R_i, 1 \leq i \leq n\}$ and suppose X_i^* is the observation corresponding to $R_{(i)}$. Then we consistently estimate a, b by a minimization procedure,

$$(\hat{a}, \hat{b}) = \arg\min_{0 \leq a \leq b \leq 1} (b - a) + \sqrt{k_n} \frac{1}{k_n} \sum_{i=1}^{k_n} \left(\frac{d^*((X_i^*, Y_i^*), \mathbb{C}_{a,b})}{R_{(k_n)}}\right)$$

The first term $b - a$ tries to keep $[a, b]$ as small as possible but the second term wants to make $[a, b]$ big in order to minimize d^*.

Fig. 5.11 XOM vs CVX diamond plots. Top row: From left to right: k=700, 400. Bottom row left to right: 300, 150

5.3.1.2 Making Use of Sliders

If you are in an exploratory mood, adding sliders to plots is helpful for examining how threshold choice affects pictorial impressions. An example is the diamond plot. Changing the threshold continuously while observing the effect on the plot, shows how sensitive conclusions are to threshold choice and gives an idea how much data should be discarded before entering the asymptotic regime.

As an example, we compare returns from Exxon (XOM) and Chevron (CVX) and unlike Fig. 5.7, this time the data is from January 3, 2005 to March 29, 2018 and is of length 3333. The diamond plots at varying thresholds indicated by the level of k, the number of upper order statistics of the L_1-norm, are given in Fig. 5.11. The version of the diamond plot with sliders was programmed by Jaakko Lehtomaa, University of Helsinki [122].

Differences in the 4 plots are apparent. Setting $k = 700$ yields points on the diamond plot in the second and 4th quadrants giving the impression it is possible that large positive and large negative values are likely. Also in the first and third quadrants the dark areas of the plots occupy almost all of the interval. As k decreases (that is, the threshold increases) the number of points in the second and 4th quadrants decrease to virtually nil and the impression of the width of the intervals in the first and 3rd quadrants decreases to a reasonably solid interval with holes only near the endpoints. Note however that at $k = 150$, we are only using a small proportion $150/3333 \approx 4\%$ of the available 3333 observations. This is frequently a feature of using extreme value methods.

5.4 Evidence Consistent with Multivariate Regular Variation from Hillish Analysis

Start with the assumptions of Theorem 5.1, that we have a sample of size n from a distribution on $\mathbb{R}_+^2$ with a regularly varying distribution on $\mathbb{R}_+^2 \setminus \{\mathbf{0}\}$ and in polar coordinates the limit measure on $[0, 1] \times (\mathbb{R}_+ \setminus \{0\})$ is $S \times \nu_\alpha$. How do we get visual evidence consistent with the assumption that this asymptotic model exists? Recall that one can get some evidence of existence of multivariate regular variation on $\mathbb{R}_+^2 \setminus \{\mathbf{0}\}$ by checking one-dimensional marginal distributions are heavy tailed and there is also the diagnostic in Sect. 3.1.1, page 95 which reduced the multivariate regular variation to an infinite number of univariate tests but the diagnostic was hard to generalize to forbidden zones beyond the origin. Hillish analysis avoids univariate tests mentioned previously but still seeks evidence of multivariate heavy tails.

The *Hillish* statistic was devised in [40] for detecting a *conditional extreme value model* [39, 41, 62, 88, 89, 166] but another use of this technique is to provide evidence consistent with the two-dimensional regular variation assumption on $\mathbb{R}_+^2 \setminus \{\mathbf{0}\}$. Using generalized polar coordinates (Sect. 2.2.2.1, page 39), the method stretches to provide evidence of hidden regular variation on $\mathbb{R}_+^2 \setminus F$ for a cone F [42].

5.4.1 Hillish Analysis for Heavy Tails on $\mathbb{R}_+^2 \setminus \{\mathbf{0}\}$

Before getting to more exotic uses of Hillish analysis for forbidden zones beyond the origin, we review the basics. The sample of observations in $[0, 1] \times (\mathbb{R}_+ \setminus \{0\})$ is $\{(\Theta_i, R_i), 1 \leq i \leq n\}$ and assume

$$tP[(\Theta_1, R_1/b(t)) \in \cdot] \to S \times \nu_\alpha, \quad \alpha > 0, \ t \to \infty, \quad (5.20)$$

is the regular variation condition. The ordered sample of Rs is

$$R_{(1)} \geq R_{(2)} \geq \ldots R_{(n)}$$

and the Θ corresponding to $R_{(i)}$ is Θ_i^* so Θ_i^* is the concomitant of $R_{(i)}$. The consistent empirical estimator of the limit measure $S \times \nu_\alpha$ is [156, p. 309ff]

$$\frac{1}{k}\sum_{j=1}^n \epsilon_{(\Theta_j, R_j/R_{(k)})} = \frac{1}{k}\sum_{j=1}^n \epsilon_{(\Theta_j^*, R_{(j)}/R_{(k)})} \Rightarrow S \times \nu_\alpha, \quad (5.21)$$

in $\mathbb{M}([0, 1] \times (\mathbb{R}_+ \setminus \{0\}))$. This yields the estimator of $S(\cdot)$ (or recall (5.8), page 161)

$$\hat{S}_n(\cdot) = \frac{1}{k}\sum_{j=1}^{k}\epsilon_{\left(\Theta_j^*, R_{(j)}/R_{(k)}\right)}\left((\cdot)\times[1,\infty)\right) \Rightarrow S(\cdot)\times\nu_\alpha([1,\infty)) = S(\cdot), \quad (5.22)$$

in $\mathbb{M}([0,1])$. Using either (5.21) or (5.22) to construct a statistical diagnostic runs into the problem of lack of knowledge of $S(\cdot)$. so we sidestep this issue with a non-parametric technique and for each $k \in \{1,\ldots,n\}$ and $1 \le i \le k$, let

$$r_i(k) = \#\{\Theta_l^* \le \Theta_i^*; l = 1,\ldots,k\}$$

which is the rank of Θ_i^* amongst $\Theta_1^*\ldots,\Theta_k^*$. Note

$$\hat{S}_n([0,\Theta_i^*]) = \frac{1}{k}\sum_{j=1}^{k}\epsilon_{\Theta_j^*}([0,\Theta_i^*]) = \frac{r_i(k)}{k}, \quad i = 1,\ldots,k. \quad (5.23)$$

Here is the information on the Hillish statistic.

Theorem 5.2 *Suppose in polar coordinates we have the regular variation described in (5.20) where the measure $S(\cdot)$ is atomless so that the function $S[0,\theta], 0 \le \theta \le 1$ is continuous. Then*

$$Hillish_{k,n} := \frac{1}{k}\sum_{i=1}^{k}\log\left(\frac{k}{r_i(k)}\right)\log\left(\frac{R_{(i)}}{R_{(k)}}\right) \xrightarrow{P} \frac{1}{\alpha} \quad (5.24)$$

as $n \to \infty, k \to \infty, n/k \to \infty$.

The use of ranks, removes the influence of $S(\cdot)$ in (5.21). The reason for the name *Hillish* is that the derivation mimics a consistency proof for the Hill estimator based on the tail empirical measure consistency (eg. [156, p. 81ff]). The statistic in (5.24) adapts to the heavy tailed case the one proposed in [40, Eqn. (2.12)] for a more general extreme value context.

Proof The proof is similar to the proof of Hill estimator consistency and some verifications are omitted.

We first verify that in $\mathbb{M}([0,1]\times(\mathbb{R}_+\setminus\{0\}))$,

$$\frac{1}{k}\sum_{i=1}^{n}\epsilon_{\hat{S}_n([0,\Theta_i^*]),R_{(i)}/R_{(k)}} \Rightarrow \mathcal{U}(\cdot)\times\nu_\alpha, \quad (5.25)$$

where $\mathcal{U}$ is uniform measure on $[0,1]$. Because S is atomless, we get from (5.22) (cf. Proposition A.1, page 225) that

$$\sup_{0\le\theta\le 1}|S([0,\theta]) - \hat{S}_n([0,\theta])| \xrightarrow{P} 0 \quad (5.26)$$

5.4 Evidence Consistent with Multivariate Regular Variation from Hillish...

as $n \to \infty$, so the distribution functions of S, $\hat{S}_n$ are uniformly close with high probability. Pick a uniformly continuous $f \in \mathbb{C}([0, 1] \times (\mathbb{R}_+ \setminus \{0\}))$ and by Theorem 1.1, (5.25) is true if

$$\frac{1}{k}\sum_{i=1}^{n} f\big(\hat{S}_n([0, \Theta_i^*]), R_{(i)}/R_{(k)}\big) \xrightarrow{P} \int_{[0,1]\times(\mathbb{R}_+\setminus\{0\})} f(u, v)du v_\alpha(dv). \quad (5.27)$$

Note if we replace $\hat{S}_n$ with S in (5.27) and set $\tilde{f}(u, v) = f(S([0, u], v) \in \mathbb{C}([0, 1] \times (\mathbb{R}_+ \setminus \{0\}))$ we get using (5.21),

$$\frac{1}{k}\sum_{i=1}^{n} f(S([0, \Theta_i^*], R_{(i)}/R_{(k)})) = \frac{1}{k}\sum_{i=1}^{n} \tilde{f}(\Theta_i^*, R_{(i)}/R_{(k)})$$

$$= \int_{[0,1]} \tilde{f}(u, v) \frac{1}{k}\sum_{i=1}^{n} \epsilon_{(\Theta_i, R_{(i)}/R_{(k)})}(du, dv) \xrightarrow{P} S \times v_\alpha(\tilde{f})$$

$$= \int_0^1 \int_{\mathbb{R}_+\setminus\{0\}} f(S[0, u], v) S(du) v_\alpha(dv) = \int_0^1 \int_{\mathbb{R}_+\setminus\{0\}} f(s, v) ds v_\alpha(dv),$$

where the last equality follows by change of variable and uses the fact $S(\cdot)$ is atomless. So it suffices to show

$$\frac{1}{k}\sum_{i=1}^{k} f(\hat{S}_n[0, \Theta_i^*], R_{(i)}/R_{(k)}) - \frac{1}{k}\sum_{i=1}^{k} f(S[0, \Theta_i^*], R_{(i)}/R_{(k)}) \xrightarrow{P} 0.$$

For any $\delta > 0$, the probability the absolute value of the difference exceeds $\epsilon > 0$ is

$$P[\frac{1}{k}\sum_{i=1}^{k} \big|f(\hat{S}_n[0, \Theta_i^*], R_{(i)}/R_{(k)}) - f(S[0, \Theta_i^*], R_{(i)}/R_{(k)})\big| > \epsilon]$$

$$= P['', \sup_{0\leq\theta\leq 1} |\hat{S}_n[0, \theta] - S[0, \theta]| > \delta]$$

$$+ P['', \sup_{0\leq\theta\leq 1} |\hat{S}_n[0, \theta] - S[0, \theta]| \leq \delta]$$

$$\leq P[\sup_{0\leq\theta\leq 1} |\hat{S}_n[0, \theta] - S[0, \theta]| > \delta] + P[\omega_f(\delta) > \epsilon].$$

Make δ so small that the modulus of continuity $\omega_f(\delta) < \epsilon$ rendering the second probability 0 and the first probability goes to 0 from (5.26). This verifies (5.25).

Apply (5.23) and the map $[0, 1] \times (\mathbb{R}_+ \setminus \{0\}) \mapsto (0, 1] \times [0, 1])$ yielding,

$$\frac{1}{k}\sum_{j=1}^{k}\epsilon_{\left(\hat{S}_n[0,\Theta_j^*],R_{(j)}/R_{(k)}\right)} = \frac{1}{k}\sum_{j=1}^{k}\epsilon_{(r_j(k)/k,R_{(j)}/R_{(k)})} \Rightarrow \mathcal{U} \times \nu_\alpha$$

in $\mathbb{M}((0,1] \times [1,\infty))$. Then apply the map $(x,y) \in (0,1] \times [1,\infty) \mapsto (1/x,y) \in [1,\infty)^2$ to the state space of the measures we get,

$$\frac{1}{k}\sum_{j=1}^{k}\epsilon_{(k/r_j(k),R_{(j)}/R_{(k)})} \Rightarrow \nu_1 \times \nu_\alpha$$

in $\mathbb{M}([1,\infty)^2)$. Evaluate the measure convergence on the set $[u,\infty) \times [v,\infty)$, and then integrate over $[1,\infty)^2$ with respect to $\frac{du}{u}\frac{dv}{v}$. Provided you justify interchange of integration and the limit, this yields the result, since

$$\int_1^\infty \int_1^\infty \nu_1[u,\infty)\nu_\alpha[v,\infty)\frac{du}{u}\frac{dv}{v} = \int_1^\infty \int_1^\infty u^{-1}v^{-\alpha}\frac{du}{u}\frac{dv}{v}$$
$$= 1 \cdot \frac{1}{\alpha}$$

while

$$\int_1^\infty \int_1^\infty \frac{1}{k}\sum_{j=1}^{k}\epsilon_{(k/r_j(k),R_{(j)}/R_{(k)})}\Big([u,\infty) \times [v,\infty)\Big)\frac{du}{u}\frac{dv}{v}$$
$$= \frac{1}{k}\sum_{j=1}^{k}\int_1^{k/r_j(k)}\frac{du}{u}\int_1^{R_{(j)}/R_{(k)}}\frac{dv}{v} = \text{Hillish}_{k,n}$$

This finishes the sketch. □

5.4.1.1 Oops! About that Continuity Assumption...

The assumption in Theorem 5.2 that the angular measure $S(\cdot)$ be atomless seems innocuous but fails with a vengeance in two important cases:

- *Asymptotic independence*, where $S(\cdot)$ is a two point distribution concentrating on $\{0,1\}$.
- *Asumptotic full dependence*, where $S(\cdot)$ concentrates all mass on an interior point of $(0,1)$.

In addition, there are estimators of the angular measure that are purely atomic. So we cannot ignore the non-atomless case. However, there are modifications of the data that allow application of Theorem 5.2.

5.4 Evidence Consistent with Multivariate Regular Variation from Hillish...

One procedure is the following: Express the regular variation assumption in polar coordinates

$$tP[(R/b(t), \Theta) \in \cdot\,] \to \nu_\alpha \times S(\cdot),$$

in $\mathbb{M}\big((\mathbb{R}_+ \setminus \{0\}) \times [0, 1]\big)$. Suppose $U \sim U[0, 1]$ is a unformly distributed random variable on $[0, 1]$ independent of (R, Θ). From Proposition 2.1,

$$tP[(R/b(t), \Theta, U) \in \cdot\,] \to \nu_\alpha \times S(\cdot) \times \mathcal{U}(\cdot),$$

in $\mathbb{M}\big((\mathbb{R}_+ \setminus \{0\}) \times [0, 1]^2\big)$ and $\mathcal{U}(\cdot)$ is uniform measure on $[0, 1]$. Define $S^c(\cdot) = S * \mathcal{U}$ as the convolution of S and $\mathcal{U}$ which is continuous. Apply the map $(r, \theta, u) \mapsto \big(r, \frac{1}{2}(\theta + u)\big)$ from $\mathbb{R}_+ \setminus \{0\} \times [0, 1]^2 \mapsto (\mathbb{R}_+ \setminus \{0\}) \times [0, 1]$ to get

$$tP\left[\left(\frac{R}{b(t)}, \frac{\Theta + U}{2}\right) \in \cdot\,\right] \to \nu_\alpha \times S^c(2\cdot),$$

in $\mathbb{M}\big((\mathbb{R}_+ \setminus \{0\}) \times [0, 1]\big)$. Now we can apply the Hillish procedure to data of the form $\{(R_i, (\Theta_i + U_i)/2), 1 \le i \le n\}$ and seek evidence of multivariate regular variation.

In the case of asymptotic independence where Θ is a Bernoulli random variable on $\{0, 1\}$, we get the angular measure corresponding to $(\Theta + U)/2$ being uniform on $[0, 1]$. For asymptotic full dependence where Θ is almost surely a fixed point, we get the angular measure corresponding to a shifted uniform.

5.4.2 Hillish Examples

Here are two simulated and two real data examples from [42] and Sect. 5.3.1. One can be skeptical about the value of simulated data since model error is often critical. However, simulations can be instructive so here goes.

5.4.2.1 Example 1: Simulation Where the Sample Size may be Too Small

We simulated a sample of size 1000 from $\Theta \sim beta(2, 10)$ and an independent sample of size 1000 from R distributed standard Pareto. Define $X = (R\Theta, R(1 - \Theta))$. Due to the large beta shape parameter of 10, we expect most Θs to be in $[0, 1/2]$ and the diamond plot confirms this expectation.

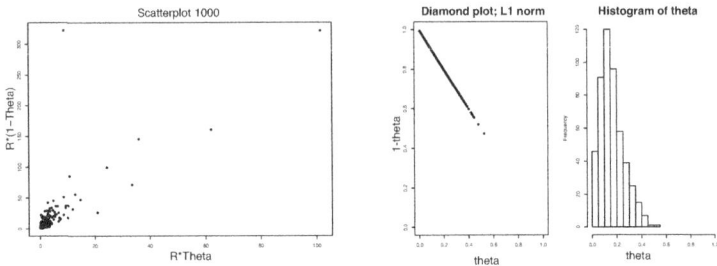

Fig. 5.12 Left: scatterplot of 1000 pairs; Right: diamond plot

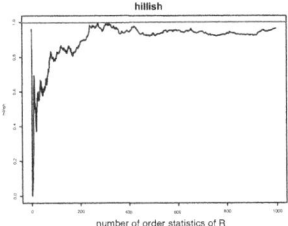

Fig. 5.13 Hillish plot $n = 1000$. Red line is at height 1

The scatterplot in Fig. 5.12 is not terribly informative but upon examination one sees some points far away from the origin and most points are above the main diagonal. The pictured diamond plot with a generous $k = 500$ would lead one to erroneously believe that the limit measure S (which in reality has a beta distribution) has support contained in $[0, 1/2]$ even though the beta has support $[0, 1]$. The sample size is just too low. (This is a hint that the asymptotic analysis requires big samples to discern the asymptotic model in difficult cases and $beta(2, 10)$ is certainly a difficult case.) The Hillish plot looks decent and correctly gives evidence of two-dimensional regular variation (Fig. 5.13).

5.4.2.2 Example 2: Simulation with Bigger Sample Size

Okay, we are fairly clear what the problems were in the previous simulation. We control the experiment so let us increase the sample size and pick the second shape parameter smaller rather than choosing a value likely to fool the diamond plot (Fig. 5.14).

This time the sample size is $n = 100,000$ for $\Theta \sim beta(2, 3)$ and again R is Pareto 1 on $[1, \infty)$ and in Cartesian coordinates we have $X = (R\Theta, R(1 - \Theta))$.

This time the scatterplot shows points more evenly distributed above and below the diagonal and the diamond plot indicates a support which is most of $[0, 1]$. The Hillish plot nails the correct value of $\alpha = 1$ and provides evidence of two-dimensional regular variation.

5.4 Evidence Consistent with Multivariate Regular Variation from Hillish...

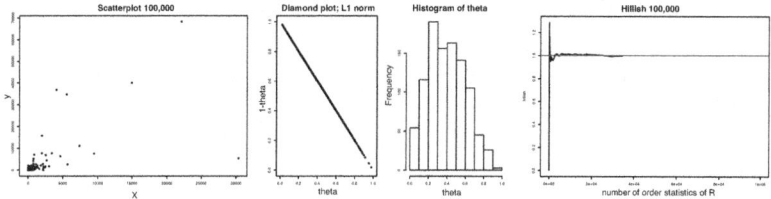

Fig. 5.14 Left: scatterplot of 100,000 pairs; Middle: diamond plot with $k = 1000$; Right: Hillish plot with red horizontal line at height 1

5.4.2.3 Example 3: Oil Returns

We return to the (CVX,XOM) stock returns discussed in Sect. 5.3.1.2. Recall the data set is size 3,333 and this length is more in the ball park of Example 1 than Example 2. Furthermore stock returns look fairly stationary but have dependence that fades slowly with increasing time lags. So we may expect difficulties.

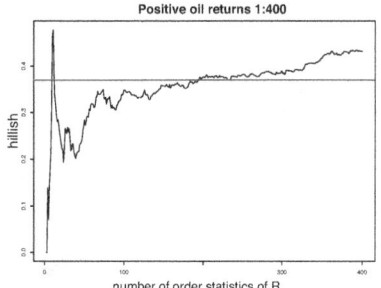

Fig. 5.15 Hillish: The red horizontal line is at height $1/\hat{\alpha} = 1/2.7 = 0.37$

We subset the data and analyze only returns which are positive for both CVX and XOM yielding 1427 values. Analysis of one-dimensional marginal data gives α estimates in both components of approximately 2.7. Making the Hillish plot for all values of R (not shown) produces a visualization that is misleading and uninformative (polite synonyms for *awful*) because the plot must be scaled to accomodate way too many order statistics.

Lessons learned from the need to correctly display information when making Hill (not Hillish) plots [61, 164] are helpful here. Using too large a value of k in Hillish uses pre-asymptotic behavior that corrupts the estimator. So to get something informative plot we plot $\{(k, \text{Hillish}_{k,n}); 2 \leq k \leq 400\}$ in Fig. 5.15 The plot also has a red horizontal line at height $1/\hat{\alpha} = 1/2.7 \approx 0.37$.

5.4.2.4 Example 4: Facebook Wall Posts

Consider the Facebook wall post data discussed in Sect. 5.3 and visualized in Fig. 5.8, page 160. The data was accessed from konect.cc [115] and consists of (out,in)-degree indexed by node for a local area. Marginal analysis yields an approximate estimate $\hat{\alpha} = 2.8$ for both margins so $1/\hat{\alpha} \approx 0.35$. We removed pairs where one or both were zero which still left 35,424 pairs. Plotting the full Hillish plot for all 35,424 R's yields an uninformative plot but plotting Hillish values for the 750 largest R's yields the reasonable picture in Fig. 5.16

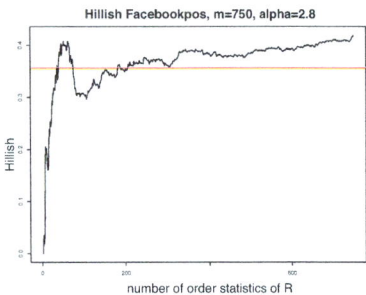

Fig. 5.16 Hillish plot for 750 values of R. Red line is at height $1/\hat{\alpha} = 0.35$

which supports the hypothesis of multivariate regular variation. Keep in mind the data is different than iid and is node-indexed.

5.5 Beyond $\mathbb{R}^2 \setminus \{0\}$: Other Cones as Forbidden Zones

There are statistical circumstances where regular variation exists on $\mathbb{R}^2 \setminus F$ where F is a cone bigger than the origin. The most common instance is where we succeed in identifying regular variation on $\mathbb{R}^2_+ \setminus \{0\}$ and the limit measure concentrates on a cone $F = \mathbb{C}_{a,b}$ corresponding to strong dependence so that in polar coordinates, the angular measure S concentrates on $[a, b] \subsetneq [0, 1]$ (Sect. 5.3.1.1, page 163). Then one can remove $\mathbb{C}_{a,b}$ and seek evidence of regular variation on $\mathbb{R}^2_+ \setminus \mathbb{C}_{a,b}$. Hillish analysis can apply to other cones and forbidden zones but, full disclosure, Hillish analysis depends on the geometry of $\mathbb{S} \setminus F$ and works best in $\mathbb{R}^2$ for the case of strong dependence (Example 2.4, page 45) [42].

Other cones

Instead of using the L_1-polar coordinate transform, we use generalized polar coordinates discussed in Sect. 2.2.2.1, page 39. Recall on a TABOF space $\mathbb{S} \setminus F$ with metric $d(\cdot, \cdot)$, the transformation to generalized polar coordinates is

$$x \mapsto \left(d(x, F), x/d(x, F)\right) = (r, \theta) \tag{5.28}$$

from $\mathbb{S} \setminus F \mapsto (\mathbb{R}_+ \setminus \{0\}) \times \aleph_F$ where $\aleph_F = \{s \in \mathbb{S} \setminus F : d(s, F) = 1\}$. If a sample of observations is $X_1, \ldots, X_n$, set $R_i = d(X_i, F)$ and $\Theta_i = X_i/R_i$ so that $\{(R_i, \Theta_i); i = 1, \ldots, n\}$ is the sample in generalized polar coordinates (Fig. 5.17).

5.5.1 Parameterizing a Cone

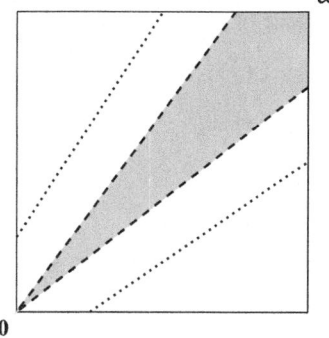

Fig. 5.17 The set [wedge] with dotted lines giving ℵ_[wedge]

The diamond plot discussed in Sect. 5.3.1 was designed to diagnose strong dependence where the limit measure $\eta(\cdot)$ in Cartesian coordinates concentrated on a subcone of the first quadrant and in L_1-polar coordinates the angular measure $S(\cdot)$ concentrated on $[a, b] \subsetneq [0, 1]$. There are two equivalent cone descriptions; one uses slopes of bounding lines to the cone and the other uses angles. As in (5.17) and (5.18), define

$$\mathbb{C}_{a,b} = [\text{wedge}] := \{x \in \mathbb{R}_+^2 : m_l x_1 \leq x_2 \leq m_u x_1\} \quad (5.29)$$

$$= \{x \in \mathbb{R}_+^2 : 0 < a \leq \frac{x_1}{x_1 + x_2} \leq b < 1\} \quad (5.30)$$

and a little algebra gives

$$0 < m_l = b^{-1} - 1 < m_u = a^{-1} - 1 < \infty. \quad (5.31)$$

Thus [wedge] is the subset of $\mathbb{R}_+^2$ bounded between the lines $y = m_l x$ and $y = m_u x$, $x > 0$ and therefore specified by the slope parameters $m_l < m_u$. The wedge is also specified by the angles $x_1/(x_1 + x_2)$ in $[a, b]$, $0 < a < b < 1$.

5.5.2 A Coordinate System for Regular Variation on $\mathbb{R}_+^2 \setminus$ [wedge]

When the forbidden zone F is [wedge], the generalized polar cooordinates reviewed in (5.28) require computation of $d(x, [\text{wedge}])$. We now use the L_2-distance $d_2(\cdot, \cdot)$. To avoid confusion, tie a string around your finger to remind yourself that [wedge] is parameterized using the L_1 metric giving angles $x_1/(x_1 + x_2)$ but $d_2(x, [\text{wedge}])$ is the L_2 distance of $x \notin [\text{wedge}]$ to [wedge].

After an easy calculation, we fine the L_2 distance of $(x_1, x_2) \in \mathbb{R}_+^2$ to the line $y = mx$, $x > 0$ is

$$\frac{|x_2 - mx_1|}{\sqrt{1+m^2}}$$

and therefore, assuming $(x_1, x_2) \notin$ [wedge], the distance of (x_1, x_2) to [wedge] is the distance to the nearest boundary line, namely,

$$d_2\big((x_1, x_2), \text{[wedge]}\big) = \begin{cases} \frac{|x_2 - m_u x_1|}{\sqrt{1+m_u^2}}, & \text{if } x_2 > m_u x_1, \\ \frac{|x_2 - m_l x_1|}{\sqrt{1+m_l^2}}, & \text{if } x_2 < m_l x_1. \end{cases} \quad (5.32)$$

(As in (5.19), page 169, one could clean up the formula in (5.32) by using a scaled distance that neglects the denominator as in (5.19).) Write

$$\mathbb{R}_+^2 \setminus \text{[wedge]} = \text{[> wedge]} \cup \text{[< wedge]}$$

where (see Fig. 5.18)

$$\text{[> wedge]} = \{(x_1, x_2) \in \mathbb{R}_+^2 \setminus \text{[wedge]} : x_2 > m_u x_1\}, \quad (5.33)$$

$$\text{[< wedge]} = \{(x_1, x_2) \in \mathbb{R}_+^2 \setminus \text{[wedge]} : x_2 < m_l x_1\}. \quad (5.34)$$

So [> wedge] consists of the points in $\mathbb{R}_+^2 \setminus$ [wedge] that are visually above the upper boundary $x_2 = m_u x_1$, $x_1 \geq 0$ and [< wedge] consists of the points in $\mathbb{R}_+^2 \setminus$ [wedge] that are below the lower boundary $x_2 = m_l x_1$ of [wedge].

Now use the discussion in Sect. 2.2.2.1, page 39 on generalized polar coordinates. Suppose X is a random element in $\mathbb{R}_+^2$ with regularly varying distribution on $\mathbb{R}_+^2 \setminus$ [wedge] and $P[X \in \cdot] \in \text{MRV}(\alpha, b(t), \eta(\cdot), \mathbb{R}_+^2 \setminus \text{[wedge]})$. Corollary 2.1, page 41 allows us to express this in generalized polar coordinates as

$$tP\bigg[\bigg(\frac{d_2(X, \text{[wedge]})}{b(t)}, \frac{X}{d_2(X, \text{[wedge]})}\bigg) \in \cdot \bigg] = tP\bigg[\bigg(\frac{R}{b(t)}, (\Theta', \Theta'')\bigg) \in \cdot \bigg]$$

$$\to (\nu_\alpha \times S)(\cdot) = \eta \circ \text{GPOLAR}^\leftarrow(\cdot) \quad (5.35)$$

in $\mathbb{M}\big((\mathbb{R}_+ \setminus \{0\}) \times \aleph_{\text{[wedge]}}\big)$ where $\nu_\alpha(x, \infty) = x^{-\alpha}$, $x > 0$, $\alpha > 0$ and $S(\cdot)$ is a probability measure on $\aleph_{\text{[wedge]}}$. The set $\aleph_{\text{[wedge]}}$ consists of two lines at distance 1 from [wedge] parallel to the boundary of [wedge] (Fig. 5.17, page 179) and these lines are

$$\aleph_{\text{[>wedge]}} = \bigg\{(x_1, x_2) \in \mathbb{R}_+^2 : 0 \leq m_u x_1 < x_2,\ x_2 = m_u x_1 + \sqrt{1+m_u^2}\bigg\},$$

$$\aleph_{\text{[<wedge]}} = \bigg\{(x_1, x_2) \in \mathbb{R}_+^2 : 0 \leq x_2 < m_l x_1,\ x_2 = m_l x_1 - \sqrt{1+m_l^2}\bigg\}.$$

5.5 Beyond $\mathbb{R}^2 \setminus \{0\}$: Other Cones as Forbidden Zones

The first line $\aleph_{[>\text{wedge}]}$ has slope m_u and intersects the y-axis at $(0, (1+m_u^2)^{1/2})$. The line $\aleph_{[<\text{wedge}]}$ has slope m_l and intersects the x-axis at $((1+m_l^2)^{1/2}/m_l, 0) = ((1+m_l^{-2})^{1/2}, 0)$. Write

$$S_>(\cdot) = S(\cdot \cap \aleph_{[>\text{wedge}]}) \text{ and } S_<(\cdot) = S(\cdot \cap \aleph_{[<\text{wedge}]}).$$

From (5.35), we see that a necessary condition for non-trivial regular variation on all of $\mathbb{R}_+^2 \setminus [\text{wedge}]$ is that both $(X_2 - m_u X_1)_+$ and $(m_l X_1 - X_2)_+$ have regularly varying distributions. This suggests the exploratory diagnostic of individually testing whether these one-dimensional variables have power laws. Then, as we will discuss, one can continue to explore with the Hillish statistic as discussed in Theorem 5.2.

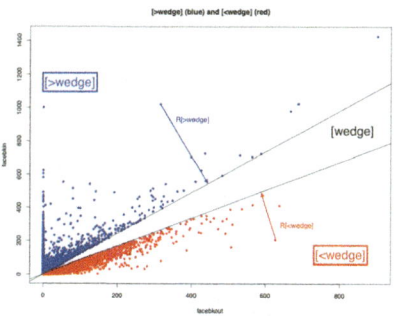

Fig. 5.18 Blue points are [>wedge] and the distance of a point in [>wedge] to [wedge] is R[>wedge]; similarly for red

In reality, there is nothing to prevent the possibility that regular variation exists on the region above [wedge] but that tails are of lower order below [wedge] or vice versa. This would happen for instance if $S(\aleph_{[<\text{wedge}]}) = 0$ but $S(\aleph_{[>\text{wedge}]}) = 1$. If this happened for a particular data set, one could search (but with no guarantee of success) for another thinner tailed regular variation on [< wedge].

In (5.35), $\Theta = (\Theta', \Theta'') \in \aleph_{[\text{wedge}]}$ which is the union of two lines. So one component of Θ determines the other since on $\aleph_{[>\text{wedge}]}$, $\Theta'' = m_u \Theta' + (1+m_u^2)^{1/2}$, $\Theta' \geq 0$ while on $\aleph_{[<\text{wedge}]}$, $\Theta'' = m_k \Theta' - (1+m_l^2)^{1/2}$, $\Theta'' \geq 0$. Write

$$T_>(r, \theta', \theta'') = (r, \theta'), \text{ on } \aleph_{[>\text{wedge}]},$$
$$T_<(r, \theta', \theta'') = (r, \theta''), \text{ on } \aleph_{[<\text{wedge}]},$$

and for $B \in \mathcal{B}(\mathbb{R}_+)$, set

$$S_>^{(1)}(B) = S_> \{(\theta', \theta'') \in \aleph_{[>\text{wedge}]} : \theta' \in B\},$$
$$S_<^{(1)}(B) = S_> \{(\theta', \theta'') \in \aleph_{[<\text{wedge}]} : \theta'' \in B\}.$$

On [> wedge], we simplify (5.35) via (5.32) using only Θ' and get

$$tP\left[X \in [>\text{wedge}], \left(\frac{X_2 - m_u X_1}{b(t)(1+m_u^2)^{1/2}}, \frac{(1+m_u^2)^{1/2}}{X_2/X_1 - m_u}\right) \in \cdot\right]$$
$$= tP\left[X \in [>\text{wedge}], \left(\frac{R}{b(t)}, \Theta'\right) \in \cdot\right] \to (\nu_\alpha \times S_>(\cdot)) \circ T_>^{-1}$$

$$=\eta \circ \text{GPOLAR}^{\leftarrow} \circ T_>^{-1}(\cdot) = \nu_\alpha \times S_>^{(1)}, \tag{5.36}$$

in $\mathbb{M}\big((\mathbb{R}_+ \setminus \{0\}) \times \mathbb{R}_+\big)$. Remember $R = d_2(X, [> \text{wedge}]) = (X_2 - m_u X_1)/(1 + m_u^2)^{1/2}$.

On $[< \text{wedge}]$ we have $R = (m_l X_1 - X_2)/(1 + m_l^2)^{1/2}$ and we reduce (5.35) by considering (R, Θ'') where $\Theta'' = X_2/d(X, [< \text{wedge}]) = (1 + m_l^2)^{1/2}/(m_l - X_2/X_1)$. We simplify (5.35) as

$$tP\left[X \in [< \text{wedge}], \left(\frac{m_l X_1 - X_2}{b(t)(1 + m_l^2)^{1/2}}, \frac{(1 + m_l^2)^{1/2}}{m_l - X_2/X_1} \right) \in \cdot \right]$$

$$= tP\left[X \in [< \text{wedge}], \left(\frac{R}{b(t)}, \Theta'' \right) \in \cdot \right] \to (\nu_\alpha \times S_<(\cdot)) \circ T_<^{-1}$$

$$= \eta \circ \text{GPOLAR}^{\leftarrow} \circ T_<^{-1}(\cdot) = \nu_\alpha \times S_<^{(1)}, \tag{5.37}$$

in $\mathbb{M}\big((\mathbb{R}_+ \setminus \{0\}) \times \mathbb{R}_+\big)$.

As in the discussion of regular variation in $\mathbb{R}_+^2 \setminus \{0\}$ in Sect. 5.3, page 158, here the Θ component in (generalized) polar coordinates for regular variation in $\mathbb{R}_+^2 \setminus [\text{wedge}]$ has been reduced to one dimension.

5.5.3 Hillish Analysis for Heavy Tails in $\mathbb{R}_+^2 \setminus [\text{wedge}]$

Now suppose we have a sample $X_1, \ldots, X_n$ of observations in $\mathbb{R}_+^2 \setminus [> \text{wedge}]$. This means we have a set of generalized polar coordinates distributed as (see (5.32) or (5.36), page 182)

$$(\Theta', R) = \left(\frac{(1 + m_u)^{1/2}}{X_2/X_1 - m_u}, \frac{|X_2 - m_u X_1|}{(1 + m_u)^{1/2}} \right)$$

where for a typical X we have $X = (X_1, X_2)$. Here $\Theta' \in \mathbb{R}_+$ instead of $[0, 1]$ as was the case when we were working on $\mathbb{R}_+^2 \setminus \{0\}$ in Sect. 5.4.1 but the Hillish procedure and the proof of Theorem 5.2, page 172 are the same. However, the value $1/\alpha$ that *Hillish* converges to corresponds to the α of the heavy tailed variable $R = d(X, [\text{wedge}])$.

5.5.4 Further Analysis of Facebook Data

In order to illustrate several techniques, this section provides more detailed heavy tail analysis of the Facebook data [193] sourced from konect.cc [115]; see also

5.5 Beyond $\mathbb{R}^2 \setminus \{0\}$: Other Cones as Forbidden Zones

[42, 123, 180]. We have discussed this dataset in Sects. 5.3.1, 5.4.2.4 and in Figs. 5.8 and 5.10 on pages 160, 161, 178.

The data format is lines of the form (node1, node2, timestamp). Each line represents that first user (node1) writes a post on the wall of second user (node2) at time given by *timestamp* so that there is one edge creation per line. This creates a directed graph and we converted the edge information to (out,in)-degree counts indexed by node (illustrated in Fig. 5.8) using the R-package *igraph* [29]. The node set of 46,952 users is a small subset of the full set of Facebook users and is restricted to a limited geographic area in New Orleans [188]. The data was collected between 09/14/2004 to 01/22/2009.

5.5.4.1 Marginal Heavy Tails

The tail indices for both out- and in-degree are both close to 2.8 which supports the assumption of multivariate regular variation for the variables (out,in). In Sect. 5.4.2.4 we adopted $\hat{\alpha} = 2.8$ and plotted the Hillish plot in Fig. 5.22 but keep in mind the information in Table 5.1, page 156 and the discussion near Fig. 5.5, page 158.

Here is more detail on the marginal analyses. Hill plots look pretty awful but altHill [156, p. 364] and [61] and *parfit* [156, p. 364], [113] which is essentially an exponential QQ plot on log-log scale measuring line slope $1/\alpha$ are effective (Figs. 5.19 and 5.20).

Out-Degree For out-degree, altHillalpha is erratic and but parfit with k=600 shows most points on the line of slope 1/2.8.

Our friends at *poweRlaw* give the optimal k, or in their lingo *ntail*, as 275 and an estimate of α for out-degree of 2.98. Note parfit for $k = 275$ gives 2.97.

In-Degree For in-degree, the limits on the x-axis of plotting for altHillalpha must be chosen judiciously and then the estimate of $\hat{\alpha} = 2.8$ is plausible. Parfit at $k = 400$

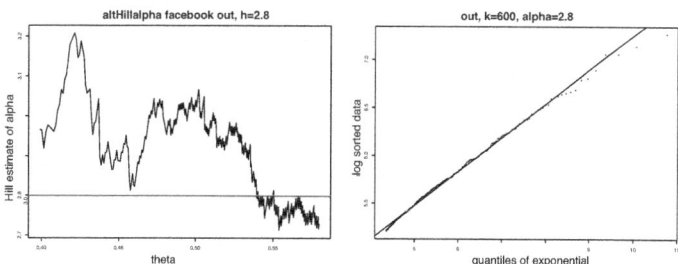

Fig. 5.19 Left: AltHillalpha for out-degree. The horizontal red line is at height 2.8. Right: parfit for out-degree using $k = 600$ largest order statistics with points hugging the line of slope 1/2.8

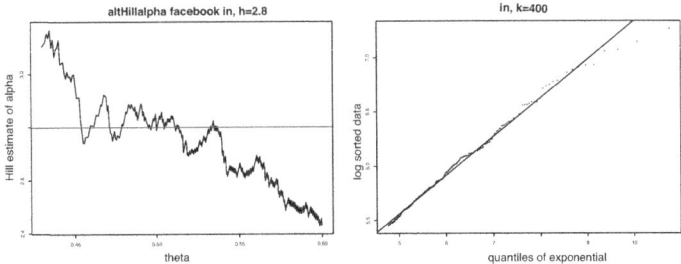

Fig. 5.20 Left: AltHillalpha for in-degree. The horizontal red line is at height 2.8. Right: parfit for out-degree using $k = 400$ largest order statistics with points hugging the line of slope $1/2.8$

gives points hugging a line of slope $1/2.8$. *PoweRlaw* gives the optimized k as 304 and the estimated α as 2.77 which is pretty close in this tail estimation world.

5.5.4.2 Dependence Analysis

The diamond plot was introduced in Sect. 5.3.1, page 161 and applied to the Facebook dataset in Fig. 5.10, page 162. We used the 200 largest values of the distance to the origin to conclude it is plausible that $S(\cdot)$ concentrates on a subinterval of $[0, 1]$. We also commented the plot was not as pristine as we would like but settled on the idea that strong asymtptotic dependence prevails.

Our experience with the beta(2,10) density for θ_1 in Fig. 5.12, page 176 is disturbing enough to make robustifying the estimates of the endpoints (a, b) of the interval of concentration an appealing idea. (Robustifying the estimates does not solve the problem that beta(2,10) is an example designed to fool the procedure.)

In Fig. 5.10, page 162, the histogram is more informative than the diamond plot and therefore, we do not use the extremes of the θ_1s suggested in Theorem 5.1 but rather quantiles and after some experimentation and appeal to paranormal forces we settled on thresholding at $k = 300$ and using the 25 and 75% of the thresholded θ_1s. This gives estimates $\theta_1 \in (0.449, 0.544) = (\hat{a}, \hat{b})$ and using these estimates, we apply (5.31), page 179, to get the wedge corresponding to slopes $m_l = 0.842$ and $m_u = 1.227$. This gives Fig. 5.18, page 181, showing subsets of the Facebook data corresponding to [wedge], [< wedge] (red) and [> wedge] (blue).

5.5.4.3 Is There Regular Variation on $\mathbb{R}_+^2 \setminus$ [wedge]?

The story so far: We possess evidence consistent with (out,in)-degree data having a regularly varying distribution on $\mathbb{R}_+^2 \setminus \{\mathbf{0}\}$ with the limit angular measure $S(\cdot)$ concentrating on $[\hat{a}, \hat{b}] = (0.449, 0.544)$ corresponding to the cone we called [wedge] bounded by lines

5.5 Beyond $\mathbb{R}^2 \setminus \{0\}$: Other Cones as Forbidden Zones

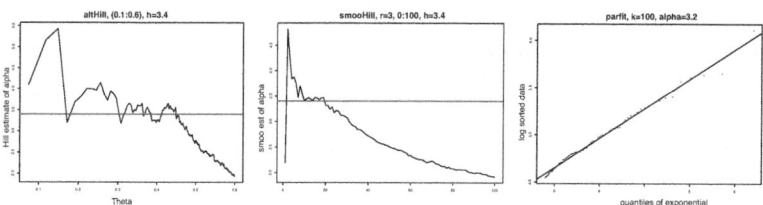

Fig. 5.21 Left: AltHill plot for distance to $y = m_u x$ of points in [> wedge]. Center: SmooHill plot for same data. Horizontal lines are at height 3.4. Right: parfit with $k = 100$

$$y = m_u x \quad y = m_l x, \, x > 0; \quad (m_l, m_u) = (0.842, 1.227).$$

So now we bravely remove [wedge] and seek a regular variation property on $\mathbb{R}_+^2 \setminus$ [wedge] using generalized polar coordinates. For reasons discussed later, this search works better on data from [> wedge] defined in (5.33), page 180 and visualized in Fig. 5.18. The generalized polar coordinates on [> wedge] are (see (5.32), page 180; (5.36), page 182)

$$(r, \theta_1) = \left(\frac{x_2 - m_u x_1}{(1 + m_u^2)^{1/2}}, \frac{(1 + m_u^2)^{1/2}}{x_2/x_1 - m_u} \right). \tag{5.38}$$

For there to be hope of multivariate regular variation, we need the univariate data representing the distance of a point in [> wedge] to the line $y = m_u x, \, x > 0$ to come from a regularly varying distribution. These values correspond to the first component in (5.28). We used *Althill* and *Parfit* [156] and because the *Althill* plot is somewhat volatile, we added the *Smoohill* plot [156, p. 365] and [164].

Consulting Table 5.1, page 156, we see that the package *poweRlaw* recommends ntail=48 with $\hat{\alpha} = 3.44$. An estimate based on an uncomfortably small $k = 48$ upper order statistics from a sample of size 12,104 seems unwise and we seek additional information from the plots in Fig. 5.21 to settle on an estimate of 3.4. Recall the title of the book contained the word *art*. Note 3.4>2.8, the latter being the index of regular variation on $\mathbb{R}_+^2 \setminus \{0\}$ so regular variation on [> wedge] gives an example of *hidden* regular variation. The *Hillish* plot in Fig. 5.22 does not provide any dramatic evidence against multivariate regular variation. The red line in Fig. 5.22 is at height $1/3.4 \approx 0.294$ and is seemingly too high; if we had drawn it at $0.26 \approx 1/3.84$ we would still estimate a regular variation index > 2.8 and hence still have HRV.

The good news: The α for $\mathbb{R}_+^2 \setminus \{\text{wedge}\}$ is 3.4 and this is greater than the $\alpha = 2.8$ associated with $\mathbb{R}_+^2 \setminus \{\mathbf{0}\}$. *Hillish* plots do not contradict the multivariate regular variation story. So we seem to have an example in a real data set consistent with *hidden* regular variation. The bad news is that both *poweRlaw* and *Parfit* suggest rather small values of k to make a convincing story which will lead to large variability in estimates. In general we see the sobering phenomena that as bigger forbidden zones are removed, more data is deleted. The length of the original Facebook vector was

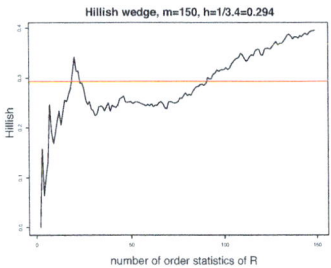

Fig. 5.22 Hillish plot for points in [> wedge]. The red line is at height $1/3.4 = 0.294$

46,953 and after deleting [wedge] the number of points in [> wedge] is reduced to 12, 104. This is still sizable but we are discarding a lot of data.

5.5.4.4 What About [< wedge]?

Now what about data in [< wedge]? The number of points in [< wedge] is 23,687 which is almost twice the number of data in [> wedge]. We have determined that data from [> wedge] could be modeled by a regularly varying distribution with an α bigger than the one used for [wedge] produced by analysis on $\mathbb{R}_+^2 \setminus \{\mathbf{0}\}$. But analysis on [< wedge] should also agree with the diamond plot in Fig. 5.10, page 162 which indicates relatively few points far from the origin in [< wedge] compared with [wedge]. So if there is regular variation on [< wedge] $= \mathbb{R}_+^2 \setminus ([\text{wedge}] \cup [> \text{wedge}])$, it should have a thinner tail than on [wedge], that is, a larger α.

Heavy tail analysis on [< wedge] proceeds in familiar steps. Subset the Facebook data and extract points below [wedge]. As in Fig. 5.18, page 181, this segments the data into those in [wedge], blue points in [> wedge] and red points in [< wedge]. Focus on red points and compute the generalized polar coordinates using the formulas in (5.32), page 180 and (5.37), page 182. This gives 23,687 data pairs (rlefacebk,thlefacebk).

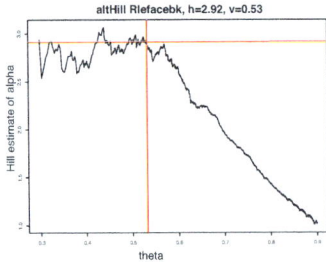

Fig. 5.23 AltHill for $R[<$ wedge] with poweRlaw estimates of $(\alpha, k) = (2.92, 214)$

First consider the distribution tail of the variable *rlefacebk*, the L_2-distance of points in [< wedge] to the boundary of [wedge]. PoweRlaw estimates summarized in Table 5.1, page 156 give $\alpha = 2.92$ and $ntail = k = 214$. Figure 5.23 is the altHill plot with red horizontal line at 2.94 and the vertical line satisfying $n^z = 2.14$ where $n = 23,687$ and so $z \approx 0.53$. Agreement between poweRlaw and altHill is apparent.

There is evidence for *multi*variate regular variation on $\mathbb{R}_+^2 \setminus ([\text{wedge}] \cup [> \text{wedge}]) = [< \text{wedge}]$, so we have coherent hidden regular variation story:

(i) The data comes from a model with MRV on $\mathbb{R}_+ \setminus \{0\}$ with $\alpha = 2.8$ and limit measure concentrating on [wedge].
(ii) Delete [wedge] from the state space and determine that the distribution has MRV on $\mathbb{R}_+^2 \setminus [\text{wedge}]$ with $\alpha = 2.94 > 2.8$ and limit measure concentrating on [> wedge].
(iii) Delete [> wedge] from the state space $\mathbb{R}_+^2 \setminus [\text{wedge}]$ and find that the distribution has MRV on $\mathbb{R}_+^2 \setminus ([\text{wedge}] \cup [> \text{wedge}])$ with $\alpha = 3.4 > 2.94 > 2.8$ and limit measure concentrating on [< wedge].

Lastly, is Hillish analysis consistent with the MRV hypothesis on [< wedge]? The Hillish plot in Fig. 5.24 for this data looks good and is consistent with MRV.

Although we have a coherent story about MRV on 3 cones, the typical imprecision in the α-estimates and the closeness of the 3 values of α should make one wary about claiming this is the only possible model for the data.

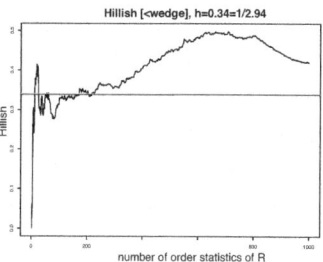

Fig. 5.24 Hillish plot for data in [< wedge] is consistent with MRV

5.6 More on Asymptotic Independence

In quantitative risk theory, *asymptotic independence* of two loss vectors refers to the fact that the two losses are unlikely to be large at the same time. The concept can be defined in arbitrary TABOF spaces by making "large" mean far from the forbidden zone. However, the most useful cases are the traditional idea of large where the forbidden zone is the origin in some Euclidean space and "large" means far from the origin. If two extreme risk events are unlikely to occur together, from a practical portfolio construction point of view, there may be little to distinguish asymptotic independence from independence. Each of the collection of asymptotically independent risk vectors can be considered in isolation, effectuating dimension reduction.

The name *asymptotic independence* is historically entrenched but usually leaves beginners to the subject scratching their heads wondering what is independent. The name originates with the problem of finding conditions on componentwise maxima of iid random vectors which makes their limit distribution a product distribution.

We have given prior basic discussions in Sect. 2.3.1.2, page 49 and in Proposition 2.2, page 49. Other treatments are [51, 79, 122, 132, 156], [28, 44, 73, 85, 104–106], and [157, Chapter 5].

5.6.1 Definitions and Basics

Return to the setting of Sect. 2.3.1.2, page 49 and suppose we have random vectors $X_i = (X_i(1), \ldots, X_i(p_i)) \in \mathbb{R}_+^{p_i}$, $i = 1, 2$ so that $X = (X_1, X_2)$ is a random vector in $\mathbb{R}_+^{p_1+p_2}$. Assume also marginal regular variation,

$$P[X_i \in \cdot] \in \text{MRV}(\alpha, b(t), \eta_i(\cdot), \mathbb{R}_+^{p_i} \setminus \{\mathbf{0}_{p_i}\}), \quad i = 1, 2. \tag{5.39}$$

Proposition 2.2(ii), page 49, gives the asymptotic independence condition (2.35) allowing the conclusion that $X = (X_1, X_2)$ has a regularly varying distribution. Conversely, if we know $P[X \in \cdot]$ is a regularly varying distribution, then by the mapping theorem and projection, we get $P[X_i \in \cdot]$ is also a regularly varying measure for $i = 1, 2$. Here are some equivalent formulations of asymptotic independence.

Proposition 5.1 (Asymptotic Independence) *[122] Suppose $X = (X_1, X_2)$ is a random vector in $\mathbb{R}_+^{p_1+p_2}$ satisfying (5.39). The following are equivalent and characterize asymptotic idependence.*

1. *The relation (2.35), page 50 holds for $b(t) = b_1(t) = b_2(t)$; that is, for any $\delta_i > 0$, $i = 1, 2$,*

$$tP[X_1/b(t) \in (\{\mathbf{0}_{p_1}\}^{\delta_1})^c, X_2/b(t) \in (\{\mathbf{0}_{p_2}\}^{\delta_2})^c] \to 0, \quad (t \to \infty).$$

2. *For any $B_i \in \mathbb{BA}(\mathbb{R}_+^{p_i} \setminus \{\mathbf{0}_{p_i}\})$, $i = 1, 2$, we have*

$$tP[X_1/b(t) \in B_1, X_2/b(t) \in B_2] \to 0, \quad (t \to \infty).$$

3. *For any $f_i \in \mathbb{C}(\mathbb{R}_+^{p_i} \setminus \{\mathbf{0}_{p_i}\})$, $i = 1, 2$, we have*

$$tE\Big(f_1\big(X_1/b(t)\big) f_2\big(X_2/b(t)\big)\Big) \to 0, \quad (t \to \infty).$$

4. *For any $v \in \{1, \ldots, p_1\}$ and $w \in \{1, \ldots, p_2\}$ and $f_l \in \mathbb{C}(\mathbb{R}_+ \setminus \{0\})$, $l = 1, 2$,*

$$tE\Big(f_1(X_1(v)/b(t)) f_2(X_2(w)/b(t))\Big) \to 0, \quad (t \to \infty).$$

5. *For each $v \in \{1, \ldots, p_1\}$ and $w \in \{1, \ldots, p_2\}$ and all $c > 0$,*

$$tP\big[X_1(v)/b(t) > c, X_2(w)/b(t) > c\big] \to 0, \quad (t \to \infty).$$

6. *We have that $P[X \in \cdot]$ is regularly varying and*

5.6 More on Asymptotic Independence

$$P[X \in \cdot] \in MRV(\alpha, b(t), \eta = \eta_1 \times \epsilon_{\mathbf{0}_{p_2}} + \epsilon_{\mathbf{0}_{p_1}} \times \eta_2, \mathbb{R}_+^{p_1+p_2} \setminus \{\mathbf{0}_{p_1+p_2}\}), \quad (5.40)$$

where $\eta_i \in \mathbb{M}(\mathbb{R}_+^{p_i} \setminus \{\mathbf{0}_{p_i}\})$, $i = 1, 2$.

Equivalently, $P[X \in \cdot]$ is regularly varying and, with respect to a norm $\|\cdot\|_{p_i}$ on $\mathbb{R}_+^{p_i}$, $i = 1, 2$, the angular measure $S(\cdot)$ of $\eta(\cdot)$ concentrates mass 1 on

$$\aleph_{\mathbf{0}_{p_1}} \times \{\mathbf{0}_{p_2}\} \cup \{\mathbf{0}_{p_1}\} \times \aleph_{\mathbf{0}_{p_2}} = \left\{ \mathbf{x} = (\mathbf{x}_1, \mathbf{0}_{p_2}) \in \mathbb{R}_+^{p_1+p_2} : \|\mathbf{x}_1\|_{p_1} = 1 \right\}$$

$$\bigcup \left\{ \mathbf{x} = (\mathbf{0}_{p_1}, \mathbf{x}_2) \in \mathbb{R}_+^{p_1+p_2} : \|\mathbf{x}_2\|_{p_2} = 1 \right\}. \quad (5.41)$$

The characterization in Item 5 is likely the most useful since it reduces a multivariate problem to checking 2-dimensional conditions. Part 6 means the angular measure concentrates on two simplices and if $p_1 = p_2 = 1$, $S(\cdot)$ concentrates on the two points $(0, 1)$ and $(1, 0)$.

Proof We show $1 \to 2 \to 3 \to 4 \to 5 \to 1$. That (5.39) and Item 1 implies Item 6 is the content of Proposition 2.2, page 49 and if Claim 6 holds then it is clear that, say, Item 5 holds.

$1 \to 2$. Suppose $B_i \in \mathbb{BA}(\mathbb{R}_+^{p_i} \setminus \{\mathbf{0}_{p_i}\})$, $i = 1, 2$ so that for a Euclidean metric $d_{p_i}(\mathbf{x}, \mathbf{y})$ on $\mathbb{R}_+^{p_i}$

$$d_{p_i}(B_i, \{\mathbf{0}_{p_i}\}) =: \epsilon_i > 0, \quad i = 1, 2.$$

Thus for $i = 1, 2$, $B_i \subset \left(\{\mathbf{0}_{p_i}\}^{\epsilon_i/2} \right)^c$ and

$$t P[X_1/b(t) \in B_1, X_2/b(t) \in B_2]$$

$$\leq t P[X_1/b(t) \in \left(\{\mathbf{0}_{p_1}\}^{\epsilon_1/2} \right)^c, X_2/b(t) \in \left(\{\mathbf{0}_{p_2}\}^{\epsilon_2/2} \right)^c] \to 0,$$

as $t \to \infty$.

$2 \to 3$. Given assumption 2, let $f_i \in \mathbb{C}(\mathbb{R}_+^{p_i} \setminus \{\mathbf{0}_{p_i}\})$, so for $i = 1, 2$, $\mathrm{supp}(f_i) \in \mathbb{BA}(\mathbb{R}_+^{p_i} \setminus \{\mathbf{0}_{p_i}\})$. Let $\|f_i\| = \sup_{\mathbf{x} \in \mathrm{supp}(f_i)} f_i(\mathbf{x}) < \infty$ and

$$tE\Big(f_1(X_1/b(t)) f_2(X_2/b(t)) \Big)$$

$$\leq \|f_1\| \|f_2\| \cdot t P[X_1/b(t) \in \mathrm{supp}(f_1), X_2/b(t) \in \mathrm{supp}(f_2)] \to 0,$$

as $t \to \infty$ from assumption 2.

$3 \to 4$. Given assumption 3. To prove claim 4, suppose for simplicity $v = w = 1$ and $f_i \in \mathbb{C}(\mathbb{R}_+ \setminus \{0\})$, $i = 1, 2$. Define $\tilde{f}_i \in \mathbb{C}(\mathbb{R}_+^{p_i} \setminus \{\mathbf{0}_{p_i}\})$ by

$$\tilde{f}_1(x(1), \ldots, x(p_1)) = f_1(x(1)), \quad \tilde{f}_2(x(1), \ldots, x(p_2)) = f_2(x(1)).$$

Then as $t \to \infty$,

$$0 \leftarrow tE\left(\tilde{f}_1(X_1/b(t))\tilde{f}_2(X_1/b(t))\right) = tE\left(f_1(X_1(1))f_2(X_2(1))\right).$$

$4 \to 5$. Suppose the assumption in 4 holds. We have to turn an expectation statement into a probability convergence; this should not require major trauma. Given a constant $c > 0$, define $f \in \mathbb{C}(\mathbb{R}_+ \setminus \{0\})$ by

$$f(x) = \begin{cases} 0, & \text{if } 0 < x < c/2, \\ \frac{2}{c}x - 1, & \text{if } c/2 \le x \le c, \\ 1, & \text{if } c \le x < \infty, \end{cases}$$

so that f is linear on $(c/2, c)$ and $f \in \mathbb{C}(\mathbb{R}_+ \setminus \{0\})$. Then since $f(x) \ge 1_{(c,\infty)}(x)$,

$$tP\left[X_1(v)/b(t) > c, X_2(w)/b(t) > c\right]$$

$$= tE\left(1_{(c,\infty)}(X_1(v)/b(t))1_{(c,\infty)}(X_2(w)/b(t))\right)$$

$$\le tE\left(f(X_1(v)/b(t))f(X_2(w)/b(t))\right) \to 0$$

as $t \to \infty$ by the assumption in Item 4.

$5 \to 1$. Assume the assumption in 5 holds and for convenience suppose the metric is L_1. Then for any $\delta_i > 0$, $i = 1, 2$,

$$tP[X_1/b(t) \in (\{\mathbf{0}_{p_1}\}^{\delta_1})^c, X_2/b(t) \in (\{\mathbf{0}_{p_2}\}^{\delta_2})^c]$$

$$\le tP\left\{\bigcup_{v=1}^{p_1}[X_1(v)/b(t) > \delta_1/p_1] \cap \bigcup_{w=1}^{p_2}[X_2(w)/b(t) > \delta_2/p_2]\right\}$$

$$\le t\sum_{v=1}^{p_1}\sum_{w=1}^{p_2} P[X_1(v)/b(t) > \delta_1/p_1, X_2(w)/b(t) > \delta_2/p_2] \to 0,$$

as $t \to \infty$. This completes the proof. □

5.6.2 Preservation of Asymptotic Independence Under Mappings: What Could Possibly Go Wrong?

Continue to suppose $X = (X_1, X_2)$ is a random vector in $\mathbb{R}_+^{p_1+p_2}$ satisfying (5.39) and that one of the equivalent conditions for asymptotic independence of the vectors in Proposition 5.1 holds. Assume for $i = 1, 2$, $q_i \leq p_i$ and we have a map $h_i : \mathbb{R}_+^{p_i} \setminus \{\mathbf{0}_{p_i}\} \mapsto \mathbb{R}_+^{q_i} \setminus \{\mathbf{0}_{q_i}\}$ and we set $Y_i = h_i(X_i)$. If we map X_i to Y_i do we still have multivariate regular variation of $Y = (Y_1, Y_2)$ and asymptotic independence? Oh the suspense! Since independence is preserved under this kind of mapping, it is fair to wonder about the *asymptotic* cousin but, of course, sharing a gene pool or a name does not guarantee similarity. Clearly we need conditions on h_i. Here is one possible formulation.

What's wrong?

Proposition 5.2 (Preservation of Asymptotic Independence) *Assume that $X = (X_1, X_2)$ is a random vector in $\mathbb{R}_+^{p_1+p_2}$ satisfying (5.39) and that X_1 and X_2 are asymptotically independent. Suppose for $i = 1, 2$ and integers $q_i \leq p_i$, we have functions $h_i : \mathbb{R}_+^{p_i} \setminus \{\mathbf{0}_{p_i}\} \mapsto \mathbb{R}_+^{q_i} \setminus \{\mathbf{0}_{q_i}\}$ and that h_i satisfies*

1. *Homogeneity: For $s > 0$ and $x \in \mathbb{R}_+^{p_i}$ we have $h_i(sx) = sh_i(x)$.*
2. *Each h_i is bounded and continuous on the unit sphere $\aleph_{\mathbf{0}_{p_i}}$. Set*

$$h_i^\vee = \sup\{h_i(a) : a \in \aleph_{\mathbf{0}_{p_i}}\}. \tag{5.42}$$

Then $Y = (Y_1, Y_2) = \big(h_1(X_1), h_2(X_2)\big)$ has a distribution which is multivariate regularly varying and Y_1 and Y_2 are asymptotically independent.

Example Suppose $\|x\|_r$ is the L_r norm on $\mathbb{R}_+^p$ for some $p \geq 1$. If $x = (x_1, x_2) \in \mathbb{R}_+^{p_1+p_2}$ we could have

$$h_1(x_1) = (\|x_1\|_2, \|x_1\|_3), \quad h_2(x_2) = \|x_2\|_1.$$

Probably the most striking case is where

$$h_1(x_1) = \sum_{l=1}^{p_1} x_1(l), \quad h_2(x_2) = \sum_{w=1}^{p_2} x_2(w).$$

Proof Assume the marginal regular variation condition (5.39). For $i = 1, 2$, we first show,

$$P[Y_i \in \cdot] = P[h_i(X_i) \in \cdot] \in \mathrm{MRV}(\alpha, b(t), \tilde{\eta}_i(\cdot), \mathbb{R}_+^{q_i} \setminus \{\mathbf{0}_{q_i}\}).$$

Let $f_i \in \mathbb{C}(\mathbb{R}_+^{q_i} \setminus \{\mathbf{0}_{q_i}\})$ so f_i has support bounded away from $\{\mathbf{0}_{q_i}\}$ and we need that $\lim_{t \to \infty} t E f_i(\mathbf{Y}_i/b(t))$ converges. To see this, apply homogeneity of h_i,

$$t E f_i\left(\frac{\mathbf{Y}_i}{b(t)}\right) = t E f_i\left(\frac{h_i(\mathbf{X}_i)}{b(t)}\right) = t E f_i \circ h_i(\mathbf{X}_i/b(t)).$$

What are the properties of $f_i \circ h_i : \mathbb{R}_+^{p_i} \mapsto \mathbb{R}_+$? Since $f_i \in \mathbb{C}(\mathbb{R}_+^{q_i} \setminus \{\mathbf{0}_{q_i}\})$ we know $f_i(\mathbf{y}_i) = 0$ if $\|\mathbf{y}_i\| = d_{p_i}(\mathbf{y}_i, \mathbf{0}_{p_i})$ is too small. However, from homogeneity again,

$$f_i \circ (h_i(\mathbf{x}_i)) = f_i(\|\mathbf{x}_i\| h_i(\mathbf{x}_i/\|\mathbf{x}_i\|))$$

and since $0 \le \|\mathbf{x}_i\| h_i(\mathbf{x}_i/\|\mathbf{x}_i\|) \le \|\mathbf{x}_i\| h_i^\vee \to 0$ as $\|\mathbf{x}_i\| \to 0$, we see that if $\|\mathbf{x}_i\|$ is too small, then $h_i(\mathbf{x}_i)$ is small so $f_i \circ h_i(\mathbf{x}_i) = 0$. We conclude $f_i \circ h_i \in \mathbb{C}(\mathbb{R}_+^{p_i} \setminus \{\mathbf{0}_{p_i}\})$. It follows from the regular variation of $P[\mathbf{X}_i \in \cdot]$ that $\lim_{t \to \infty} t E f_i(\mathbf{Y}_i/b(t))$ converges.

We have now verified that $\mathbf{Y}_1$ and $\mathbf{Y}_2$ each have regularly varying distributions and it remains to show $\mathbf{Y}_1$ and $\mathbf{Y}_2$ have asymptotic independence. We check Condition 3 in Proposition 5.1 so let $f_i \in \mathbb{C}(\mathbb{R}_+^{q_i} \setminus \{\mathbf{0}_{q_i}\})$, $i = 1, 2$ and we must prove,

$$t E f_1(\mathbf{Y}_1/b(t)) f_2(\mathbf{Y}_2/b(t)) = t E f_1 \circ h_1(\mathbf{X}_1/b(t)) f_2 \circ h_2(\mathbf{X}_2/b(t)) \to 0.$$

This follows from part 3 of Proposition 5.1 because $f_i \circ h_i \in \mathbb{C}(\mathbb{R}_+^{p_i} \setminus \{\mathbf{0}_{p_i}\})$ and $\mathbf{X}_1, \mathbf{X}_2$ are asymptotically independent. □

5.6.2.1 A Partial Converse Giving Asymptotic Independence

A partial converse where asymptotic independence of $\mathbf{Y}_1$ and $\mathbf{Y}_2$ implies that of $\mathbf{X}_1$ and $\mathbf{X}_2$ seems too good to be true and would be surprising.[2] Careful formulation is required. We discuss one simple and striking case.

Proposition 5.3 ([122]) *Suppose $\mathbf{X} = (\mathbf{X}_1, \mathbf{X}_2)$ is a random vector in $\mathbb{R}_+^{p_1+p_2}$ with a regularly varying distribution:*

$$P[\mathbf{X} \in \cdot] \in MRV(\alpha, b(t), \eta(\cdot), \mathbb{R}_+^{p_1+p_2} \setminus \{\mathbf{0}_{p_1+p_2}\}). \tag{5.43}$$

Then $\mathbf{X}_1$ and $\mathbf{X}_2$ are asymptotically independent iff the two random variables $\Sigma_1 := \sum_{l=1}^{p_1} X_1(l)$ and $\Sigma_2 := \sum_{w=1}^{p_2} X_2(w)$ are asymptotically independent.

---

[2] A graduate school highlight was when I commented to a faculty advisor that some result was "surprising". The advisor warmly responded: "It is surprising only if you don't understand what is going on." Comeuppance hopefully leads to wisdom.

5.7 Measuring Extremal Dependence with the EDM

Claims of implausibility should be tempered by the understanding that we start with $P[X \in \cdot]$ being regularly varying; this is a stronger starting assumption than (5.39) used in Proposition 5.1.

Proof For the easy direction: Suppose X has a regularly varying distribution and therefore X_1 and X_2 each have distributions that are regularly varying by the mapping theorem. Assume X_1 and X_2 are asymptotically independent and then we apply Proposition 5.2. The functions $h_i : \mathbb{R}_+^{p_i} \mapsto \mathbb{R}_+$, $i = 1, 2$, defined by $h_i(\boldsymbol{x}_i) = \sum_{l=1}^{p_i} x_i(l)$ satisfy the assumptions 1,2 in Proposition 5.2 where the unit sphere is defined by the L_1-norm. So Proposition 5.2 yields that the two sums Σ_1, Σ_2 are asymptotically independent.

We construct a proof of the converse using mapping arguments. To prevent confusion between the limit measure and its angular measure, use the convention that for a vector $\boldsymbol{\xi}$ whose distribution is regularly varying, the limit distribution and angular measure with respect to the L_1-norm are denoted by $\eta_{\boldsymbol{\xi}}(\cdot)$ and $S_{\boldsymbol{\xi}}(\cdot)$. For $\boldsymbol{x} \in \mathbb{R}_+^p$, write $\|\boldsymbol{x}\| = \sum_{i=1}^p x_i$ and recall

$$S_{\boldsymbol{\xi}}(\cdot) = \eta_{\boldsymbol{\xi}}\{\boldsymbol{x} : \boldsymbol{x}/\|\boldsymbol{x}\| \in \cdot\}.$$

So now suppose (5.43) so that it follows that Σ_1, Σ_2 marginally have distributions that are regularly varying. Further suppose that Σ_1, Σ_2 are asymptotically independent. From Proposition 5.1, part 6, page 188, we have

$$\begin{aligned}1 &= S_{\Sigma_1,\Sigma_2}\{(0,1),(1,0)\} \\ &= \eta_{\Sigma_1,\Sigma_2}\left\{(y_1, y_2) \in \mathbb{R}_+^2 \setminus \{\boldsymbol{0}_2\} : \frac{(y_1, y_2)}{y_1 + y_2} = (0,1) \text{ or } (1,0)\right\} \\ &= \eta_{\Sigma_1,\Sigma_2}\left\{(y_1, y_2) \in \mathbb{R}_+^2 \setminus \{\boldsymbol{0}_2\} : (y_1, y_2) = (0,1) \text{ or } (1,0)\right\} \\ &= \eta_X\left\{(\boldsymbol{x}_1, \boldsymbol{x}_2) \in \mathbb{R}_+^{p_1+p_2} : (\|\boldsymbol{x}_1\|, \|\boldsymbol{x}_2\|) = (0,1) \text{ or } (1,0)\right\} \\ &= \eta_X\left\{(\boldsymbol{x}_1, \boldsymbol{x}_2) : \boldsymbol{x}_1 = \boldsymbol{0}_{p_1}, \boldsymbol{x}_2 \in \aleph_{\boldsymbol{0}_{p_2}} \text{ or } \boldsymbol{x}_1 \in \aleph_{\boldsymbol{0}_{p_1}}, \boldsymbol{x}_2 = \boldsymbol{0}_{p_2}\right\} \\ &= S_X\left\{\boldsymbol{a} = (\boldsymbol{a}_1, \boldsymbol{a}_2) \in \aleph_{\boldsymbol{0}_{p_1+p_2}} : \boldsymbol{a} \in (\{\boldsymbol{0}_{p_1}\} \times \aleph_{\boldsymbol{0}_{p_2}}) \cup (\aleph_{\boldsymbol{0}_{p_1}} \times \{\boldsymbol{0}_{p_2}\})\right\}.\end{aligned}$$

So by Part 6 of Proposition 5.1, the angular measure of X_1, X_2 concentrates where it should to get asymptotic independence; see (5.41). □

5.7 Measuring Extremal Dependence with the EDM

The *extremal dependence measure* or EDM should be thought of as the summarization of the diamond plot with a single number. Though the EDM is useful for exploration, this phrasing hints that we should not have strong expectations that this concept will cure cancer or bring world peace.

Numerical summaries such as the EDM help explore for asymptotic independence. So if, for instance, we are trying to construct two baskets of securities that are unlikely to incur large losses together, the EDM belongs in the toolbox. Proposition 5.1, page 188, informs us that asymptotic independence between random vectors can be reduced to asymptotic independence between random variables. This provides justification for concentrating on multivariate regularly varying measures in two dimensions.

The EDM definition has drifted a bit over the years and those with a yen for history can consult [28, 43, 91, 121, 122, 133, 155] and [114, page 48]. Despite the drift, the philosophy behind the idea has stayed firm: Summarize how concentrated the two-dimensional angular measure is about a point. There is a competing notion with similar goals advanced by Davis and Mikosch called the *extremogram* [44, 46, 121] and [114, page 49] which is oriented toward time series; it has some advantages and disadvantages compared to EDM [121].

Suppose $X = (X, Y)$ is an $\mathbb{R}_+^2$ random vector and

$$P[X \in \cdot\,] \in MRV(\alpha, b(t), \eta(\cdot), \mathbb{R}_+^2 \setminus \{\mathbf{0}_2\}).$$

It is convenient to use the L_1 metric to define the unit sphere

$$\aleph_{\mathbf{0}_2} = \{(\theta, 1 - \theta) : 0 \leq \theta \leq 1\}$$

and this choice plays nicely with the diamond plot. The angular measure of $\eta(\cdot)$ corresponding to the measure of $\theta \in [0, 1]$ is $S(\cdot)$.

The definition of the EDM is

$$\text{EDM}(X) = \text{EDM}(X, Y) = 4 \int_{[0,1]} \theta(1 - \theta) S(d\theta). \tag{5.44}$$

Of course in $[0, 1]$ the function $\theta(1-\theta)$ achieves its max of $1/4$ at $1/2$ so since $S(\cdot)$ is a probability measure, the formulation in (5.44) means

$$0 \leq \text{EDM}(X) \leq 1.$$

However, keep in mind we have restricted attention to $\mathbb{R}_+^2$ and if 4 quadrant regular variation on $\mathbb{R}^2$ for the distribution of X were considered, the definition (5.44), would be modified to have integration over $[-1, 1]$ and then we could have negative values for EDM.

If $S(\cdot)$ concentrates mass at $1/2$, then $\text{EDM}(X) = 1$. This is the case of full dependence with asymptotically equal marginals. For the case of asymptotic independence where $S(\cdot)$ concentrates on $\{0, 1\}$, the integrand is zero and $\text{EDM}(X) = 0$.

5.7 Measuring Extremal Dependence with the EDM

The EDM is scaled to feel like correlation. Observe,

$$\text{EDM}(X) = 4\int_{[0,1]} \theta(1-\theta) S(d\theta) = 4\int_{[0,1]} (\theta - \theta^2) S(d\theta).$$

$$= 4\int_{[0,1]} \left(\frac{1}{4} - \theta^2 + \theta - \frac{1}{4}\right) S(d\theta)$$

$$= 1 - 4\int_{[0,1]} \left(\theta - \frac{1}{2}\right)^2 S(d\theta) = 1 - 4\,\text{Int}. \quad (5.45)$$

The original definition [155], apart from a different parameterization, was (5.45). The expression (5.45) makes it clear that EDM is a moment designed to smell like correlation and that it measures the spread of mass of $S(\cdot)$ about 1/2. If EDM is relatively large, then the integral in (5.45) is relatively small and mass of $S(\cdot)$ is relatively concentrated near 1/2.

We have used the L_1-metric to define $\aleph_{0_2}$ and $S(\cdot)$. What would happen if some other norm was used? It is possible [121] to define an equivalence relation between dependence measures and show that switching norms produces an equivalent dependence measure. See also Problem 5.7, page 224.

Recall the consistent estimator $\hat{S}_n(\cdot)$ of $S(\cdot)$ in (5.8), page 161. If in (5.44) we replace $S(\cdot)$ by $\hat{S}_n(\cdot)$, then because the integrand is bounded and continuous we get a consistent estimator $\widehat{\text{EDM}(X)}$ of $\text{EDM}(X)$. Likewise, if we replace $S(\cdot)$ by $\hat{S}_n(\cdot)$ in the integral in (5.45), we get a consistent estimator of the integral. Remember the notation from page 161: Given iid observations $X, X_1, \ldots, X_n$ with polar coordinate versions $\{(R_i, \Theta_i), 1 \le i \le n\}$, where $\Theta_i = (\Theta_i, 1 - \Theta_i)$, we then have the estimators

$$\widehat{\text{EDM}(X)} = \frac{4}{k}\sum_{j=1}^{n} \Theta_j(1-\Theta_j) 1_{[R_j \ge R_{(k)}]}, \quad (5.46)$$

$$\widehat{\text{Int}} = \frac{1}{k}\sum_{j=1}^{n} (\Theta_j - \frac{1}{2})^2 1_{[R_j \ge R_{(k)}]}. \quad (5.47)$$

The expression $1 - 4\,\widehat{\text{Int}}$ is similar to the one used on page 274 of [122].

5.7.1 Exploring Pairwise Extremal Dependence with the Dependence Graph

Portfolio managers doing portfolio construction seek to assign assets to different baskets such that two different baskets have low probability of simultaneously experiencing big losses.

Assuming that regular variation with tail equivalent marginals holds, there are two facts to keep in mind:

1. Asymptotic independence of vectors can be reduced to pairwise asymptotic independence as explained in Proposition 5.1, page 188.
2. A relatively large EDM means there is a tendency for large values in $\mathbb{R}_+^2$-valued data to sit near the main diagonal and hence are extremally dependent.

Given multivariate $\mathbb{R}_+^p$-valued data that plausibly comes from a regularly varying distribution with tail equivalent marginals, here is an exploratory method [122] that examines asymptotic independence of subvectors by doing pairwise computation of the EDM.

- Start by thinking of each of the p components of the random vector as nodes in an undirected graph. This can be visualized compellingly, for instance, using the R-package *igraph* [29]. On the right are 5 nodes representing 5-dimensional data giving absolute log-returns of adjusted stock prices of (Google, Microsoft, Apple, Exxon, Chevron, BP) between January 3, 2005 and March 29, 2018. One edge is drawn.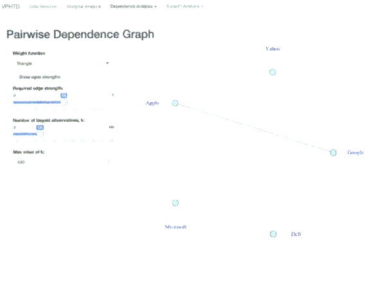

- For each pair of nodes, draw a connecting edge with weight equal to the estimated EDM between that pair. One edge is drawn in the example above.
- Decide on a threshold EDM value. If the EDM of a pair is below the threshold, delete the weighted edge between the pair.
- This resolves the nodes into connected subgroups. Within each subgroup are components that are highly extremally dependent. Across subgroups are node pairs that are weakly extremally dependent since their EDM is below the selected threshold.
- Sliders linked to the threshold EDM value allow for exploration of stability of the graph components as a function of the threshold EDM value.

5.7.1.1 Example of How to Ferret Out the Strongest Dependencies

We continue [122] with the example started in the first bullet above but add one security from the reinsurance industry. The securities are: Google, Microsoft, Apple, Chevron, Exxon, British Petroleum and the additional security is CATCo Reinsurance Opportunities Fund (CAT.L).

5.7 Measuring Extremal Dependence with the EDM

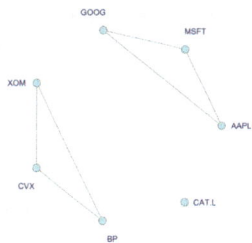

Fig. 5.25 An exploratory graph illustrating the strongest preliminary pairwise asymptotic dependencies

So three securities are from the tech sector, three from energy industries and one from reinsurance. Potentially three different sectors of economic activity are represented and we will see we can discern differences from the data. All observations range from December 20, 2010 to July 10, 2018.

Absolute log returns from all securities except CAT.L have very similar tail indices. Since CAT.L's tail index was quite different, we rank transformed [156, page 310], [87, 91, 94] each security to establish equal tails. Using a value of $k = 200$ for (5.46) and progressively sliding the required edge strength threshold down produces Fig. 5.25. As one might expect, energy stocks are in one group, tech securities in a second and the reinsurance stock stands apart.

Here the phrase *strongest dependencies* really means *strongest dependencies as measured by the estimated EDM–whatever the EDM measures*. Of course a legitimate criticism of this analysis is that time dependencies between data observation have been ignored in favor of a simpler theory consistent with iid assumptions. If you come from a time series culture that is interested in dependence modeling, think of the analysis presented here as producing suggestive pictures.

5.7.2 Hang on! Is the Estimated EDM Asymptotically Normal?

Asymptotic normality of a statistic is often used to justify (asymptotic) confidence levels of confidence intervals or confidence sets. Since multivariate regular variation is only an asymptotic model, care must be exercised. Diligence is required to make sure only observables are present in the final result and that the variance of the asymptotically normal distribution is not 0; unfortunately an asymptotic variance of 0 is not uncommon but there are workarounds for this awkward problem which are only moderately clunky.

Hang on!

We use the notation around (5.46) and (5.47) and we aim to impose conditions that imply weak convergence in $\mathbb{R}$,

$$\sqrt{k}\left(\widehat{\mathrm{EDM}}(X) - \mathrm{EDM}(X)\right) \Rightarrow N(0, \sigma^2), \tag{5.48}$$

for some asymptotic variance $\sigma^2 > 0$. When asymptotic independence is present, $\mathrm{EDM}(X) = 0$ so (5.48) cannot hold without modification.

The asymptotic normality of the EDM and related functions has been treated in [121, 122, 155] and [114, Chapter 9] and is related to the large literature on asymptotic normality of the tail empirical measure; see [156, p. 292] and [30–33, 60, 65, 70, 127, 128].

5.7.2.1 The Clunk Function $g(\cdot)$

To be able to sidestep the issue of an asymptotic variance of 0, we take a function of the data using a function $g : [0, 1] \mapsto \mathbb{R}_+$ which is bounded and continuous. The argument θ of g, we think of as the first component of $(\theta, 1 - \theta) \in \aleph_0$. Consider a test statistic

$$\widehat{\mathrm{EDM}(g)} := \frac{1}{k} \sum_{i=1}^{n} g(\Theta_i) 1_{[R_i \geq R_{(k)}]} \tag{5.49}$$

and note if $g(\theta) = 4\theta(1 - \theta)$ then $\widehat{\mathrm{EDM}(g)} = \widehat{\mathrm{EDM}(X)}$. The kth largest in the sample $R_1, \ldots, R_n$ is $R_{(k)}$ and required moments of the distribution $S(\cdot)$ with respect to g are

$$\mu_g := \int_{[0,1]} g(\theta) S(d\theta) =: S(g), \tag{5.50}$$

$$\sigma_g^2 = \int_{[0,1]} (g(\theta) - \mu_g)^2 S(d\theta). \tag{5.51}$$

Theorem 5.3 *Continue to use the notation for two-dimensional multivariate regular variation from page 161 and suppose* $g : [0, 1] \mapsto \mathbb{R}_+$ *is bounded and continuous. Define,*

$$B_n(t) := \begin{cases} \sqrt{k} \left(\frac{n}{k} E \left(g(\Theta_1) 1_{[R_1/b(\frac{n}{k}) \geq t^{-1/\alpha}]} \right) - \mu_g \frac{n}{k} P \left[\frac{R_1}{b(n/k)} > t^{-1/\alpha} \right] \right), & \text{if } t > 0, \\ 0, & \text{if } t = 0. \end{cases}$$

Suppose,

1. *As* $n \to \infty$, $k = k(n) \to \infty$, $n/k \to \infty$ *that*

$$\lim_{n \to \infty} B_n(t) = 0 \tag{5.52}$$

   locally uniformly for $t \in [0, \infty)$;
2. $\sigma_g^2 > 0$.

Then

5.7 Measuring Extremal Dependence with the EDM

$$T =: \frac{\sqrt{k}}{\sigma_g} \left(\widehat{EDM(g)} - \mu_g \right) = \frac{\sqrt{k}}{\sigma_g} \left(\frac{1}{k} \sum_{i=1}^{n} g(\Theta_i) 1_{[R_i \geq R_{(k)}]} - \mu_g \right) \Rightarrow N(0, 1).$$
(5.53)

The condition $\sigma_g > 0$ is required by the Lindeberg condition for the central limit theorem [76, page 284, Remark (e)], though a slightly weaker condition would suffice. If $S(\cdot)$ is the angular measure corresponding to asymptotic independence so that $S(\{0, 1\}) = 1$ and the function g is chosen to give the EDM, $g(\theta) = 4\theta(1-\theta)$, then $\mu_g = 0$ and (worse) $\sigma_g^2 = 0$. For asymptotic independence and similar cases where the limit variance is zero, if you want meaningful asymptotic normality, you must sagely pick a different g. Suggestions for some choices of g are in Sect. 5.7.3.

The result in (5.53) appears to be related to the asymptotic normality of the tail empirical process except we have exceedance counts weighted by $\{g(\Theta_i), 1 \leq i \leq n\}$; see [156, p. 292] for what is essentially the $g \equiv 1$ case but keep in mind $g \equiv 1$ violates the condition that $\sigma_g > 0$.

About condition (5.52): The assumption (5.52) is a second-order condition [51] that allows replacement of a pre-asymptotic centering by the natural and simpler asymptotic condition. Intuitively, this condition guarantees that the dependence between Θ_i and R_i decays sufficiently fast with n, as R_i is conditioned to lie above $b(n/k)$. This can also be viewed as a rate condition on $k = k(n)$. Note that if R_1 and Θ_1 are independent to begin with, the left side of (5.52) is zero for all n and the condition is thus satisfied in this case. Except in special cases, second order conditions such as (5.52) are difficult to verify from data and since you are the model builder, barring evidence to the contrary, you might as well assume your model satisfies this condition.

A sufficient condition for (5.52) is that as $n \to \infty$,

$$\sqrt{k} \left(\frac{n}{k} P \left[\left(\frac{R_1}{b(n/k)}, \Theta_1 \right) \in \cdot \right] - \frac{n}{k} P \left[\frac{R_1}{b(n/k)} \in \cdot \right] \times S \right) \to 0$$

in $M((0, \infty) \times [0, 1])$. This condition only involves the distribution of (R_1, Θ_1), and not the particular function g. Other uses of this and similar conditions are in [190].

Proof We follow the proof in [121]. Consider the process on $\mathbb{R}_+$,

$$W_n(t) = \frac{1}{\sigma_g \sqrt{k}} \sum_{i=1}^{n} \left(g(\Theta_i) - \mu_g \right) 1_{[R_i/b(\frac{n}{k}) \geq t^{-1/\alpha}]}, \qquad t > 0,$$

with $W_n(0) = 0$. Suppose we can prove that $W_n \Rightarrow W$ in $D[0, \infty)$, where W is standard Brownian motion. It is an established fact (eg. [156, p. 81]) that in $\mathbb{R}_+$,

$$\frac{R_{(k)}}{b(n/k)} \xrightarrow{P} 1,$$

where recall $R_{(k)}$ is the kth largest of $R_1, \ldots, R_n$. Since the limit is a constant, W_n and $R_{(k)}/b(n/k)$ converge jointly [13, p. 37], and thus we may apply the composition map $D[0,\infty) \times \mathbb{R}_+ \mapsto \mathbb{R}$ defined by $(f, c) \mapsto f(c)$ to deduce

$$W_n\left(\left(\frac{R_{(k)}}{b(n/k)}\right)^{-\alpha}\right) \Rightarrow W(1).$$

The result then follows by unpacking notation and observing,

$$\sigma_g W_n\left(\left(\frac{R_{(k)}}{b(n/k)}\right)^{-\alpha}\right) = \frac{1}{\sqrt{k}} \sum_{i=1}^n \left(g(\Theta_i) - \mu_g\right) 1_{[\frac{R_i}{b(n/k)} \geq ((\frac{R_{(k)}}{b(n/k)})^{-\alpha})^{-1/\alpha}]}$$

$$= \frac{1}{\sqrt{k}} \sum_{i=1}^n \left(g(\Theta_i) - \mu_g\right) 1_{[R_i \geq R_{(k)}]} = \frac{1}{\sqrt{k}} \left(\sum_{i=1}^n g(\Theta_i) 1_{[R_i \geq R_{(k)}]} - k\mu_g\right)$$

$$= \sqrt{k}\left(\frac{1}{k} \sum_{i=1}^n g(\Theta_i) 1_{[R_i \geq R_{(k)}]} - \mu_g\right).$$

Now we focus on proving $W_n \Rightarrow W$ in $D[0, \infty)$. First, we show finite dimensional distributions converge. As a warmup, start by fixing $t > 0$ and let $K_j(n)$ be the jth index i for which $R_i/b(n/k) \geq t^{-1/\alpha}$. By the *Découpage de Lévy*, (eg. [157, page 212]) the sequence $\{\Theta_{K_j(n)}\}$ is iid with common distribution

$$P[\Theta_1 \in \cdot | R_1/b(n/k) \geq t^{-1/\alpha}].$$

For $t > 0$, define

$$N_n = \sum_{i=1}^n \epsilon_{R_i/b(\frac{n}{k})}[t^{-1/\alpha}, \infty).$$

Write

$$W_n(t) = \frac{1}{\sigma_g \sqrt{k}} \sum_{j=1}^{N_n} \left(g(\Theta_{K_j(n)}) - \mu_g\right)$$

$$= \frac{1}{\sigma_g \sqrt{k}} \sum_{j=1}^{N_n} \left(g(\Theta_{K_j(n)}) - Eg(\Theta_{K_j(n)})\right) + \frac{1}{\sigma_g \sqrt{k}} N_n \left(Eg(\Theta_{K_j(n)}) - \mu_g\right)$$

$$= C_n + D_n.$$

First consider D_n. We write D_n in terms of the process $B_n(t)$ defined before (5.52). We have,

5.7 Measuring Extremal Dependence with the EDM

$$\sigma_g D_n = \frac{N_n}{k}\sqrt{k}\left(E(g(\Theta_1) \mid R_1/b(n/k) \geq t^{-1/\alpha}) - \mu_g\right)$$

$$= \frac{N_n/k}{\frac{n}{k}P[R_1 > b(\frac{n}{k})t^{-1/\alpha}]}$$

$$\times \sqrt{k}\left(\frac{n}{k}E(g(\Theta_1)1_{[R_1/b(\frac{n}{k}) \geq t^{-1/\alpha}]}) - \mu_g\frac{n}{k}P[R_1 > b(\frac{n}{k})t^{-1/\alpha}]\right)$$

$$= \frac{N_n/k}{\frac{n}{k}P[R_1 > b(n/k)t^{-1/\alpha}]} B_n(t).$$

It follows from [156, Theorem 6.2(9), page 180] that $N_n/k \xrightarrow{P} \nu_\alpha[t^{-1/\alpha}, \infty) = t$, and regular variation of $P[R_1 > r]$ similarly implies that $\frac{n}{k}P[R_1 \geq b(\frac{n}{k})t^{-1/\alpha}] \to t$. Assumption (5.52) says that $B_n \to 0$ locally uniformly, so $D_n \xrightarrow{P} 0$.

To deal with C_n, let $\sigma_{gn}^2 = \text{Var}(g(\Theta_{K_1(n)})) = \text{Var}(g(\Theta_1) \mid R_1/b(n/k) \geq t^{-1/\alpha})$. Remember t is fixed. Define the process Y_n by $Y_n(0) = 0$ and

$$Y_n(s) = \frac{1}{\sigma_{gn}\sqrt{k}}\sum_{j=1}^{ks}\left(g(\Theta_{K_j(n)}) - Eg(\Theta_{K_j(n)})\right), \quad s > 0. \tag{5.54}$$

Since the *Découpage de Lévy* [157, page 212] guarantees that the sequence $\{\Theta_{K_j(n)}, j \geq 1\}$ is iid, the functional central limit theorem for triangular arrays [13, page 94] and [143] implies that $Y_n \Rightarrow W$ in $D[0, \infty)$, where W is a Brownian motion. As before, $N_n/k \xrightarrow{P} t$, a deterministic limit, so we have joint convergence and may apply composition to obtain in $\mathbb{R}$,

$$Y_n(N_n/k) = \frac{1}{\sigma_{gn}\sqrt{k}}\sum_{j=1}^{N_n}\left(g(\Theta_{K_j(n)}) - Eg(\Theta_{K_j(n)})\right) \Rightarrow W(t).$$

The left-hand side equals $\frac{\sigma_g}{\sigma_{gn}}C_n$. Regular variation and $g(\cdot)$ being bounded and continuous implies $\lim_{n\to\infty}\sigma_{gn} = \sigma_g$, and since we assume $\sigma_g > 0$, we obtain $C_n \Rightarrow W(t) \sim N(0, t)$ in $\mathbb{R}$. Since $D_n \xrightarrow{P} 0$, Slutsky's Theorem gives that $W_n(t) \Rightarrow W(t)$ in $\mathbb{R}$ for fixed t. So ends the warmup.

We sneak up on convergence of finite dimensional distributions. Fix $0 < s < t < \infty$ and re-define $K_j(n)$ as the jth index i for which $R_i/b(n/k) \in [t^{-1/\alpha}, s^{-1/\alpha})$. Also re-define

$$N_n = \sum_{i=1}^{n}\epsilon_{R_i/b(\frac{n}{k})}[t^{-1/\alpha}, s^{-1/\alpha}),$$

and write

$$W_n(t) - W_n(s) = \frac{1}{\sigma_g \sqrt{k}} \sum_{j=1}^{N_n} \left(g(\Theta_{K_j(n)}) - \mu_g \right)$$

$$= \frac{1}{\sigma_g \sqrt{k}} \sum_{j=1}^{N_n} \left(g(\Theta_{K_j(n)}) - Eg(\Theta_{K_1(n)}) \right) + \frac{1}{\sigma_g \sqrt{k}} N_n \left(Eg(\Theta_{K_1(n)}) - \mu_g \right)$$

$$= C_n + D_n.$$

Mimic the steps in the warmup to show $D_n \xrightarrow{P} 0$ using the re-defined N_n, which now satisfies $N_n/k \xrightarrow{P} \nu_\alpha[t^{-1/\alpha}, s^{-1/\alpha}) = t - s$ [156, Theorem 6.2(9)]. To deal with C_n, re-define $\sigma_{gn}^2 = \text{Var}(g(\Theta_{K_1(n)})) = \text{Var}(g(\Theta_1) \mid R_1/b(n/k) \in [t^{-1/\alpha}, s^{-1/\alpha}))$ and with the re-definitions the definition of the process Y_n in (5.54) stays the same. and $Y_n \Rightarrow W$ in $D[0, \infty)$. After the composition step $Y_n(N_n/k) \Rightarrow W(t - s)$ and we get $W_n(t) - W_n(s) = C_n + o_p(1) \Rightarrow W(t - s) \sim N(0, t - s)$.

Consider now an arbitrary number of *disjoint* intervals $(s_m, t_m]$, $m = 1, \ldots, M$ with $0 < s_m < t_m < \infty$. As before, define $K_j^m(n)$ to be the jth index i for which $R_i/b(n/k) \in [t_m^{-1/\alpha}, s_m^{-1/\alpha})$, and set

$$N_n^m = \sum_{i=1}^n \epsilon_{R_i/b(\frac{n}{k})}[t_m^{-1/\alpha}, s_m^{-1/\alpha}).$$

As in the previous steps, using obvious notation, decompose each of the M increments as

$$W_n(t_m) - W_n(s_m) = C_n^m + D_n^m$$

and verify that $D_n^m \xrightarrow{P} 0$ for each m. Next, for each m, define processes Y_n^m as in (5.54) but with $K_j^m(n)$ instead of $K_j(n)$. The Découpage de Lévy implies that the M sequences $\{\Theta_{K_j^m(n)} : j \geq 1\}$ are independent for fixed n, and hence the processes Y_n^m are also independent. The previously established convergence result, which was proven for one single sequence of processes $\{Y_n\}$, thus holds jointly:

$$(Y_n^1, \ldots, Y_n^M) \Rightarrow (W^1, \ldots, W^M),$$

where the limit is M-dimensional Brownian motion consisting of independent standard Brownian motions. Composition with $N_n^m/k \xrightarrow{P} t_m - s_m$ for $m = 1, \ldots, M$, lets us conclude that in $\mathbb{R}^M$,

$$\left(W_n(t_m) - W_n(s_m), 1 \leq m \leq M \right) = (C_n^1, \ldots, C_n^M) + \left(o_p(1), \ldots, o_p(1) \right)$$
$$\Rightarrow N(0, \text{diag}(t_1 - s_1, \ldots, t_M - s_M)).$$

5.7 Measuring Extremal Dependence with the EDM

Since $W_n(0) = 0$, we may apply CUMSUM to get finite-dimensional convergence. □

5.7.2.2 Proof of Tightness: How to Avoid Crushing a Ninja

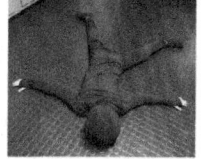

Now to prove tightness, the elephant in the room capable of crushing even a ninja. Since the limit process W has continuous paths, it suffices to prove [13, Theorem 13.5]

$$\limsup_{n \to \infty} E(|W_n(t) - W_n(s)|^2 |W_n(s) - W_n(r)|^2) \leq (t-r)^2 \tag{5.55}$$

for all $0 \leq r \leq s \leq t$. Fix r, s, t and write

$$\alpha_i = \left(g(\Theta_i) - \mu_g\right) 1_{[R_i/b(\frac{n}{k}) \in [t^{-1/\alpha}, s^{-1/\alpha})]}$$

$$\beta_i = \left(g(\Theta_i) - \mu_g\right) 1_{[R_i/b(\frac{n}{k}) \in [s^{-1/\alpha}, r^{-1/\alpha})]}.$$

Using that $\alpha_i \beta_i = 0$, check that

$$\sigma_g^4 E(|W_n(t) - W_n(s)|^2 |W_n(s) - W_n(r)|^2) = \frac{1}{k^2} E \left(\sum_i \alpha_i\right)^2 \left(\sum_i \beta_i\right)^2$$

$$= \frac{n(n-1)}{k^2} E(\alpha_1^2 \beta_2^2) + \frac{n(n-1)(n-2)}{k^2} E(\alpha_1^2 \beta_2 \beta_3)$$

$$+ \frac{n(n-1)(n-2)}{k^2} E(\alpha_1 \alpha_2 \beta_3^2) + \frac{n(n-1)(n-2)(n-3)}{k^2} E(\alpha_1 \alpha_2 \beta_3 \beta_4).$$

All expectations factorize by independence, and using simple bounds for the coefficients we get

$$\sigma_g^4 E(|W_n(t) - W_n(s)|^2 |W_n(s) - W_n(r)|^2)$$

$$\leq \frac{n^2}{k^2} E(\alpha_1^2) E(\beta_1^2) + \frac{n^3}{k^2} E(\alpha_1^2)(E(\beta_1))^2$$

$$+ \frac{n^3}{k^2} (E(\alpha_1))^2 E(\beta_1^2) + \frac{n^4}{k^2} (E(\alpha_1))^2 (E(\beta_1))^2.$$

Define the function $h : (0, \infty) \times [0, 1] \to \mathbb{R}$ by

$$h(r, \theta) = \left(g(\theta) - \mu_g\right)^2 1_{[r \in [t^{-1/\alpha}, s^{-1/\alpha})]}$$

and this function is $\nu_\alpha \times S$-a.e. continuous and compactly supported in $(0, \infty) \times [0, 1]$. Regular variation, and the fact that $E(\alpha_1^2) = Eg(R_1/b(n/k), \Theta_1)$ gives

$$\frac{n}{k} E(\alpha_1^2) \to \nu_\alpha[t^{-1/\alpha}, s^{-1/\alpha}) \int_{[0,1]} (g(\theta) - \mu_g)^2 S(d\theta) = \sigma_g^2 (t - s).$$

Similarly, $\frac{n}{k} E(\beta_1^2) \to \sigma_g^2(s - r)$. Moreover,

$$\sqrt{k} \frac{n}{k} E(\alpha_1) = \sqrt{k} \left(E(g(\Theta_1) 1_{[t^{-1/\alpha} \le \frac{R_1}{b(\frac{n}{k})} < s^{-1/\alpha}]}) - \mu_g \frac{n}{k} P\left[\frac{R_1}{b(\frac{n}{k})} \in [t^{-1/\alpha}, s^{-1/\alpha}) \right] \right)$$

$$\to 0 \quad (n \to \infty)$$

by (5.52). Thus $\frac{n^2}{k}(E(\alpha_1))^2 = (\sqrt{k}\frac{n}{k}E(\alpha_1))^2 \to 0$, and similarly we also get $\frac{n^2}{k}(E(\beta_1))^2 \to 0$. Combining these results yields

$$\sigma_g^4 \limsup_n E(|W_n(t) - W_n(s)|^2 |W_n(s) - W_n(r)|^2)$$

$$\le \sigma_g^2(t-s)\sigma_g^2(s-r) \le \sigma_g^4(t-r)^2.$$

We conclude that (5.55) holds and we have conquered tightness.

Wahoo!

5.7.3 Unclunking the Clunk Function $g(\cdot)$

Here are several cases where judicious choice of the function g avoids the problem that the asymptotic variance is zero.

5.7.3.1 Asymptotic Independence

Section 5.7.1 has a discussion of an exploratory method to partition a vector into sub-vectors that are approximately asymptotically independent. So suppose exploration has yielded two components X_1, X_2 which are candidates for asymptotically independent random variables. Can we base a test for this on asymptotic normality using results of Sect. 5.7.2?

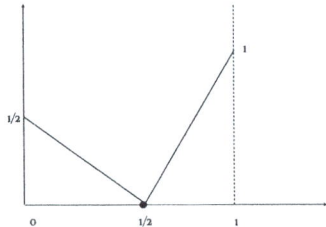

Fig. 5.26 A simple fix

5.7 Measuring Extremal Dependence with the EDM

We cannot use the EDM as a test statistic since the function $g(\theta) = 4\theta(1-\theta)$ fails the $\sigma_g > 0$ condition for asymptotic independence. A quick reminder why: For asymptotic independence the angular measure $S(\cdot)$ concentrates on the end points of $[0, 1]$ and is of the form

$$S(\cdot) = p_0 \epsilon_{\{0\}}(\cdot) + p_1 \epsilon_{\{1\}}(\cdot), \quad p_0 + p_1 = 1.$$

So $S(\cdot)$ gives mass p_0 to $\{0\}$ and p_1 to $\{1\}$. If $g(\cdot)$ is the clunk function of Theorem 5.3, a quick calculation shows

$$\sigma_g^2 = \int_{[0,1]} (g(\theta) - \mu_g)^2 S(d\theta)$$
$$= p_1^2 p_0 \big(g(0) - g(1)\big)^2 + p_1 p_0^2 \big(g(1) - g(0)\big)^2$$
$$= \big(g(0) - g(1)\big)^2 (p_1^2 p_0 + p_1 p_0^2) = \big(g(0) - g(1)\big)^2 p_0 p_1.$$

The obvious problem is that if $g(0) = g(1)$ then $\sigma_g^2 = 0$; this is the case for the EDM $g(\theta) = 4\theta(1-\theta)$, An obvious solution is to pick a different $g(\cdot)$ [122] such that $g(0) \neq g(1)$. For asymptotic independence, Fig. 5.26 gives a choice of g where $g(0) = 1/2$, $g(1) = 1$ and g is still bounded and continuous. This choice of g is somewhat ad hoc but c'est la vie.

For the g beautifully drawn in Fig. 5.26, we have

$$g(x) = \begin{cases} \frac{1}{2} - x, & \text{if } 0 \leq x \leq \frac{1}{2}, \\ 2x - 1, & \text{if } \frac{1}{2} \leq x \leq 1. \end{cases}$$

If $p_0 = 1/2 = p_1$, then $\mu_g = 3/4$ and $\sigma_g^2 = 1/16$ and asymptotic normality in (5.53) gives

$$4 \cdot \sqrt{k}(\widehat{\mathrm{EDM}}(g) - 3/4) \Rightarrow N(0, 1), \quad (n \to \infty, k \to \infty, k/n \to 0). \tag{5.56}$$

If you have no reason to optimistically believe $p_0 = 1/2 = p_1$, you have to estimate p_0, p_1 and calculate μ_g and σ_g from these estimates.

5.7.3.2 Strong Dependence

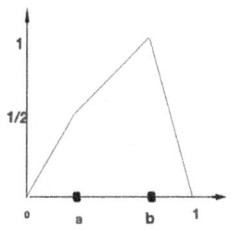

Fig. 5.27 Strong dependence

Consider the case of two variables exhibiting strong dependence so the angular measure S concentrates on $[a, b] \subset [0, 1]$ and $0 < a < b < 1$. In Problem 5.9, page 224, you are asked to verify that if g is constant on $[a, b]$, then $\sigma_g = 0$. This embarrassment is removed if you make $g(a) = 1/2$ and $g(b) = 1$ and for simplicity [122] make g linear between a, b as is illustrated in the dazzling Fig. 5.27. This makes

$$g(x) = \begin{cases} \frac{x}{2a}, & \text{if } x \in [0, a), \\ \frac{1}{2}\left(\frac{x-a}{b-a}\right) + \frac{1}{2}, & \text{if } x \in [a, b], \\ 1 - \frac{x-b}{1-b}, & \text{if } x \in (b, 1]. \end{cases} \quad (5.57)$$

Since S does not charge points outside $[a, b]$, the definition of g on $[0, 1] \setminus [a, b]$ is immaterial for μ_g, σ_g (except we must make sure g is bounded and continuous) but the pre-asymptotic computation of $\widehat{\text{EDM}}(g)$ needs a definition on all of $[0, 1]$. Theorem 5.1, page 163, discusses ways to estimate support endpoints a and b (probably not the last word on this subject) and application of (5.53) requires also knowledge of μ_g, σ_g which will typically require some knowledge of S beyond just its support. So either we must estimate S or try the work-around discussed next.

If $S(\cdot)$ is uniform on $[a, b]$, with $0 < a < b < 1$, you are asked in Problem 5.10 to confirm the superficially surprising fact that

$$(\mu_g, \sigma_g^2) = (3/4, 1/48). \quad (5.58)$$

5.7.3.3 The Angular Measure S Concentrates on Two Disjoint Subintervals

For $0 < a < b < c < d < 1$, suppose S concentrates on $[a, b] \cup [c, d]$. Set $p_0 = S[a, b]$ and $p_1 = S[c, d]$ and make g constant on each interval, say

$$g(x) = \begin{cases} c_0, & \text{if } x \in [a, b], \\ c_1, & \text{if } x \in [c, d], \end{cases}$$

with linear interpolation elsewhere as suggested by Fig. 5.28. By now familiar calculations we have

$$\mu_g = c_0 p_0 + c_1 p_1, \quad \sigma_g^2 = p_0 p_1 (c_0 - c_1)^2,$$

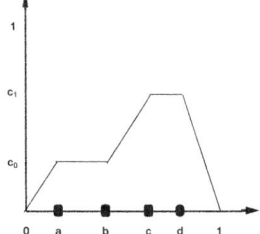

Fig. 5.28 Two intervals

so as long as $c_0 \neq c_1$, we avoid the dreaded $\sigma_g = 0$ case. You do not have to know detailed information about S on the two intervals but just the amount of mass p_0, p_1 plopped into each. This is one reason to prefer this g rather than the traditional EDM with $g(\theta) = 4\theta(1 - \theta)$.

The circumstance of the angular measure concentrating on two disjoint intervals would arise when seeking hidden regular variation where a forbidden zone in polar coordinates (r, θ) of a θ-interval is removed and regular variation sought on what is left. See the discussion in Sect. 5.5.4.3, page 184. For example, the model could be

$$(\Theta, R) = B(\Theta_1, R_1) + (1 - B)(\Theta_2, R_2)$$

5.7 Measuring Extremal Dependence with the EDM

where $B, R_i, \Theta_i, i = 1, 2$ are independent and B is Bernoulli with values $\{0, 1\}$, Θ_1 has range $[.4, .6]$, Θ_2 has range $[.25, .3] \cup [.85, .9]$ and $P[R_1 > r] = r^{-1}, r \geq 1$ and $P[R_2 > r] = r^{-2}, r \geq 1$.

The other circumstance where this case arises is if we have strong dependence where the support of S, $[a, b] \subset [0, 1]$, is a proper subset of $[0, 1]$; see [122, page 271]. We know we cannot use asymptotic normality of $\widehat{\text{EDM}}(g)$ with g that is constant on $[a, b]$ because $\sigma_g = 0$ but here is a work-around. Suppose in $\mathbb{R}_+^2$ we have an MRV sample $\{(X_i, Y_i); 1 \leq i \leq n\}$ in Cartesian coordinates with polar coordinate

$$\Theta_i = \frac{X_i}{X_i + Y_i}.$$

Let $\{B_i, 1 \leq i \leq n\}$ be iid symmetric Bernoulli random variables with values $\{0, 1\}$ and independent of $\{(X_i, Y_i)\}$. For $i = 1, \ldots, n$, define

$$(\tilde{X}_i, \tilde{Y}_i) = B_i \left(\frac{1}{2}X_i, \frac{1}{2}X_i + Y_i\right) + (1 - B_i)\left(X_i + \frac{1}{2}Y_i, \frac{1}{2}Y_i\right).$$

Apply the map $(x, y) \mapsto x/(x+y)$ to get the first coordinate of the polar component of $(\tilde{X}_i, \tilde{Y}_i)$ on $\aleph_{\{0\}}$,

$$\tilde{\Theta}_i = B_i \left(\frac{X_i}{2X_i + 2Y_i}\right) + (1 - B_i)\left(\frac{2X_i + Y_i}{2X_i + 2Y_i}\right) = B_i \frac{\Theta_i}{2} + (1 - B_i)\frac{1 + \Theta_i}{2}.$$

If the original angular measure of the MRV distribution of $\{(X_i, Y_i)\}$ concentrates on $[a, b]$, then the angular measure of the transformed MRV variables $\{(\tilde{X}_i, \tilde{Y}_i)\}$ concentrates on $[a/2, b/2] \cup [(a+1)/2, (b+1)/2]$ and the transformed data is an example of data where the angular measure concentrates on the union of two disjoint intervals. For asymptotic normality, the transformed data qualifies for a $g(\cdot)$ which is constant on the two intervals as in Fig. 5.28.

5.7.3.4 Asymptotic Full Dependence

Asymptotic full dependence means S concentrates at a point of $[0, 1]$ which for specificity we assume is $1/2$. Under this assumption, in force throughout this section, no matter what clunk function g we choose, $\sigma_g = 0$. To salvage methodology based on asymptotic normality, we take a cue from Sect. 5.7.3.3 and modify the data. Remember, if you know a model, you also know a model derived from the model.

So, as in Sect. 5.7.3.3, suppose in $\mathbb{R}_+^2$ we have an MRV sample $\{(X_i, Y_i); 1 \leq i \leq n\}$ in Cartesian coordinates with polar coordinates

$$\Theta_i = \frac{X_i}{X_i + Y_i}, \quad R_i = X_i + Y_i$$

and

$$P[(X_i, Y_i) \in \cdot] \in \text{MRV}(\alpha, b(t), \eta(\cdot), \mathbb{R}_+^2 \setminus \{\mathbf{0}_2\}) \tag{5.59}$$

where the limit measure $\eta(\cdot)$ has angular measure $S(\cdot) = \epsilon_{1/2}(\cdot)$ concentrating on $\{1/2\}$. Let $\{U_i, i \geq 1\}$ be iid $U(0, 1/8)$ independent of $\{(X_i, Y_i)\}$ and define

$$(\tilde{X}_i, \tilde{Y}_i) = \Big(R_i(\Theta_i + U_i), R_i(1 - \Theta_i - U_i) \Big)$$

so that

$$\tilde{R}_i = \tilde{X}_i + \tilde{Y}_i = R_i \quad \tilde{\Theta}_i = \frac{\tilde{X}_i}{\tilde{X}_i + \tilde{Y}_i} = \Theta_i + U_i. \tag{5.60}$$

Note the definition of $\tilde{Y}_i$ includes the small possibility that Θ_i is close to 1 and U_i is close to 1/8, thus making $\tilde{Y}_i$ negative. Also, the support $\tilde{\Theta} \in [0, 1 + 1/8]$ could be scaled to the unit interval but, to paraphrase Alfred E. Neuman,[3] we will not worry. Forgoing the index i, write $(\tilde{R}, \tilde{\Theta}) = (R, \Theta) + (0, U)$, use the polar coordinate version of (5.59) and apply the First Binding Proposition 2.1, page 47, to get as $t \to \infty$,

$$tP\Big[\Big(\Big(\frac{R}{b(t)}, \Theta\Big), (0, U)\Big) \in \cdot\Big] \to (\nu_\alpha \times \epsilon_{1/2}(\cdot)) \times P[(0, U) \in \cdot] \tag{5.61}$$

in $\mathbb{M}\big((0, \infty) \times [0, 1] \times [\frac{1}{2}, \frac{5}{8}]\big) = \mathbb{M}\Big(\big(\mathbb{R}_+ \times [0, 1] \times [\frac{1}{2}, \frac{5}{8}]\big) \setminus \big(\{0\} \times [0, 1] \times [\frac{1}{2}, \frac{5}{8}]\big)\Big)$. If you have trouble applying the First Binding Proposition 2.1, match

$$\mathbb{S}_1 = \mathbb{R}_+ \times [0, 1], \quad \mathbb{F}_1 = \{0\} \times [0, 1], \quad \mathbb{S}_2 = \{0\} \times \Big[\frac{1}{2}, \frac{5}{8}\Big].$$

Now apply the map $T : (r, \theta, 0, u) \mapsto (r, \theta + u)$ to (5.61) which yields,

$$tP\Big[\Big(\Big(\frac{\tilde{R}}{b(t)}, \tilde{\Theta}\Big)\Big) \in \cdot\Big] = tP\Big[\Big(\Big(\frac{R}{b(t)}, \Theta + U\Big)\Big) \in \cdot\Big] \to \nu_\alpha \times P\Big[\frac{1}{2} + U \in \cdot\Big]$$

so the limiting angular measure of $(\tilde{R}, \tilde{\Theta})$ is uniform on $[1/2, 5/8]$.

So we conclude unsurprisingly that starting from an asymptotically fully dependent sample, our transformation using uniform random variables produces a sample

---

[3] Mad Magazine: "What me worry?"

5.7 Measuring Extremal Dependence with the EDM

with strong dependence. Testing methods from Sects. 5.7.3.2 and 5.7.3.3 can be applied.

5.7.4 Is Relying on Asymptotic Normality Wise?

We began this chapter urging caution, modest goals and emphasizing exploration rather than confirmation. Now with asymptotic normality, we seem to embrace the hubris of testing hypotheses. So perhaps we should pause and experiment to see how helpful asymptotic normality can be.

We focus initially on full dependence since according to the outline in Sect. 5.7.3.4 there is no need to estimate $S(\cdot)$. In Sect. 5.7.5, starting page 214, we outline a preferential attachment network growth model

What do Yoda would?

that incorporates the feature of *reciprocity*. This model leads to asymptotically fully dependent data [23, 189] But this data is not iid and we find the test for asymptotic full dependence does not work well. Generally it is not easy to find confirmed samples of real data that are asymptotically fully dependent so we begin with a simple simulation example giving asymptotic full dependence and where the iid assumption is assured. The idea is that if your method does *not* work on simulated data where there is no model error, you should quietly slip out of town and/or abandon the method. Recall the moments given in (5.58), page 206.

5.7.4.1 Simple Simulation Example

We simulate 30,000 random samples from the model $(X, X + E)$ where the independent random variables (X, E) are standard Pareto and standard exponential. Form the variables $R = X + X + E$ and $\Theta = X/R$ and then $\tilde{\Theta} = \Theta + U(0, 1/8)$ as outlined in (5.60), page 208. For the function $g(\cdot)$ in (5.57) and Fig. 5.27, we then have $a = 1/2$ and $b = 1/2 + 1/8 = 5/8$. Sort the R's in decreasing order filter the $\tilde{\Theta}$s, retaining only those $\tilde{\Theta}$s such that the corresponding R is bigger than a designated threshold, taken to be an upper order statistic of the R's indexed by k. Pump the filtered $\tilde{\Theta}$s into the g of (5.57) and the approximately normal test statistic called T in (5.53) has sample values indicated in the following Table 5.2 as t along with $P[|Z| > |t|]$ for a standard normal variable Z.

Table 5.2 Approximately normal statistic for several thresholds

k	100	200	300	600	1000	1500
t	0.0671568	0.22732	−0.578519	−1.53528	−3.29499	−6.49845
$P[|Z| > |t|]$	0.946456	0.820169	0.562913	0.124713	0.00098423	$8.1147e{-}11$

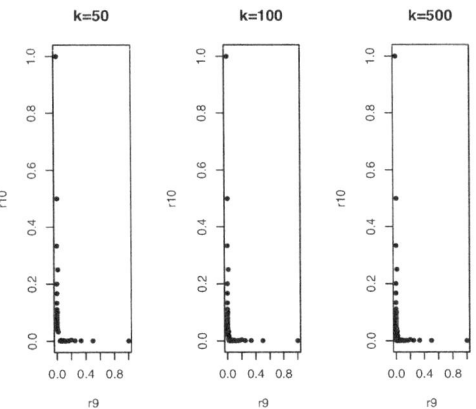

Fig. 5.29 Horizontal axis is Sodankylä and vertical axis is Kouvola

Values seem sensitive to choice of k. For smaller values of k, say $k = 100, 200, 300$, there is no strong evidence against full dependence.

5.7.4.2 What About Real Data Mr. Smarty Pants?

Estimation and testing procedures tend to behave well with simulated data since there are large sample sizes and no model error unless it is purposely introduced into the simulation. Does the model verification based on asymptotic normality of the EDM provide comfort when fitting real data? The subsequent Sect. 5.7.5 demonstrably produces data from an MRV model whose limit measure shows asymptotic full dependence but the attempt to use EDM-asymptotic normality is not a resounding success, presumably because the data is not remotely iid.

Ok smarty pants!

So lesson learned and we stay away from network data (sigh) when exploring the usefulness of EDM-normality.

Rainfall Data For the project in [122], we downloaded daily rainfall data from separate locations in Finland from the Finnish Meteorological Institute. To reduce seasonal effects, we only used observations from summer months June, July and August and the total number of observations is $n = 3864$. One expects sites that are geographically relatively close to each other to exhibit dependence so we selected two measuring sites, Kouvola and Sodankylä, that are relatively far apart. (Sodankylä is in the north and Kouvola is in the extreme south of Finland and Google maps helpfully informs us that these sites are 899 km apart which takes 10 hours and 41 minutes to drive by car, something, I imagine, few Europeans would

5.7 Measuring Extremal Dependence with the EDM

do.) While it does not appear these two locations show independence, there does seem to be extremal independence of the largest observations.

Hill plots (mercifully not shown) struggle to reveal anything so since the data is not likely to be very heavy tailed, we resorted to bargaining with the devil and using the rank transformation [87, 174] and [156, p. 310 ff]. The rank transformation standardizes marginal distributions to $\alpha = 1$ while preserving the asymptotic dependence structure. However, even if the original data pairs are close to iid, the rank-transformed data may not be.

After rank-transforming, we display scatterplots in Fig. 5.29 of the 2-dimensional data, restricted to the k-biggest sums, for $k = 50, 100, 500$. There is little visual evidence against asymptotic independence as data seems to hug the axes quite satisfactorily. Apply the L_1 polar coordinate transformation

$$(x_1, x_2) \mapsto \left(x_1 + x_2, \frac{x_1}{x_1 + x_2}\right) = (r, \theta_1)$$

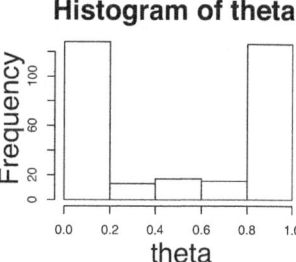

and make a histogram of $\{\theta_1 : r \geq r_0\}$ where r_0 is the 300th largest of the r's. The plot confirms the impression of mass concentrating at the endpoints of $[0, 1]$ consistent with asymptotic independence and also suggests $p_0 = p_1 = 0.5$ and (5.56), page 205 should be appropriate. So proceeding with the function g in Fig. 5.26, page 204, we produce the following Table 5.3 for observed values t of T in (5.53), page 199.

For $k = 25, 50, 75$ we would not reject the hypothesis of asymptotic independence but for k larger than 75 the *meh* results are not supportive. However, remember the rank transformation destroys the iid assumption so it is not surprising the test statistic struggles a bit. There is some evidence supporting asymptotic independence but the fragility of the result for differing thresholds leaves us nervous.

Stock Data [122] We downloaded daily stock data using the R-package [144] *Quantmod* [169]. Observations range from December 20, 2010 to July 10, 2018. Observations were converted to log-returns by taking the logarithm of the price and calculating differences. The resulting returns have similar tail indices for positive and negative tails and, for simplicity, we used absolute returns. Converting from prices to returns produces data that seems stationary but it is well known the data is not iid and in fact can exhibit long range dependence. However, we proceeded anyway.

Table 5.3 Approximately normal statistic for several thresholds

k	25	50	75	100	200	300
t	−0.4879064	−1.079969	−1.461097	−2.122587	−6.627537	−9.832442
$P[|Z| > |t|]$	0.6256161	0.2801559	0.1439887	0.03378847	3.413336e−11	8.161509e−23

We investigate two questions:

1. Are returns in the tech-sector asymptotically independent of those in the oil industry? Their spheres of economic activity do not directly overlap. We do not expect independence since both depend on the general state of the economy but why should large changes track each other? We tested Exxon (XOM) against Microsoft (MSFT).
2. Will returns within the same sector, say two companies in tech, exhibit asymptotic full dependence? We would *expect* large relative changes in one company to be mirrored by the other. We tested Google (GOOG) versus Microsoft (MSFT).

Read on for details but spoiler alert for those who peek at the last page of a mystery novel: Let's just say it is not easy to be comfortable that either of the extreme cases, asymptotic independence or asymptotic full dependence, exist here.

1. Asymptotic Independence The scatterplot of absolute log-returns in Fig. 5.30 for MSFT vs XOM is not encouraging. While there are some points close to the axes, the effect is not pronounced. Hoping to salvage something from (MSFT,XOM) absolute returns we mapped

$$(\text{MSFT},\text{XOM}) = (x_1, x_2) \mapsto \big(x_1 + x_2, x_1/(x_1 + x_2)\big) = (r, \theta)$$

and made histograms of θ's. Even accounting for the scrunched up x-axis, there is no evidence of migration to the endpoints as a function of threshold. With so little visual encouragement, we abandon the idea that these tech and oil stocks are asymptotically independent.

2. Asymptotic Full Dependence We looked at absolute returns from Google (GOOG) vs Microsoft (MSFT) to see if the intuition that two companies in the same sector should have large relative price movements in common. The scatterplot in the left panel of Fig. 5.31 of absolute returns of GOOG vs MSFT shows some tendency

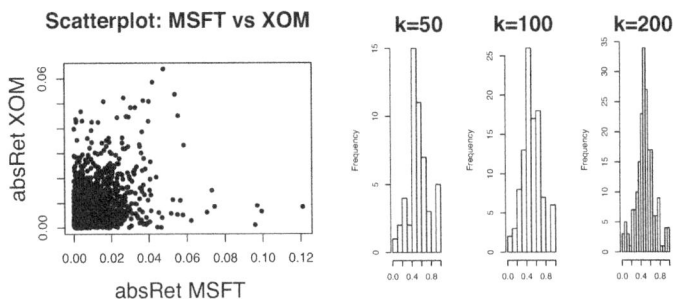

Fig. 5.30 Left: scatterplot of absolute returns from MSFT vs XOM. Right: Three histograms of MSFT/(MSFT+XOM) thresholded according to sum of components at levels k=50, 100, 200 largest of the sums

5.7 Measuring Extremal Dependence with the EDM

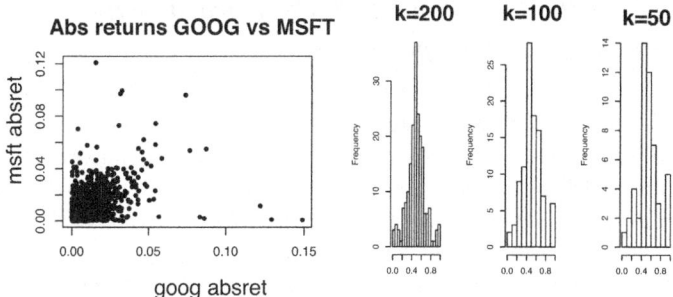

Fig. 5.31 Left: scatterplot of absolute returns from GOOG vs MSFT. Right: three histograms of GOOG/(GOOG+MSFT) thresholded according to sum of components at levels k=200, 100, 50 largest of the sums

to funnel towards the diagonal but the tendency is weak so maybe we should not have high hopes. On the other hand, the histograms of

$$\{GOOG/(GOOG+MSFT): GOOG+MSFT \geq r\}$$

for r taken as the 200, 100, 50th order statistic of the sums does seem visually to be consistent with full dependence. Despite being slightly wary due to the scatterplot Fig. 5.31 we proceed to test for full dependence using EDM-asymptotic normality.

A quick calculation of the mean of GOOG/(GOOG+MSFT) for varying degrees of thresholding shows a mean always close to 0.5. So we employ the techniques of Sect. 5.7.3.4, page 207 which builds on Sect. 5.7.3.2, page 205 with the clunk function g in (5.57) with $a = 0.5$, $b = 1/2 + 1/8 = 5/8$.

Form R, Θ and then remember $\tilde{\Theta} = \Theta + U(0, 1/8)$. Sort R's in decreasing order, filter the $\tilde{\Theta}$s according to R being bigger than the kth largest R. Pump the filtered $\tilde{\Theta}$s through g, form the asymptotically normal statistic T with observed values t as in Table 5.2, page 209. A comparable Table 5.4 for this (real Smarty Pants) data follows.

Relying on EDM-normality and using Table 5.4, it seems pretty safe to reject the hypothesis of asymptotic full dependence. For the two stock data examples, neither asymptotic independence for (MSFT, XOM) nor asymptotic full dependence for (GOOG,MSFT) seem likely based on the techniques of EDM-normality. Just keep in mind the data is not iid.

Table 5.4 Approximately normal statistic for several thresholds

k	25	50	75	100	200	300
t	-9.5267	-9.40756	-11.17193	-12.03624	-15.58042	-19.39744
$P[|Z| > |t|]$	1.623e–21	5.078e–21	5.595e–29	2.292e–33	9.891e–55	8.112e–84

5.7.5 Preferential Attachment with Reciprocity Gives Asymptotic Full Dependence

In network science, models of preferential attachment with reciprocity lead to heavy tailed models with asymptotic full dependence and therefore it makes sense to simulate from this model and see how well asymptotic normality holds up. Here is a brief description of the model and relevent results [19, 23, 103, 186, 187]. Statisticians note: (i) the model depends on the following parameter vector $(\alpha, \beta, \gamma, \delta_{in}, \delta_{out}, \rho)$ and after re-parameterization two parameters may be replaced by tail indices ι_{in}, ι_{out} of degree distributions. (ii) The model produces data that is not iid.

Before describing the model, some comments about reciprocity. For modeling social network behavior this is a natural concept and describes mutual interaction among users: you link to a friend and the friend replies back to you. For a directed graph $G = (V, E)$, there is a quantitative measure of reciprocity called the *reciprocity coefficient* [80, 116] defined as the proportion of edges in G that are reciprocated:

$$r(G) := \frac{|\{(w, v) \in E : (v, w) \in E\}|}{|E|}. \tag{5.62}$$

Empirical studies [103, 115] show social networks may have particularly high values for $r(G)$, and these values are significantly higher than what would be predicted from just preferential attachment. Hence there is value to adding reciprocation to the social network modeling.

5.7.5.1 Description of the Model

The model describes a growing sequence of directed graphs

$$\{G(n) = (V(n), E(n)), n \geq 0\}$$

where $G(n)$ is the graph after n steps of development, $V(n)$ is the set of nodes in $G(n)$ and $E(n)$ is the set of directed edges in $G(n)$. The edge set $E(n)$ consists of ordered pairs $(w_1, w_2) \in E(n)$, where $w_1, w_2 \in V(n)$, and the ordered pair (w_1, w_2) represents a directed edge $w_1 \mapsto w_2$. When $n = 0$, we have $V(0) = \{1\}$ and $E(0) = \{(1, 1)\}$ since we imagine the model sequence initialized with one node (labeled node 1) and a self-loop. The cardinality of a subset A of either $V(n)$ or $E(n)$ is denoted by $|A|$ so $|V(n)|$ and $|E(n)|$ are the cardinalities of the node set and edge set. We write $V(n) = \{1, \ldots, |V(n)|\} = [V(n)]$.

For $w \in V(n)$, set $\left(D_w^{in}(n), D_w^{out}(n)\right)$ to be the in- and out-degrees of node w so

$$D_w^{in}(n) = |\{(u, w) : (u, w) \in E(n)\}|, \quad D_w^{out}(n) = |\{(w, v) : (w, v) \in E(n)\}|.$$

5.7 Measuring Extremal Dependence with the EDM

We use the convention that $D_w^{\text{in}}(n) = D_w^{\text{out}}(n) = 0$ if $w \notin V(n)$. For $n \geq 0$, given $G(n)$ we get $G(n+1)$ by flipping a trinomial coin (Fig. 5.32 is a visual aid) with outcome probabilities α, β, γ in $(0, 1)$ and observing which scenario occurs along with whether or not reciprocal edges are created.

(i) With probability α, add a new node $|V(n)| + 1$ with a directed edge $(|V(n)| + 1, w)$, where $w \in V(n)$ is chosen with probability

$$\frac{D_w^{\text{in}}(n) + \delta_{\text{in}}}{\sum_{w \in V(n)}(D_w^{\text{in}}(n) + \delta_{\text{in}})} = \frac{D_w^{\text{in}}(n) + \delta_{\text{in}}}{|E(n)| + \delta_{\text{in}}|V(n)|}, \quad (5.63)$$

and update the node set $V(n+1) = V(n) \cup \{|V(n)| + 1\}$. (The equivalence in the denominators is readily proven using induction on n.) With probability $\rho \in (0, 1)$, a reciprocal edge $(w, |V(n)| + 1)$ is added and we update the edge set as $E(n+1) = E(n) \cup \{(|V(n)| + 1, w), (w, |V(n)| + 1)\}$. With probability $1 - \rho$, no reciprocation takes place and $E(n+1) = E(n) \cup \{(|V(n)| + 1, w)\}$.

(ii) With probability γ, add a new node $|V(n)| + 1$ with a directed edge $(w, |V(n)| + 1)$, where $w \in V(n)$ is chosen with probability

$$\frac{D_w^{\text{out}}(n) + \delta_{\text{out}}}{\sum_{w \in V(n)}(D_w^{\text{out}}(n) + \delta_{\text{out}})} = \frac{D_w^{\text{out}}(n) + \delta_{\text{out}}}{|E(n)| + \delta_{\text{out}}|V(n)|}, \quad (5.64)$$

and update the node set $V(n+1) = V(n) \cup \{|V(n)| + 1\}$. If, with probability $\rho \in (0, 1)$, a reciprocal edge $(|V(n)| + 1, w)$ is added, update the edge set as $E(n+1) = E(n) \cup \{(|V(n)| + 1, w), (w, |V(n)| + 1)\}$. If the reciprocal edge is not created, set $E(n+1) = E(n) \cup \{(w, |V(n)| + 1)\}$.

(iii) With probability $\beta \in (0, 1)$ we do not add a new node but only add a new edge (v, w) between two existing nodes $v, w \in V(n)$, with probability

$$\frac{D_w^{\text{in}}(n) + \delta_{\text{in}}}{\sum_{w \in V(n)}(D_w^{\text{in}}(n) + \delta_{\text{in}})} \frac{D_v^{\text{out}}(n) + \delta}{\sum_{v \in V(n)}(D_v^{\text{out}}(n) + \delta)}$$

$$= \frac{D_w^{\text{in}}(n) + \delta_{\text{in}}}{|E(n)| + \delta_{\text{in}}|V(n)|} \frac{D_v^{\text{out}}(n) + \delta_{\text{out}}}{|E(n)| + \delta_{\text{out}}|V(n)|}.$$

Then with probability $\rho \in (0, 1)$, we add a reciprocal edge (w, v) and update the edge set as $E(n+1) = E(n) \cup \{(v, w), (w, v)\}$. If with probability $1 - \rho$ the reciprocal edge is not created, then $E(n+1) = E(n) \cup \{(v, w)\}$.

Note with probability $\alpha + \gamma$ we add a new node.

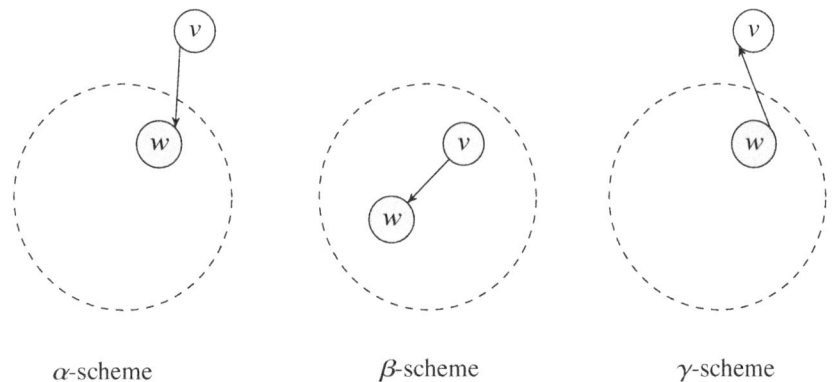

Fig. 5.32 Results from trinomial coin flip

5.7.5.2 Where Are the Heavy Tails?

Heavy tails are elusive in this model and often buried under jargon about *power laws*. Here is the precise story. First consider the limiting empirical frequency of joint degree counts. Define the count variable $N_{k,l}(n)$ of the number of nodes $w \in V(n)$ for which the in- and out-degrees are (k, l),

$$N_{k,l}(n) := \sum_{w=1}^{|V(n)|} 1_{\left[(D_w^{\text{in}}(n), D_w^{\text{out}}(n))=(k,l)\right]} \qquad (5.65)$$

and the empirical frequency of nodes having in- and out-degrees (k, l) is $N_{k,l}(n)/|V(n)|$. If $\delta_{\text{in}} = \delta_{\text{out}}$, as $n \to \infty$, for $1 \leq k, l < \infty$, the frequency converges to a probability mass function limit $p_{k,l}$,

$$\frac{N_{k,l}(n)}{|V(n)|} \xrightarrow{P} p_{k,l} \qquad (5.66)$$

and this limit mass function is a discrete multivariate regularly varying distribution. (The condition $\delta_{\text{in}} = \delta_{\text{out}}$ may be a function of the current method of proof in [23].) Manufacture fictitious random variables (I, O) such that

$$P\left[(I, O) = (k, l)\right] = p_{k,l}, \quad 1 \leq k, l < \infty,$$

and marginally there are positive indices $\iota_{\text{in}}, \iota_{\text{out}}$ such that

$$P[I > x] \in RV_{-\iota_{\text{in}}}, \quad P[O > x] \in RV_{-\iota_{\text{out}}}.$$

In fact, $P[(I, O) \in \cdot\,] \in \text{MRV}$. Define

5.7 Measuring Extremal Dependence with the EDM

$$\lambda_1 = \frac{1}{2}\left(1 + \beta + \sqrt{(\alpha - \gamma)^2 + 4(\alpha + \beta)(\beta + \gamma)\rho^2}\right) =: \frac{1}{2}\left(1 + \beta + \sqrt{D_0}\right).$$

If $\lambda_1 \geq \log 2$, then $P[(\mathcal{I}, \mathcal{O}) \in \cdot]$ is in

$$\text{MRV}\left(\frac{1 + \rho + \delta(1 - \beta)}{\lambda_1}, b(t) = t^{\lambda_1/(1+\rho+\delta(1-\beta))}, \eta(\cdot), \mathbb{R}_+^2 \setminus \{\mathbf{0}\}\right), \qquad (5.67)$$

and

$$\iota_{\text{in}} = \iota_{\text{out}} = \iota = \frac{1 + \rho + \delta(1 - \beta)}{\lambda_1}. \qquad (5.68)$$

The limit measure $\eta(\cdot)$ in (5.67) concentrates Pareto mass on the ray $y = mx$, $x > 0$ where

$$m = \frac{\gamma - \alpha + \sqrt{D_0}}{2(\beta + \gamma)\rho},$$

and the angular measure S on $[0, 1]$ concentrates mass 1 at the point

$$\theta_0 := \frac{2(\beta + \gamma)\rho}{2(\beta + \gamma)\rho + \gamma - \alpha + \sqrt{D_0}}. \qquad (5.69)$$

From the definitions, note $\gamma - \alpha + \sqrt{D_0} \geq 0$ so $m \geq 0$ and the point of concentration of $S(\cdot)$ is in $[0, 1]$.

While these formulas seem complicated, the point is simple: The pair $(\mathcal{I}, \mathcal{O})$ has an MRV distribution that is fully asymptotically dependent and explicit formulas in terms of the parameters of the model yield all indices, slopes and points of concentration. These explicit formulas can be easily programmed.

5.7.5.3 But What Is the Data? The Slippery Issue of the Double Limit

It is time to review, breath deeply and assess. Define $F(\cdot)$ as the multivariate regularly varying distribution with mass function $\{p_{k,l}\}$ of the mythical pair $(\mathcal{I}, \mathcal{O})$; that is,

$$F(\cdot) := \sum_{k,l} p_{k,l} \epsilon_{(k,l)}(\cdot)$$

$$= \sum_{k,l} P[(\mathcal{I}, \mathcal{O}) = (k, l)] \epsilon_{(k,l)}(\cdot). \qquad (5.70)$$

What data?

Write (5.66) as convergence of measures on $\mathbb{N}_+ \times \mathbb{N}_+$ and as $n \to \infty$ we have,

$$F_n(\cdot) := \frac{1}{|V(n)|} \sum_{w=1}^{|V(n)|} \epsilon_{(D_w^{\text{in}}(n), D_w^{\text{out}}(n))}(\cdot)$$

$$= \sum_{(k,l) \in \mathbb{N}_+ \times \mathbb{N}_+} \frac{N_{k,l}(n)}{|V(n)|} \epsilon_{(k,l)}(\cdot) \Rightarrow F(\cdot). \tag{5.71}$$

Owing to (5.67),

$$tF(b(t) \cdot) = tP\left[\left(\frac{I}{b(t)}, \frac{O}{b(t)}\right) \in \cdot\right] \to \eta(\cdot),$$

and in polar form

$$tP\left[\left(\frac{R}{b(t)}, \Theta\right) \in \cdot\right] = tP\left[\left(\frac{I+O}{b(t)}, \frac{I}{I+O}\right) \in \cdot\right] \to \nu_\iota \times S(\cdot)$$

where ι is read from (5.68) and $S(\cdot)$ concentrates at the point in (5.69). So first we have to let $n \to \infty$ to reach the first stop in our journey to asymptopia which yields the multivariate regularly varying distribution $F(\cdot)$ and then, in the second stop on the yellow brick road we have to scale $F(\cdot)$ and let $t \to \infty$ to get the limit measure $\eta(\cdot)$ or angular measure $S(\cdot)$.

So what is the data? Will asymptotic normality of the EDM be a useful diagnostic? What does *useful* mean? Inquiring minds want to know!

Here is the optimistic path that we *hope* is effective.

(a) For large n, F_n approximates F.
(b) Use the observables $\{(D_w^{\text{in}}(n), D_w^{\text{out}}(n)), w \in V(n)\}$ to estimate $\eta(\cdot)$.
(c) Therefore, use the observables $\{\Theta_w(n) := D_w^{\text{in}}(n)/(D_w^{\text{in}}(n) + D_w^{\text{out}}(n)), w \in V(n)\}$ to estimate $S(\cdot)$.
(d) Asymptotic normality with clunk function $g(\cdot)$ as discussed in Theorem 5.3, page 198, approximately holds when applied to $\{\Theta_w(n) : (D_w^{\text{in}}(n) + D_w^{\text{out}}(n)) > r\}$ for some chosen threshold r that is selected judiciously, psychically or perhaps using MDSP.

Thus, the data is $\{\Theta_j(n), j \leq |V(n)|\}$, where usually reality has fixed the value of n. For estimation, we need to threshold the data and use only $\{\Theta_w(n) : D_w^{\text{in}}(n) + D_w^{\text{out}}(n) > r\}$ for a chosen threshold r.

For preferential attachment models, sometimes the steps in this path are investigated (eg. [10, 63, 184]) but often just assumed and crossing porous academic and funding boundaries between statistics, operations research, computer science, applied probability and network science can produce strange reactions from different communities with different cultures.

5.7 Measuring Extremal Dependence with the EDM

So what about data formats? A common data format for directed networks (eg. [112, page 17] is simply an edge list consisting of two columns of nodes where on each line the first node1 is the originating node and the second node2 on the same line is the receiving node of the directed edge. The is represented by the ordered pair (node1,node2). Sometimes each line is accompanied by a time stamp representing when the link was created. For instance on the Stanford University repository SNAP [118, 123], you find the citation network data *cit-HepPh* which is the Arxiv High Energy Physics paper citation network from a limited time span early in Arxiv's life. The first few lines looks like this:

```
# Directed graph
# Paper citation network of Arxiv High Energy Physics
# Nodes: 34546 Edges: 421578
# FromNodeId      ToNodeId
9907233           9301253
9907233           9504304
9907233           9505235
9907233           9506257
9907233           9606402
```

Papers are coded with identifier numbers and the paper in the first column referenced the paper in the second column thus forming a directed edge of a graph.

Software such as *igraph* [29] takes the edge list as input and outputs the degrees $\{(D_w^{in}(n), D_w^{out}(n)), w \in V(n)\}$. Typically it is the degree list that is used as "the data".

5.7.5.4 A Controlled Experiment: Simulation of the Network

The best way to run a controlled experiment where you know the parameters and there is no model error is to simu late a model. Using software by Daniel Cirkovic of Texas A&M prepared for [23], we simulated the model of network growth with preferential attachment and reciprocity using the following parameters:

Control?

α	β	γ	$\delta_{in} = \delta_{out}$	ρ	n
0.2	0.7	0.1	0.25	0.2	100,000

Based on thi's input parameter vector, using formulas from Sect. 5.7.5.2, some arithmetic produces the following values of the derived parameters:

D_0	λ_1	m	θ_0	$\iota_{in} = \iota_{out} = \iota$
0.1252	1.02693	0.7933156	0.5576263	1.285383

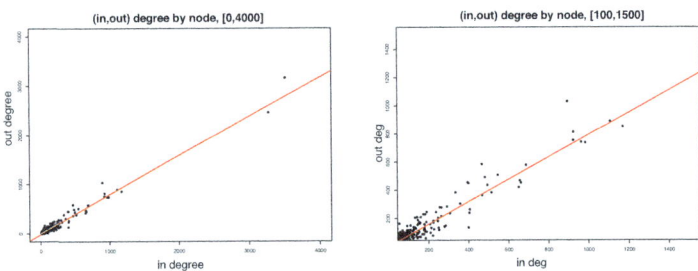

Fig. 5.33 Left: scatterplot of in- and out-degrees for first 4000 nodes. Right: scatterplot of degrees for nodes 100 to 1500. In both cases, the red line is at slope $m = 0.79$

After 100,000 steps we expect roughly $(\alpha + \gamma = 0.3) \times 100{,}000 = 30{,}000$ nodes to be created and indeed the simulation produced 30,082 nodes. With $n = 10^5$, we calculated $\{(D_w^{in}(n), D_w^{out}(n)), w = 1, \ldots, 30{,}082\}$ and this is displayed in the two panels of Fig. 5.33. With 30,082 points care must be taken to produce informative scatterplots. One potential problem is that plentiful points relatively near the origin mush together and hide any revealing structure and a second potential problem is that accommodating extreme points distorts the plot. The maximum in-degree is 14,977 and the max out-degree is 11,901 and these are omitted from the plotting to avoid distortion. Both plots seem to show points adhering to a neighborhood of the line $y = mx$ where $m = 0.79$.

It is difficult to confirm this visual assessment using the diamond plot and histogram in Fig. 5.34 where the left plot for $k = 100$ produces an uncomfortably broad interval $[0.245, 0.758]$ making it difficult to guess $S(\cdot)$ concentrates at a point. However the right panel histogram has a mode at a value plausibly close to the correct $\theta_0 \approx 0.557$ which is encouraging. Clearly, for the non-clairvoyant, additional tools that help to distinguish between strong dependence and full dependence would be welcome.

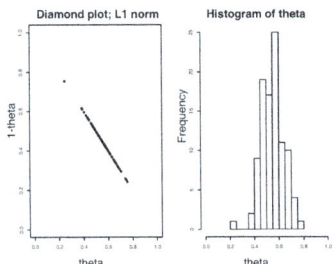

Fig. 5.34 Diamond plot: (in,out)-degree, $k = 100$

Another responsibility of heavy tail statistical analysis is to estimate tail indices. The panels in Fig. 5.35 show the qqest plots [156, page 367] for both in-degree (left) and out-degree (right) as a function of k, the number of upper order statistics

5.7 Measuring Extremal Dependence with the EDM

Table 5.5 Approximately normal statistic for several thresholds

k	25	50	75	100	200	300
t	-1.069014	-2.450439	-4.040233	-5.164809	-10.07177	-14.71127
$P[|Z|>|t|]$	0.2850635	0.0142682	5.339802e–05	2.40685e–07	7.363732e–24	5.457725e–49

used in estimation. Both commendably show the plots close to $1.2 \approx \iota = \iota_{\text{in}} = \iota_{\text{out}}$ over a range of k-values. For integer valued data, the Hill plot or the altHill plot [61] and [156, page 364] are often volatile and hard to interpret and frequently techniques based on the QQ-estimator produce more stable and therefore more easily interpretable plots.

Now what about the test statistic for full dependence discussed in Sect. 5.7.3.4 which modifies the Θ's and converts the angular measure from one that concentrates at a point θ_0 to a uniform distribution on $[\theta_0, \theta_0 + 1/8]$. This transforms Θ's to $\tilde{\Theta}$'s. In (5.57) we have $a = \theta_0 = 0.5576$, $b = \theta_0 + 1/8 = 0.68262$. Sort the $\{R_i\}$ in decreasing order with $R_{(k)}$ as the kth largest. Then thin the $\tilde{\Theta}$'s defined in (5.60), retaining only those $\tilde{\Theta}_i$ corresponding to $R_i \geq R_{(k)}$. The retained modified angles are then put into the g function to get the test statistic

$$T := \sqrt{k}\sqrt{48}\left(\frac{1}{k}\sum_{i=1}^{n} g(\tilde{\Theta}_i) 1_{[R_i \geq R_{(k)}]} - 3/4\right)$$

where the sum has k non-zero summands. We mimic Table 5.2, page 209, to obtain the preceding Table 5.5. Again, sample values of T at various thresholds of k are indicated by t along with $P[|Z| > |t|]$ for a standard normal variable Z.

Except for rather small choices of k, the test ineffectively indicates it is unlikely to see something bigger in absolute value than what you saw. So we conclude, the asymptotic normality test designed for randomly sampled data is not brilliant for this network data unless the goal is to reject a hypothesis of iid data from a distribution with asymptotic full dependence. Techniques depending on slopes such as Hill and QQ plotting work acceptably and diamond plotting works passably but do not use

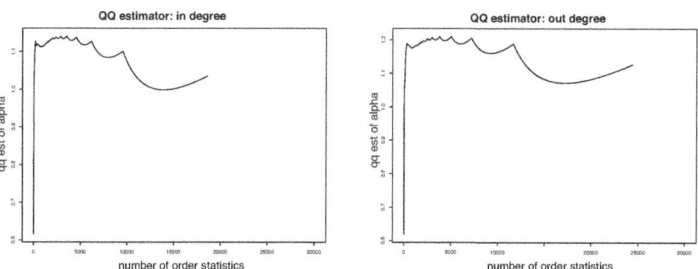

Fig. 5.35 QQ-estimates of ι_{in} (left) and ι_{out} (right) as a function of k, the number of upper order statistics used in estimation. In both cases a reasonable estimate is approximately 1.2

EDM-asymptotic normality with degree data to discern full dependence. (Sigh! You read it here first.) Other problems include having to take a double limit (first $n \to \infty$ to get a limiting MRV frequency distribution and then $t \to \infty$ to obtain the limit measure of regular variation) and the lack of guarantees that for fixed n, we are close to an asymptotic model.

Skepticism hinted at in Sect. 5.7.4 is warranted for network data.

5.8 Problems

5.1 [122] Suppose $\mathbb{S} = [0, 1]$ and consider the measures in $\mathbb{M}([0, 1])$,

$$m_n = (1 - \frac{1}{n})\epsilon_0 + \frac{1}{n}\epsilon_1, \, n \geq 1; \quad m_0 = \epsilon_0.$$

Check that $m_n \to m_0$ in $\mathbb{M}([0, 1])$ but the supports fail to converge since $\mathrm{supp}(m_n) = \{0, 1\}$ but $\mathrm{supp}(m_0) = \{0\}$.

5.2 Suppose $X_1, \ldots, X_n$ is a sample of $\mathbb{R}_+$-valued random variables from a distribution with a regularly varying tail. Let $X_{(k)}$ be the kth largest in the sample. Show $X_{(k)} \xrightarrow{P} \infty$ requires $k/n \to 0$. (Hint: One way to do this is to start by proving the assertion assuming the underlying distribution is exponential using the Renyi representation (eg. [76, 146], [152, p. 439ff], [156, p. 114]) and then transforming to the distribution of X_1.)

5.3 (The Rast Operator) [122]. Consider $\aleph_+ = \{x \in \mathbb{R}_+^p : \|x\| = 1\}$. Define the box $[\mathbf{0}, \mathbf{0} + \frac{1}{m}\mathbf{1})$ as the box with sides length $1/m$ with lower left corner $\mathbf{0}$ so that for $x \in \mathbb{R}_+^p$, $[x, x + \frac{1}{m}\mathbf{1}) = x + [\mathbf{0}, \mathbf{0} + \frac{1}{m}\mathbf{1})$. Let

$$G_m = \left\{ x = (x_1, \ldots, x_p) \in \mathbb{R}_+^p : x_i \in \{0, \frac{1}{m}, \ldots, \frac{m-1}{m}\} \right\} = \{0, \frac{1}{m}, \ldots, \frac{m-1}{m}\}^p$$

be *gridpoints*. For a compact $K \in \mathcal{K}(\aleph_+)$ define $\mathrm{Rast}(K)$, the rasterized version of K, as the closure of

$$\bigcup_{x \in G_m} [x, x + \frac{1}{m}\mathbf{1}) : [x, x + \frac{1}{m}\mathbf{1}) \cap K \neq \emptyset\}$$

the boxes that intersect K. Assume K satisfies

$$\forall x \in G_m : \quad K \cap [x, x + \frac{1}{m}\mathbf{1}] \neq \emptyset \text{ implies } K \cap (x, x + \frac{1}{m}\mathbf{1}) \neq \emptyset. \qquad (5.72)$$

(Condition (5.72) says that if K intersects a cell, it does not do so only on the boundary of the cell. Prove $\mathrm{Rast}(\cdot, m)$ is continuous at K; that is, if $K_n \to K$ in the Hausdorff metric, then also for fixed m, $\text{Rast}(K_n, m) \to \text{Rast}(K, m)$. (Hint: Condition 1 and 2, page 163 could be helpful.)

5.4 (Asymptotic Full Dependence) [42] Suppose $Z_1 \sim \text{Pareto}(1.5)$ and $Z_2 \sim \text{Pareto}(2.5)$ and independent of each other. Let B_1, B_2 be iid Bernoulli (0.5) random variables also independent of Z_1 and Z_2. Now define the vector $X = (X_1, X_2)$ as

$$X_1 = B_1 Z_1 + (1 - B_1) Z_2,$$
$$X_2 = B_1 Z_1 + B_2(1 - B_1)(1.5 Z_2) + (1 - B_2)(1 - B_1)(0.5 Z_2).$$

Verify:

1. By construction

$$X = \begin{cases} (Z_1, Z_1), & \text{with probability } P[B_1 = 1] = \frac{1}{2}, \\ (Z_2, 1.5 Z_2), & \text{with probability } P[B_1 = 0, B_2 = 1] = \frac{1}{4}, \\ (Z_2, 0.5 Z_2), & \text{with probability } P[B_1 = 0, B_2 = 0] = \frac{1}{4}, \end{cases}$$

and $X = (X_1, X_2)$ lies on the line $y = x$ with probability 0.5. With probability 0.25 each it is either on the line $y = 0.5x$ or on $y = 1.5x$.

2. With these definitions, $P[X \in \cdot] \in \text{MRV}(\alpha = 1.5, b(t) = t^{2/3}, \eta(\cdot), \mathbb{R}_+^2 \setminus \{\mathbf{0}\})$ and η concentrates on the diagonal $[\text{diag}] := \{(x, y) \in \mathbb{R}_+^2 : y = x\}$. So $P[X \in \cdot]$ has asymptotic full dependence.

3. Consider $\mathbb{R}_+^2 \setminus [\text{diag}]$. What is $\aleph_{[\text{diag}]}$? What are the generalized polar coordinates? Show there is also regular variation with tail parameter $\alpha_0 = 2.5$ on $\mathbb{R}_+^2 \setminus [\text{diag}]$ and find the angular measure of the generalized polar coordinates.

5.5 [84] and Balkema and Nolde [2] Suppose $\{Z_i := (X_i, Y_i), i \geq 1\}$ are iid random vectors from a continuous distribution. To improve context, suppose

$$P[Z_1 \in \cdot] \in \text{MRV}(\alpha, b(t), \eta(\cdot), \mathbb{R}_+^2 \setminus \{\mathbf{0}_2\}).$$

For $n \geq 1$, define $M_n = (\vee_{i=1}^n X_i, \vee_{i=1}^n Y_i)$, and let p_n be the probability of a simultaneous record in both coordinates by index n; that is,

$$p_n = P[M_n = Z_i, \text{ for some } i = 1, \ldots, n].$$

Then $p_n \to 0$ as $n \to \infty$ iff the components of Z_1 have asymptotic independence. What is the behavior of p_n under the assumption of asymptotic full dependence?

5.6 (EDM and Product Moments) [121] Suppose $X = (X(1), X(2))$ is an $\mathbb{R}_+^2$-valued random vector and

$$P[X \in \cdot] \in MRV(\alpha, b(t), \eta(\cdot), \mathbb{R}_+^2 \setminus \{\mathbf{0}_2\}).$$

Show

$$\text{EDM}(X) = \lim_{x \to \infty} E\left(\frac{X(1)X(2)}{(X(1)+X(2))^2} \bigg| X(1)+X(2) > x\right).$$

Consequently, given iid observations $X_1, \ldots, X_n$ with polar coordinate versions $\{(R_i, \Theta_i), 1 \le i \le n\}$, where $\Theta_i = (\Theta_i, 1 - \Theta_i)$, rederive the EDM estimator in (5.46).

5.7 [121] Suppose we have two norms $\|x\|_i$, $i = 1, 2$ on $\mathbb{R}_+^2$ and $\aleph_i = \{x \in \mathbb{R}_+^2 : \|x\|_i = 1\}$ is the unit *sphere* with respect to $\|x\|_i$. Suppose

$$P[X \in \cdot] \in MRV(\alpha, b(t), \eta(\cdot), \mathbb{R}_+^2 \setminus \{\mathbf{0}_2\}).$$

Depending on which norm is specified we get two angular measures $S_i(\cdot)$, $i = 1, 2$ on $\aleph_i$. This gives two different possible definitions of EDM, namely for $i = 1, 2$

$$\text{EDM}_i(X) = \int_{\aleph_i} a_1 a_2 S_i(d\mathbf{a}).$$

Show $\text{EDM}_1(X)$, $\text{EDM}_2(X)$ are equivalent in the sense that there exist two constants m_l, m_r (possibly depending on α) such that

$$m_l \text{EDM}_1(X) \le \text{EDM}_2(X) \le m_r \text{EDM}_1(X)$$

for any X satisfying $P[X \in \cdot]$ is regularly varying with index α.

5.8 For $0 < a < b < 1$, suppose the angular measure on $[0, 1]$ concentrates on the endpoints of $[a, b]$ so that

$$S = p_0 \epsilon_{\{a\}} + p_1 \epsilon_{\{b\}}$$

with $p_0 + p_1 = 1$. Compute μ_g, σ_g^2 and verify $\sigma_g = 0$ if $g(a) = g(b)$.

5.9 For $0 < a < b < 1$, suppose the angular measure S on $[0, 1]$ concentrates on $[a, b]$ and g is constant on $[a, b]$. Verify $\sigma_g = 0$.

5.10 For $0 < a < b < 1$, suppose the angular measure $S(\cdot)$ on $[0, 1]$ concentrates on $[a, b]$ and $S(\cdot)$ is uniform on $[a, b]$.

1. If g gives the EDM so $g(\theta) = 4\theta(1 - \theta)$, $0 \le \theta \le 1$, compute μ_g and σ_g^2.
2. If g is given by (5.57) and Fig. 5.27, compute μ_g and σ_g^2. Verify $\mu_g = 3/4$ and $\sigma_g^2 = 1/48$.

5.11 Assume (5.66) to prove (5.71).

Appendix A
A Crash Course on Regularly Varying Functions

This is an overview of the theory of regularly varying functions designed to rapidly get the reader functional. Further references and more detail can be found in [49, 172, 15, 81, 150, 51]. This account was shamelessly cribbed from [156] which was mostly cribbed from [150] which is now called [157] and the latter relied heavily on [49].

A.1 Preliminaries from Analysis

A.1.1 Uniform Convergence

If $\{f_n, n \geq 0\}$ are real valued functions on $\mathbb{R}$ (or, in fact, any metric space) then f_n converges uniformly on $A \subset \mathbb{R}$ to f_0 if

$$\sup_{x \in A} |f_0(x) - f_n(x)| \to 0 \tag{A.1}$$

as $n \to \infty$. The definition would still make sense if the range of $f_n, n \geq 0$ were a metric space but then $|f_0(x) - f_n(x)|$ would need to be replaced by $d(f_0(x), f_n(x))$, where $d(\cdot, \cdot)$ is the metric. For functions on $\mathbb{R}$, the phrase *local uniform convergence* means that (A.1) holds for any compact interval A.

A very useful fact is that monotone functions converging pointwise to a continuous limit converge locally uniformly. (See [150, page 1] for additional material.)

Proposition A.1 *Suppose $U_n, n \geq 0$ are nondecreasing, real valued functions on $\mathbb{R}$, and that U_0 is continuous. If for all x,*

$$U_n(x) \to U_0(x), \quad (n \to \infty)$$

then $U_n \to U_0$ locally uniformly; i.e. for any $a < b$

$$\sup_{x \in [a,b]} |U_n(x) - U_0(x)| \to 0.$$

Proof One proof of this fact is outlined as follows: If U_0 is continuous on $[a, b]$, then it is uniformly continuous. From the uniform continuity, for any x, there is an interval-neighborhood O_x on which $U_0(\cdot)$ oscillates by less than a given ϵ. This gives an open cover of $[a, b]$. Compactness of $[a, b]$ allows us to prune $\{O_x, x \in [a, b]\}$ to obtain a finite subcover $\{(a_i, b_i), i = 1, \ldots, K\}$. Using this finite collection and the monotonicity of the functions leads to the result: Given $\epsilon > 0$, there exists some large N such that if $n \geq N$ then

$$\max_{1 \leq i \leq K} \left(|U_n(a_i) - U_0(a_i)| \bigvee |U_n(b_i) - U_0(b_i)| \right) < \epsilon, \tag{A.2}$$

(by pointwise convergence). Observe that

$$\sup_{x \in [a,b]} |U_n(x) - U_0(x)| \leq \max_{1 \leq i \leq K} \sup_{x \in [a_i,b_i]} |U_n(x) - U_0(x)|. \tag{A.3}$$

For any $x \in [a_i, b_i]$, we have by monotonicity

$$U_n(x) - U_0(x) \leq U_n(b_i) - U_0(a_i)$$
$$\leq U_0(b_i) + \epsilon - U_0(a_i), \quad \text{(by (A.2))}$$
$$\leq 2\epsilon,$$

with a similar lower bound. This is true for all i and hence we get uniform convergence on $[a, b]$.

$\square$

A.1.2 Inverses of Monotone Functions

Suppose $H : \mathbb{R} \mapsto (a, b)$ is a nondecreasing function on $\mathbb{R}$ with range (a, b), $-\infty \leq a < b \leq \infty$. With the convention that the infimum of an empty set is $+\infty$, we define the (left continuous) inverse $H^{\leftarrow} : (a, b) \mapsto \mathbb{R}$ of H as

$$H^{\leftarrow}(y) = \inf\{s : H(s) \geq y\}.$$

See Fig. A.1.

A A Crash Course on Regularly Varying Functions

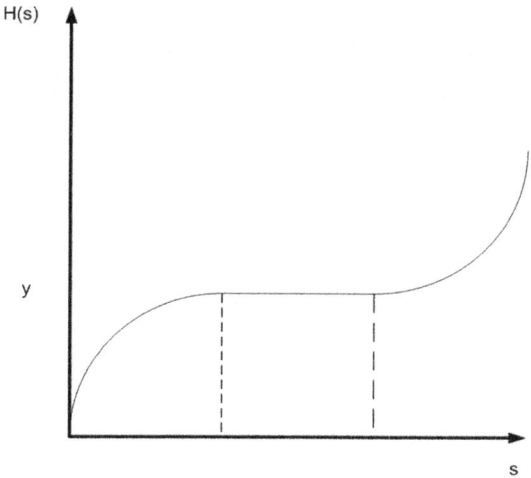

Fig. A.1 The inverse at y is the foot of the left dotted perpendicular

In case the function H is right continuous we have the following desirable properties:

$$A(y) := \{s : H(s) \geq y\} \text{ is closed}, \tag{A.4}$$

$$H(H^{\leftarrow}(y)) \geq y \tag{A.5}$$

$$H^{\leftarrow}(y) \leq t \text{ iff } y \leq H(t). \tag{A.6}$$

For (A.4), observe that if $s_n \in A(y)$ and $s_n \downarrow s$, then $y \leq H(s_n) \downarrow H(s)$ so $H(s) \geq y$ and $s \in A(y)$. If $s_n \uparrow s$ and $s_n \in A(y)$, then $y \leq H(s_n) \uparrow H(s-) \leq H(s)$ and $H(s) \geq y$ so $s \in A(y)$ again and $A(y)$ is closed. Since $A(y)$ is closed, $\inf A(y) \in A(y)$; that is, $H^{\leftarrow}(y) \in A(y)$ which means $H(H^{\leftarrow}(y)) \geq y$. This gives (A.5). Lastly, (A.6) follows from the definition of $H^{\leftarrow}$.

A.1.3 Convergence of Monotone Functions

For any function H denote

$$C(H) = \{x \in \mathbb{R} : H \text{ is finite and continuous at } x\}.$$

A sequence $\{H_n, n \geq 0\}$ of nondecreasing functions on $\mathbb{R}$ converges weakly to H_0 if as $n \to \infty$ we have

$$H_n(x) \to H_0(x),$$

for all $x \in C(H_0)$. We will denote this by $H_n \to H_0$. No other form of convergence for monotone functions will be relevant. If $F_n, n \geq 0$ are probability distributions on $\mathbb{R}$, then a myriad of names give equivalent concepts: complete convergence, vague convergence, weak* convergence, narrow convergence. If $X_n, n \geq 0$ are random variables and X_n has distribution function $F_n, n \geq 0$, then $X_n \Rightarrow X_0$ means $F_n \to F_0$. For the proof of the following, see [14], [150, page 5], and [153, page 259].

Proposition A.2 *If $H_n, n \geq 0$ are nondecreasing functions on $\mathbb{R}$ with range (a, b) and $H_n \to H_0$, then $H_n^{\leftarrow} \to H_0^{\leftarrow}$ in the sense that for $t \in (a, b) \cap C(H_0^{\leftarrow})$*

$$H_n^{\leftarrow}(t) \to H_0^{\leftarrow}(t).$$

A.1.4 Cauchy's Functional Equation

Let $k(x), x \in \mathbb{R}$ be a function which satisfies

$$k(x + y) = k(x) + k(y), x, y \in R.$$

If k is measurable and bounded on a set of positive measure, then $k(x) = cx$ for some $c \in \mathbb{R}$. (See [172], [15, page 4].)

A.2 Regular Variation: Definition and First Properties

The theory of regularly varying functions is an essential analytical tool for dealing with heavy tails, long range dependence and domains of attraction. Roughly speaking, *regularly varying functions* are those functions which behave asymptotically like power functions. We will deal currently only with real functions of a real variable.

Definition A.1 A measurable function $U : \mathbb{R}_+ \mapsto \mathbb{R}_+$ is regularly varying at ∞ with index $\rho \in \mathbb{R}$ (written $U \in RV_\rho$) if for $x > 0$

$$\lim_{t \to \infty} \frac{U(tx)}{U(t)} = x^\rho.$$

We call ρ the *exponent of variation*.

If $\rho = 0$ we call U *slowly varying*. Slowly varying functions are generically denoted by $L(x)$. If $U \in RV_\rho$, then $U(x)/x^\rho \in RV_0$ and setting $L(x) = U(x)/x^\rho$ we see it is always possible to represent a ρ-varying function as $x^\rho L(x)$.

A A Crash Course on Regularly Varying Functions

Example A.1 The canonical ρ-varying function is x^ρ. The functions $\log(1 + x)$, $\log\log(e + x)$ are slowly varying, as is $\exp\{(\log x)^\alpha\}$, $0 < \alpha < 1$. Any function U such that $\lim_{x\to\infty} U(x) =: U(\infty)$ exists positive and finite is slowly varying. The following functions are not regularly varying: e^x, $\sin(x + 2)$. Note $[\log x]$ is slowly varying, but $\exp\{[\log x]\}$ is not regularly varying.

In probability applications we are concerned with distributions whose tails are regularly varying. Examples are

$$1 - F(x) = x^{-\alpha}, \quad x \geq 1, \quad \alpha > 0,$$

and the extreme value distribution

$$\Phi_\alpha(x) = \exp\{-x^{-\alpha}\}, \quad x \geq 0.$$

$\Phi_\alpha(x)$ has the property

$$1 - \Phi_\alpha(x) \sim x^{-\alpha} \text{ as } x \to \infty.$$

A stable law with index α, $0 < \alpha < 2$ has the property

$$1 - G(x) \sim c x^{-\alpha}, \quad x \to \infty, \quad c > 0.$$

The Cauchy density $f(x) = (\pi(1 + x^2))^{-1}$ has a distribution function F with the property

$$1 - F(x) \sim (\pi x)^{-1}.$$

If $N(x)$ is the standard normal distribution function then $1 - N(x)$ is not regularly varying nor is the tail of the Gumbel extreme value distribution $1 - \exp\{-e^{-x}\}$.

The definition of regular variation can be weakened slightly (cf [49, 76, 150]).

Proposition A.3

(i) *A measurable function $U : \mathbb{R}_+ \mapsto \mathbb{R}_+$ varies regularly if there exists a function h such that for all $x > 0$*

$$\lim_{t\to\infty} U(tx)/U(t) = h(x).$$

In this case $h(x) = x^\rho$ for some $\rho \in \mathbb{R}$ and $U \in RV_\rho$.

(ii) *A monotone function $U : \mathbb{R}_+ \mapsto \mathbb{R}_+$ varies regularly provided there are two sequences $\{\lambda_n\}, \{b_n\}$ of positive numbers satisfying*

$$b_n \to \infty, \quad \lambda_n \sim \lambda_{n+1}, \quad n \to \infty, \quad (A.7)$$

and for all $x > 0$

$$\lim_{n\to\infty} \lambda_n U(b_n x) =: \chi(x) \text{ exists positive and finite.} \tag{A.8}$$

In this case $\chi(x)/\chi(1) = x^\rho$ and $U \in RV_\rho$ for some $\rho \in \mathbb{R}$.

We frequently refer to (A.8) as the *sequential form of regular variation*. For probability purposes, it is most useful. Typically U is a distribution tail, $\lambda_n = n$ and b_n is a distribution quantile.

Proof

(i) The function h is measurable since it is a limit of a family of measurable functions. Then for $x > 0$, $y > 0$

$$\frac{U(txy)}{U(t)} = \frac{U(txy)}{U(tx)} \cdot \frac{U(tx)}{U(t)}$$

and letting $t \to \infty$ gives

$$h(xy) = h(y)h(x).$$

So h satisfies the Hamel equation, which by change of variable can be converted to the Cauchy equation. Therefore, the form of h is $h(x) = x^\rho$ for some $\rho \in \mathbb{R}$.

(ii) For concreteness assume U is nondecreasing. Assume (A.7) and (A.8) and we show regular variation. Since $b_n \to \infty$, for each t there is a finite $n(t)$ defined by

$$n(t) = \inf\{m : b_{m+1} > t\}$$

so that

$$b_{n(t)} \leq t < b_{n(t)+1}.$$

Therefore by monotonicity for $x > 0$

$$\left(\frac{\lambda_{n(t)+1}}{\lambda_{n(t)}}\right)\left(\frac{\lambda_{n(t)}U(b_{n(t)}x)}{\lambda_{n(t)+1}U(b_{n(t)+1})}\right)$$
$$\leq \frac{U(tx)}{U(t)} \leq \left(\frac{\lambda_{n(t)}}{\lambda_{n(t)+1}}\right)\left(\frac{\lambda_{n(t)+1}U(b_{n(t)+1}x)}{\lambda_{n(t)}U(b_{n(t)})}\right).$$

Now let $t \to \infty$ and use (A.7) and (A.8) to get $\lim_{t\to\infty} \frac{U(tx)}{U(t)} = 1\frac{\chi(x)}{\chi(1)}$. Regular variation follows from part (i).

$\square$

Remark A.1 Proposition A.3 (ii) remains true if we only assume (A.8) holds on a dense set. This is relevant to the case where U is nondecreasing and $\lambda_n U(b_n x)$ converges weakly.

A.3 A Maximal Domain of Attraction

Suppose $\{X_n, n \geq 1\}$ are iid with common distribution function $F(x)$. The extreme is

$$M_n = \bigvee_{i=1}^{n} X_i = \max\{X_1, \ldots, X_n\}.$$

One of the extreme value distributions is

$$\Phi_\alpha(x) := \exp\{-x^{-\alpha}\}, \quad x > 0, \ \alpha > 0.$$

What are conditions on F, called *domain of attraction conditions*, so that there exists $b_n > 0$ such that

$$P[b_n^{-1} M_n \leq x] = F^n(b_n x) \to \Phi_\alpha(x) \tag{A.9}$$

weakly. How do you characterize the normalization sequence $\{b_n\}$?

Set $x_0 = \sup\{x : F(x) < 1\}$ which is called the right end point of F. We first check (A.9) implies $x_0 = \infty$. Otherwise if $x_0 < \infty$ we get from (A.9) that for $x > 0, b_n x \to x_0$; i.e. $b_n \to x_0 x^{-1}$. Since $x > 0$ is arbitrary we get $b_n \to 0$ whence $x_0 = 0$. But then for $x > 0$, $F^n(b_n x) = 1$, which violates (A.9). Hence $x_0 = \infty$.

Furthermore $b_n \to \infty$ since otherwise on a subsequence $n', b_{n'} \leq K$ for some $K < \infty$. Then, since $F(K) < 1$,

$$0 < \Phi_\alpha(1) = \lim_{n' \to \infty} F^{n'}(b_{n'}) \leq \lim_{n' \to \infty} F^{n'}(K) = 0,$$

which is a contradiction.

In (A.9), take logarithms to get for $x > 0$, $\lim_{n \to \infty} n(-\log F(b_n x)) = x^{-\alpha}$. Now use the relation $-\log(1-z) \sim z$ as $z \to 0$ and (A.9) is equivalent to

$$\lim_{n \to \infty} n(1 - F(b_n x)) = x^{-\alpha}, \quad x > 0. \tag{A.10}$$

From (A.10) and Proposition A.3 we get

$$1 - F(x) \sim x^{-\alpha} L(x), \quad x \to \infty, \tag{A.11}$$

for some $\alpha > 0$. To characterize $\{b_n\}$ set $U(x) = 1/(1 - F(x))$ and (A.10) is the same as

$$U(b_n x)/n \to x^\alpha, \quad x > 0,$$

and inverting, we find via Proposition A.2 that

$$\frac{U^\leftarrow(ny)}{b_n} \to y^{1/\alpha}, \quad y > 0. \tag{A.12}$$

So $U^\leftarrow(n) = (1/(1-F))^\leftarrow(n) \sim b_n$ and this determines b_n by the convergence to types theorem. See [76, 153, 150].

Conversely if (A.11) holds, define $b_n = U^\leftarrow(n)$ as previously. Then

$$\lim_{n \to \infty} \frac{1 - F(b_n x)}{1 - F(b_n)} = x^{-\alpha}$$

and we recover (A.10) provided $1 - F(b_n) \sim n^{-1}$ or what is the same provided $U(b_n) \sim n$ i.e., $U(U^\leftarrow(n)) \sim n$. Recall from (A.6), that $z < U^\leftarrow(n)$ iff $U(z) < n$ and setting $z = U^\leftarrow(n)(1 - \varepsilon)$ and then $z = U^\leftarrow(n)(1 + \varepsilon)$ we get

$$\frac{U(U^\leftarrow(n))}{U(U^\leftarrow(n)(1+\varepsilon))} \leq \frac{U(U^\leftarrow(n))}{n} \leq \frac{U(U^\leftarrow(n))}{U(U^\leftarrow(n)(1-\varepsilon))}.$$

Let $n \to \infty$, remembering $U = 1/(1 - F) \in RV_\alpha$. Then

$$(1+\varepsilon)^{-\alpha} \leq \liminf_{n \to \infty} n^{-1} U(U^\leftarrow(n)) \leq \limsup_{n \to \infty} U(U^\leftarrow(n)) \leq (1-\varepsilon)^{-\alpha}$$

and since $\varepsilon > 0$ is arbitrary the desired result follows.

A.4 Regular Variation: Deeper Results; Karamata's Theorem

There are several deeper results which give the theory power and utility: uniform convergence, Karamata's theorem which says a regularly varying function integrates the way you expect a power function to integrate, and finally the Karamata representation theorem.

A.4.1 Uniform Convergence

The first useful result is the uniform convergence theorem.

Proposition A.4 *If $U \in RV_\rho$ for $\rho \in \mathbb{R}$, then*

$$\lim_{t \to \infty} U(tx)/U(t) = x^\rho$$

locally uniformly in x on $(0, \infty)$. If $\rho < 0$, then uniform convergence holds on intervals of the form (b, ∞), $b > 0$. If $\rho > 0$ uniform convergence holds on intervals $(0, b]$ provided U is bounded on $(0, b]$ for all $b > 0$.

If U is monotone the result already follows from the discussion in Sect. A.1.1, since we have a family of monotone functions converging to a continuous limit. For detailed discussion see [172, 49, 81, 15].

A.4.2 Integration and Karamata's Theorem

The next set of results examines the integral properties of regularly varying functions [108, 109, 49, 172, 15]. For purposes of integration, a ρ-varying function behaves roughly like x^ρ. We assume all functions are locally integrable and since we are interested in behavior at ∞ we assume integrability on intervals including 0 as well.

Theorem A.1 (Karamata's Theorem)

(a) *Suppose $\rho \geq -1$ and $U \in RV_\rho$. Then $\int_0^x U(t)dt \in RV_{\rho+1}$ and*

$$\lim_{x \to \infty} \frac{xU(x)}{\int_0^x U(t)dt} = \rho + 1. \tag{A.13}$$

If $\rho < -1$ (or if $\rho = -1$ and $\int_x^\infty U(s)ds < \infty$) then $U \in RV_\rho$ implies $\int_x^\infty U(t)dt$ is finite, $\int_x^\infty U(t)dt \in RV_{\rho+1}$ and

$$\lim_{x \to \infty} \frac{xU(x)}{\int_x^\infty U(t)dt} = -\rho - 1. \tag{A.14}$$

(b) *If U satisfies*

$$\lim_{x \to \infty} \frac{xU(x)}{\int_0^x U(t)dt} = \lambda \in (0, \infty) \tag{A.15}$$

then $U \in RV_{\lambda-1}$. If $\int_x^\infty U(t)dt < \infty$ and

$$\lim_{x \to \infty} \frac{xU(x)}{\int_x^\infty U(t)dt} = \lambda \in (0, \infty) \tag{A.16}$$

then $U \in RV_{-\lambda-1}$.

What Theorem A.1 emphasizes is that for the purposes of integration, the slowly varying function can be passed from inside to outside the integral. For example the way to remember and interpret (A.13) is to write $U(x) = x^\rho L(x)$ and then observe

$$\int_0^x U(t)dt = \int_0^x t^\rho L(t)dt$$

and pass the $L(t)$ in the integrand outside as a factor $L(x)$ to get

$$\sim L(x)\int_0^x t^\rho dt = L(x)x^{\rho+1}/(\rho+1)$$
$$= xx^\rho L(x)/(\rho+1) = xU(x)/(\rho+1),$$

which is equivalent to the assertion (A.13).

Proof

(a) For certain values of ρ, uniform convergence suffices after writing, for instance,

$$\frac{\int_0^x U(s)ds}{xU(x)} = \int_0^1 \frac{U(sx)}{U(x)} ds.$$

If we wish to proceed, using elemmaentary concepts, consider the following approach, which follows [49].

If $\rho > -1$ we show $\int_0^\infty U(t)dt = \infty$. From $U \in RV_\rho$ we have

$$\lim_{s \to \infty} U(2s)/U(s) = 2^\rho > 2^{-1}$$

since $\rho > -1$. Therefore there exists s_0 such that $s > s_0$ necessitates $U(2s) > 2^{-1}U(s)$. For n with $2^n > s_0$ we have

$$\int_{2^{n+1}}^{2^{n+2}} U(s)ds = 2\int_{2^n}^{2^{n+1}} U(2s)ds > \int_{2^n}^{2^{n+1}} U(s)ds$$

and so setting $n_0 = \inf\{n : 2^n > s_0\}$ gives

$$\int_{s_0}^\infty U(s)ds \geq \sum_{n:2^n > s_0} \int_{2^{n+1}}^{2^{n+2}} U(s)ds > \sum_{n \geq n_0} \int_{2^{n_0+1}}^{2^{n_0+2}} U(s)ds = \infty.$$

Thus for $\rho > -1, x > 0$, and any $N < \infty$ we have

$$\int_0^t U(sx)ds \sim \int_N^t U(sx)ds, t \to \infty,$$

since $U(sx)$ is a ρ-varying function of s. For fixed x and given ε, there exists N such that for $s > N$

$$(1-\varepsilon)x^\rho U(s) \leq U(sx) \leq (1+\varepsilon)x^\rho U(s)$$

and thus

$$\limsup_{t\to\infty} \frac{\int_0^{tx} U(s)ds}{\int_0^t U(s)ds} = \limsup_{t\to\infty} \frac{x\int_0^t U(sx)ds}{\int_0^t U(s)ds}$$

$$= \limsup_{t\to\infty} \frac{x\int_N^t U(sx)ds}{\int_N^t U(s)ds}$$

$$\leq \limsup_{t\to\infty} x^{\rho+1}(1+\varepsilon)\frac{\int_N^t U(s)ds}{\int_N^t U(s)ds}$$

$$= (1+\varepsilon)x^{\rho+1}.$$

An analogous argument applies for lim inf and thus we have proved

$$\int_0^x U(s)ds \in RV_{\rho+1}$$

when $\rho > -1$.

In case $\rho = -1$ then either $\int_0^\infty U(s)ds < \infty$ in which case $\int_0^x U(s)ds \in RV_{-1+1} = RV_0$ or else $\int_0^\infty U(s)ds = \infty$ and the previous argument is applicable. So we have checked that for $\rho \geq -1$, $\int_0^x U(s)ds \in RV_{\rho+1}$.

We now focus on proving (A.13) when $U \in RV_\rho$, $\rho \geq -1$. Define the function

$$b(x) := xU(x)/\int_0^x U(t)dt, \qquad (A.17)$$

so that integrating $b(x)/x$ leads to the representations

$$\int_0^x U(s)ds = c\exp\left\{\int_1^x t^{-1}b(t)dt\right\}$$

$$U(x) = cx^{-1}b(x)\exp\left\{\int_1^x t^{-1}b(t)dt\right\}. \qquad (A.18)$$

We must show $b(x) \to \rho + 1$. Observe first that

$$\liminf_{x\to\infty} 1/b(x) = \liminf_{x\to\infty} \frac{\int_0^x U(t)dt}{xU(x)}$$

$$= \liminf_{x\to\infty} \int_0^1 \frac{U(sx)}{U(x)}ds.$$

Now make a change of variable $s = x^{-1}t$ and by Fatou's lemma this is

$$\geq \int_0^1 \liminf_{x\to\infty}(U(sx)/U(x))ds$$

$$= \int_0^1 s^\rho ds = \frac{1}{\rho+1}$$

and we conclude

$$\limsup_{x\to\infty} b(x) \leq \rho + 1. \qquad (A.19)$$

If $\rho = 1$ then $b(x) \to 0$ as desired, so now suppose $\rho > -1$.
We observe the following properties of $b(x)$:

(i) $b(x)$ is bounded on a semi-infinite neighborhood of ∞ (by (A.19)).
(ii) b is slowly varying since $xU(x) \in RV_{\rho+1}$ and $\int_0^x U(s)ds \in RV_{\rho+1}$.
(iii) We have

$$b(xt) - b(x) \to 0$$

as $x \to \infty$ and the convergence is uniformly bounded for t in finite intervals.

The last statement follows since by slow variation

$$\lim_{x\to\infty} (b(xt) - b(x))/b(x) = 0$$

and the denominator is ultimately bounded.
From (iii) and dominated convergence

$$\lim_{x\to\infty} \int_1^s t^{-1}(b(xt) - b(x))dt = 0$$

and the left side may be rewritten to obtain

$$\lim_{x\to\infty} \left\{ \int_1^s t^{-1} b(xt) dt - b(x) \log s \right\} = 0. \qquad (A.20)$$

From (A.18)

$$c \exp\left\{\int_1^x t^{-1}b(t)dt\right\} = \int_0^x U(s)ds \in RV_{\rho+1}$$

and from the regular variation property

$$(\rho+1)\log s = \lim_{x\to\infty} \log\left\{\frac{\int_0^{xs} U(t)dt}{\int_0^x U(t)dt}\right\}$$

$$= \lim_{x\to\infty} \int_x^{xs} t^{-1}b(t)dt = \lim_{x\to\infty} \int_1^s t^{-1}b(xt)dt$$

and combining this with (A.20) leads to the desired conclusion that $b(x) \to \rho+1$.

(b) We suppose (A.15) holds and check $U \in RV_{\lambda-1}$. Set

$$b(x) = xU(x)/\int_0^x U(t)dt$$

so that $b(x) \to \lambda$. From (A.18)

$$U(x) = cx^{-1}b(x)\exp\left\{\int_1^x t^{-1}b(t)dt\right\}$$

$$= cb(x)\exp\left\{\int_1^x t^{-1}(b(t)-1)dt\right\}$$

and since $b(t) - 1 \to \lambda - 1$, U satisfies the representation of a $(\lambda - 1)$-varying function. □

A.4.3 Karamata's Representation

Theorem A.1 leads in a straightforward way to what has been called the *Karamata representation* of a regularly varying function.

Corollary A.1 (The Karamata Representation)

(i) *The function L is slowly varying iff L can be represented as*

$$L(x) = c(x)\exp\left\{\int_1^x t^{-1}\varepsilon(t)dt\right\}, \quad x > 0, \tag{A.21}$$

where $c : \mathbb{R}_+ \mapsto \mathbb{R}_+$, $\varepsilon : \mathbb{R}_+ \mapsto \mathbb{R}_+$ and

$$\lim_{x \to \infty} c(x) = c \in (0, \infty), \tag{A.22}$$

$$\lim_{t \to \infty} \varepsilon(t) = 0. \tag{A.23}$$

(ii) *A function* $U : \mathbb{R}_+ \mapsto \mathbb{R}_+$ *is regularly varying with index* ρ *iff* U *has the representation*

$$U(x) = c(x) \exp\left\{ \int_1^x t^{-1} \rho(t) dt \right\} \tag{A.24}$$

where $c(\cdot)$ *satisfies* (A.22) *and* $\lim_{t \to \infty} \rho(t) = \rho$. *(This is obtained from (i) by writing* $U(x) = x^\rho L(x)$ *and using the representation for* L.*)*

Proof If L has a representation (A.21) then it must be slowly varying since for $x > 1$

$$\lim_{t \to \infty} L(tx)/L(t) = \lim_{t \to \infty} (c(tx)/c(t)) \exp\left\{ \int_t^{tx} s^{-1} \varepsilon(s) ds \right\}.$$

Given ε, there exists t_0 by (A.23) such that

$$-\varepsilon < \varepsilon(t) < \varepsilon, \quad t \geq t_0,$$

so that

$$-\varepsilon \log x = -\varepsilon \int_t^{tx} s^{-1} ds \leq \int_t^{tx} s^{-1} \varepsilon(s) ds \leq \varepsilon \int_t^{tx} s^{-1} ds = \varepsilon \log x.$$

Therefore $\lim_{t \to \infty} \int_t^{tx} s^{-1} \varepsilon(s) ds = 0$ and $\lim_{t \to \infty} L(tx)/L(t) = 1$.

Conversely suppose $L \in RV_0$. In a matter similar to (A.17), define

$$b(x) := xL(x) / \int_0^x L(s) ds$$

and by Karamata's theorem, $b(x) \to 1$, as $x \to \infty$. Note

$$L(x) = x^{-1} b(x) \int_0^x L(s) ds.$$

Set $\varepsilon(x) = b(x) - 1$ so $\varepsilon(x) \to 0$ and

$$\int_1^x t^{-1} \varepsilon(t) dt = \int_1^x \left(L(t) / \int_0^t L(s) ds \right) dt - \log x$$

$$= \int_1^x d\left(\log \int_0^t L(s) ds \right) - \log x$$

A A Crash Course on Regularly Varying Functions

$$= \log\left(x^{-1}\int_0^x L(s)ds \Big/ \int_0^1 L(s)ds\right)$$

whence

$$\exp\left\{\int_1^x t^{-1}\varepsilon(t)dt\right\} = x^{-1}\int_0^x L(s)ds \Big/ \int_0^1 L(s)ds$$

$$= L(x)\Big/\left(b(x)\int_0^1 L(s)ds\right), \tag{A.25}$$

and the representation follows with

$$c(x) = b(x)\int_0^1 L(s)ds.$$

Example A.2 The Cauchy density

$$F'(x) = \frac{1}{2\pi}\left(\frac{1}{1+x^2}\right), \quad x \in \mathbb{R},$$

satisfies

$$F'(x) \sim \frac{1}{2\pi}x^{-2}, \quad x \to \infty,$$

and hence

$$1 - F(x) \sim \frac{1}{2\pi}x^{-1}, \quad x \to \infty.$$

A.4.4 Differentiation

The previous results describe the asymptotic properties of the indefinite integral of a regularly varying function. We now describe what happens when a ρ-varying function is differentiated.

Proposition A.5 Suppose $U : \mathbb{R}_+ \mapsto \mathbb{R}_+$ is absolutely continuous with density u so that

$$U(x) = \int_0^x u(t)dt.$$

(a) (von Mises) If

$$\lim_{x\to\infty} xu(x)/U(x) = \rho, \qquad (A.26)$$

then $U \in RV_\rho$. (See [181].)

(b) (Landau) If $U \in RV_\rho$, $\rho \in \mathbb{R}$, and u is monotone then (A.26) holds and if $\rho \neq 0$ then $|u|(x) \in RV_{\rho-1}$. (See [120, 172, 150] and [49, page 23, 109].)

Proof

(a) Set

$$b(x) = xu(x)/U(x)$$

and as before we find

$$U(x) = U(1) \exp\left\{\int_1^x t^{-1} b(t) dt\right\}$$

so that U satisfies the representation theorem for a ρ-varying function.

(b) Suppose u is nondecreasing. An analogous proof works in the case u is nonincreasing. Let $0 < a < b$ and observe

$$(U(xb) - U(xa))/U(x) = \int_{xa}^{xb} u(y) dy / U(x).$$

By monotonicity we get

$$u(xb)x(b-a)/U(x) \geq (U(xb) - U(xa))/U(x) \geq u(xa)x(b-a)/U(x). \qquad (A.27)$$

From (A.27) and the fact that $U \in RV_\rho$ we conclude

$$\limsup_{x\to\infty} xu(xa)/U(x) \leq (b^\rho - a^\rho)/(b-a) \qquad (A.28)$$

for any $b > a > 0$. So let $b \downarrow a$, which is tantamount to taking a derivative. Then (A.28) becomes

$$\limsup_{x\to\infty} xu(xa)/U(x) \leq \rho a^{\rho-1} \qquad (A.29)$$

for any $a > 0$. Similarly from the left-hand equality in (A.27) after letting $a \uparrow b$ we get

$$\liminf_{x\to\infty} xu(xb)/U(x) \geq \rho b^{\rho-1} \qquad (A.30)$$

for any $b > 0$. Then (A.26) results by setting $a = 1$ in (A.29) and $b = 1$ in (A.30). □

A.5 Regular Variation: Further Properties

For the following list of properties, it is convenient to define *rapid variation* or regular variation with index ∞. We say $U : \mathbb{R}_+ \mapsto \mathbb{R}_+$ is regularly varying with index ∞ ($U \in RV_\infty$) if for every $x > 0$

$$\lim_{t \to \infty} \frac{U(tx)}{U(t)} = x^\infty := \begin{cases} 0, & \text{if } x < 1, \\ 1, & \text{if } x = 1, \\ \infty, & \text{if } x > 1. \end{cases}$$

Similarly $U \in RV_{-\infty}$ if

$$\lim_{t \to \infty} \frac{U(tx)}{U(t)} = x^{-\infty} := \begin{cases} \infty, & \text{if } x < 1, \\ 1, & \text{if } x = 1, \\ 0, & \text{if } x > 1. \end{cases}$$

The following proposition, modelled after [49] (see also [51]), collects useful properties of regularly varying functions.

Proposition A.6

(i) *If $U \in RV_\rho$, $-\infty \leq \rho \leq \infty$, then*

$$\lim_{x \to \infty} \log U(x) / \log x = \rho$$

so that

$$\lim_{x \to \infty} U(x) = \begin{cases} 0, & \text{if } \rho < 0, \\ \infty, & \text{if } \rho > 0. \end{cases}$$

(ii) *(Potter bounds.) Suppose $U \in RV_\rho$, $\rho \in \mathbb{R}$. Take $\varepsilon > 0$. Then there exists t_0 such that for $x \geq 1$ and $t \geq t_0$*

$$(1 - \varepsilon) x^{\rho - \varepsilon} < \frac{U(tx)}{U(t)} < (1 + \varepsilon) x^{\rho + \varepsilon}. \tag{A.31}$$

(iii) *If $U \in RV_\rho$, $\rho \in \mathbb{R}$, and $\{a_n\}$, $\{b_n\}$ satisfy $0 < b_n \to \infty$, $0 < a_n \to \infty$, and $b_n \sim c a_n$, as $n \to \infty$ for $0 < c < \infty$, then $U(b_n) \sim c^\rho U(a_n)$. If $\rho \neq 0$ the result also holds for $c = 0$ or ∞. Analogous results hold with sequences replaced by functions.*

(iv) If $U_1 \in RV_{\rho_1}$ and $U_2 \in RV_{\rho_2}$, $\rho_2 < \infty$, and $\lim_{x \to \infty} U_2(x) = \infty$ then

$$U_1 \circ U_2 \in RV_{\rho_1 \rho_2}.$$

(v) Suppose U is nondecreasing, $U(\infty) = \infty$, and $U \in RV_\rho$, $0 \leq \rho \leq \infty$. Then

$$U^\leftarrow \in RV_{\rho^{-1}}.$$

(vi) Suppose U_1, U_2 are nondecreasing and ρ-varying, $0 < \rho < \infty$. Then for $0 \leq c \leq \infty$

$$U_1(x) \sim c U_2(x), \quad x \to \infty$$

iff

$$U_1^\leftarrow(x) \sim c^{-\rho^{-1}} U_2^\leftarrow(x), \quad x \to \infty.$$

(vii) If $U \in RV_\rho$, $\rho \neq 0$, then there exists a function U^* which is absolutely continuous, strictly monotone, and

$$U(x) \sim U(x)^*, \quad x \to \infty.$$

Proof

(i) We give the proof for the case $0 < \rho < \infty$. Suppose U has Karamata representation

$$U(x) = c(x) \exp\left\{ \int_1^x t^{-1} \rho(t) dt \right\}$$

where $c(x) \to c > 0$ and $\rho(t) \to \rho$. Then

$$\log U(x)/\log x = o(1) + \int_1^x t^{-1} \rho(t) dt / \int_1^x t^{-1} dt \to \rho.$$

(ii) Using the Karamata representation

$$U(tx)/U(t) = (c(tx)/c(t)) \exp\left\{ \int_1^x s^{-1} \rho(ts) ds \right\}$$

and the result is apparent since we may pick t_0 so that $t > t_0$ implies $\rho - \varepsilon < \rho(ts) < \rho + \varepsilon$ for $s > 1$.

(iii) If $c > 0$ then from the uniform convergence property in Proposition A.4

$$\lim_{n \to \infty} \frac{U(b_n)}{U(a_n)} = \lim_{n \to \infty} \frac{U(a_n(b_n/a_n))}{U(a_n)} = \lim_{t \to \infty} \frac{U(tc)}{U(t)} = c^\rho.$$

(iv) Again by uniform convergence, for $x > 0$

$$\lim_{t\to\infty} \frac{U_1(U_2(tx))}{U_1(U_2(t))} = \lim_{t\to\infty} \frac{U_1(U_2(t)(U_2(tx)/U_2(t)))}{U_1(U_2(t))}$$

$$= \lim_{y\to\infty} \frac{U_1(yx^{\rho_2})}{U_1(y)} = x^{\rho_2\rho_1}.$$

(v) Let $U_t(x) = U(tx)/U(t)$ so that if $U \in RV_\rho$ and U is nondecreasing then $(0 < \rho < \infty)$

$$U_t(x) \to x^\rho, \quad t \to \infty,$$

which implies by Proposition A.2

$$U_t^\leftarrow(x) \to x^{\rho^{-1}}, \quad t \to \infty;$$

that it,

$$\lim_{t\to\infty} U^\leftarrow(xU(t))/t = x^{\rho^{-1}}.$$

Therefore

$$\lim_{t\to\infty} U^\leftarrow(xU(U^\leftarrow(t)))/U^\leftarrow(t) = x^{\rho^{-1}}.$$

This limit holds locally uniformly since monotone functions are converging to a continuous limit. Now $U \circ U^\leftarrow(t) \sim t$ as $t \to \infty$, and if we replace x by $xt/U \circ U^\leftarrow(t)$ and use uniform convergence we get

$$\lim_{t\to\infty} \frac{U^\leftarrow(tx)}{U^\leftarrow(t)} = \lim_{t\to\infty} \frac{U^\leftarrow((xt/U \circ U^\leftarrow(t))U \circ U^\leftarrow(t))}{U^\leftarrow(t)}$$

$$= \lim_{t\to\infty} \frac{U^\leftarrow(xU \circ U^\leftarrow(t))}{U^\leftarrow(t)} = x^{\rho^{-1}}$$

which makes $U^\leftarrow \in RV_{\rho^{-1}}$.

(vi) If $c > 0, 0 < \rho < \infty$ we have for $x > 0$

$$\lim_{t\to\infty} \frac{U_1(tx)}{U_2(t)} = \lim_{t\to\infty} \frac{U_1(tx)U_2(tx)}{U_2(tx)U_2(t)} = cx^\rho.$$

Inverting we find for $y > 0$

$$\lim_{t\to\infty} U_1^\leftarrow(yU_2(t))/t = (c^{-1}y)^{\rho^{-1}}$$

and so
$$\lim_{t\to\infty} U_1^\leftarrow(yU_2\circ U_2^\leftarrow(t))/U_2^\leftarrow(t) = (c^{-1}y)^{\rho^{-1}}$$

and since $U_2 \circ U_2^\leftarrow(t) \sim t$
$$\lim_{t\to\infty} U_1^\leftarrow(yt)/U_2^\leftarrow(t) = (c^{-1}y)^{\rho^{-1}}.$$

Set $y=1$ to obtain the result.

(vii) For instance if $U \in RV_\rho, \rho > 0$ define
$$U^*(t) = \int_1^t s^{-1} U(s)\,ds.$$

Then $s^{-1}U(s) \in RV_{\rho-1}$ and by Karamata's theorem
$$U(x)/U^*(x) \to \rho.$$

U^* is absolutely continuous and since $U(x) \to \infty$ when $\rho > 0$, U^* is ultimately strictly increasing. □

Appendix B
Notation Summary

Notation tends to accumulate so just in case you forget something, here is a partial list of terms. Consult as needed.

- Metric spaces: Measures are defined on $\mathbb{S}$ and many random elements live in $\mathbb{S}$. The space $\mathbb{S}$ is always a complete separable metric space (CSMS). This is sufficient generality to cover intended applications.

  - $d(x, y)$ = metric on $\mathbb{S}$ giving distance from $x \in \mathbb{S}$ to $y \in \mathbb{S}$.
  - For $A \subset \mathbb{S}$ and $x \in \mathbb{S}$, the distance from x to A is
    $$d(x, A) := \inf_{y \in A} d(x, y).$$
    If $A = \emptyset$, we use the convention that $d(x, \emptyset) = \infty$.
  - If A, B are two subsets of $\mathbb{S}$, write
    $$d(A, B) = \inf_{\substack{a \in A, \\ b \in B}} d(a, b).$$
    If $B = \emptyset$, we use the convention, $d(A, \emptyset) = \infty$.
  - $\mathcal{F}(\mathbb{S})$ = closed subsets of $\mathbb{S}$.
  - $\mathcal{G}(\mathbb{S})$ = open subsets of $\mathbb{S}$.
  - $\mathcal{S}$ = Borel σ-algebra generated by $\mathcal{G}(\mathbb{S})$; that is, by open sets or open balls of $\mathbb{S}$.
  - $\mathcal{K}(\mathbb{S})$ = compact subsets of $\mathcal{S}$.
  - $B(x, r) = \{y \in \mathbb{S} : d(x, y) < r\}$; for $r > 0$, open ball of radius r about center $x \in \mathbb{S}$.
  - A^δ = δ-swelling of subset $A \subset \mathbb{S}$; more precisely, for $\delta > 0$,
    $$A^\delta = \{x \in \mathbb{S} : d(x, A) < \delta\} = \cup_{x \in A} B(x, \delta).$$

- A^0 = interior of the subset $A \subset \mathbb{S}$.
- $\bar{A}$ = closure of the subset $A \subset \mathbb{S}$.
- $\partial A = \bar{A} \setminus A^0$ = boundary of $A \subset \mathbb{S}$.

- TABOF spaces: $\mathbb{S}_F \equiv \mathbb{S}(F) := \mathbb{S} \setminus F$; the TABOF space $\mathbb{S}$ with the closed subset $F \in \mathcal{F}(\mathbb{S})$ removed. When discussing regular variation of measures, F must be a closed cone.

  - $\mathbb{BA}(\mathbb{S}_F)$ is the class of Borel sets bounded away from the forbidden zone F.
  - Support: For $f : \mathbb{S} \mapsto \mathbb{R}_+$, the support of f, denoted $\mathrm{supp}(f)$, is the closure of $\{x \in \mathbb{S} : f(x) > 0\}$. For a measure $\mu(\cdot)$, the support of μ is $\mathrm{supp}(\mu)$, the smallest closed set F such that $\mu(F^c) = 0$.
  - $\mathbb{C}(\mathbb{S}_F)$ is the class of bounded, non-negative, continuous functions on $\mathbb{S}_F$ whose support $\mathrm{supp}(f)$ is bounded away from F. Not to be confused with the venerable $\mathbb{C}[0, 1]$, the continuous functions on $[0, 1]$.
  - For a bounded, uniformly continuous function $f : \mathbb{S} \mapsto \mathbb{R}_+$, the modulus of continuity is

  $$\omega_f(\delta) = \sup\{|f(x) - f(y)| : d(x, y) < \delta\}$$

  and $\lim_{\delta \to 0} \omega_f(\delta) = 0$.
  - For a non-negative measure $\mu(\cdot)$ defined on Borel sets of $\mathbb{S}$ and a measurable function $f : \mathbb{S} \mapsto \mathbb{R}_+$, set $\mu(f) = \int_{\mathbb{S}} f d\mu$.
  - $\mathbb{M}(\mathbb{S}_F)$ is the class of measures $\mu(\cdot)$ on $\mathbb{S}_F$ satisfying $\mu(D) < \infty$ for all $D \in \mathbb{BA}(\mathbb{S}_F)$.
  - Unit sphere: When $F \in \mathcal{F}(\mathbb{S})$, write

  $$\aleph_F := \{x \in \mathbb{S}_F : d(x, F) = 1\}.$$

  Special cases:

  · $\aleph_{0_2} = \{x \in \mathbb{R}^2 : \|x\| = 1\}$,
  · $\aleph_{[\mathrm{axes}]} = \{x \in \mathbb{R}_+^2 : x_1 \wedge x_2 = 1\}$,

- For $x \in \mathbb{S}$ and $A \subset \mathbb{S}$, set $\epsilon_x(A)$ = the Dirac probability measure putting all mass at point x so that

$$\epsilon_x(A) = \begin{cases} 1, & \text{if } x \in A, \\ 0, & \text{if } x \notin A. \end{cases} \quad (\text{B.1})$$

- Vectors in $\mathbb{R}^p$ and $\mathbb{R}^\infty$:

  - For integer $p \geq 1$, $\mathbf{0}_p = (0, \ldots, 0) \in \mathbb{R}^p$. We do not always attach the pedantic subscript p to $\mathbf{0}_p$ if the meaning is clear. Similarly, $\mathbf{0}_\infty \in \mathbb{R}^\infty$ is $\mathbf{0}_\infty = (0, 0, \ldots)$. In Chap. 4, $\mathbf{0}$ is the function on $[0, 1]$ which is identically 0.
  - For $x \in \mathbb{R}^p$, set $(x, \mathbf{0}_\infty) = (x_1, \ldots, x_p, 0, 0, \ldots)$.

B Notation Summary

- Interpret relations and operations on vectors componentwise. For example:
    - $x \leq y$ means $x_i \leq y_i$ for $i = 1, \ldots, p$.
    - $x/y = (x_i/y_i, i = 1, \ldots, p)$ and $1/x = (1/x_1, \ldots, 1/x_p)$.
    - $x > 0$ means $x_i > 0$ for $i = 1, \ldots, p$.
    - Intervals: $[a, b] = \{y : a \leq y \leq b\}$.
- Set $e_i \in \mathbb{R}^p$ to be the basis vector $e_i = (0, \ldots, 0, 1, 0 \ldots, 0)$ where the "1" is in the ith place. We allow $p = \infty$.
- Projection operators on vectors in $\mathbb{R}_+^p$ and sequences $\mathbb{R}_+^\infty$:
    - $\mathrm{PROJ}_p : \mathbb{R}_+^\infty \mapsto \mathbb{R}_+^p$ by $\mathrm{PROJ}_p(x) = x_{|p} = (x_1, \ldots, x_p)$. See Sect. 2.3.3, page 65.
    - $T_p : \mathbb{R}_+^p \mapsto \mathbb{R}_+$ by $T_p(x_1, \ldots, x_p) = x_p$. See page 71 and Sect. 2.3.3.
- Subspaces of $\mathbb{R}_+^\infty$ (see page 100 and 104):
    - $F_{\leq j} = \{x \in \mathbb{R}_+^\infty : \sum_{i=1}^\infty \epsilon_{x_i}(0, \infty) \leq j\}$,
    - $F_{=j} = \{x \in \mathbb{R}_+^\infty : \sum_{i=1}^\infty \epsilon_{x_i}(0, \infty) = j\}$.
    - $\mathbb{R}_+^{\infty \downarrow} = \{x \in \mathbb{R}_+^\infty : x_1 \geq x_2 \geq \ldots\}$.
    - $\mathbb{H}_{=j} = \{x \in \mathbb{R}_+^{\infty \downarrow} : x_j > 0, x_{j+1} = 0\}$, $j \geq 1$.
    - $\mathbb{H}_{\leq j} = \{x \in \mathbb{R}_+^{\infty \downarrow} : x_{j+1} = 0\}$, $j \geq 1$; $\mathbb{H}_{\leq 0} = \{0_\infty\}$.

- The class of regularly varying functions on $\mathbb{R}_+$ with index $\rho \in \mathbb{R}$ is denoted RV_ρ. This is the class of functions $U(x) : \mathbb{R}_+ \mapsto \mathbb{R}_+$ satisfying for $x > 0$,

$$\lim_{t \to \infty} \frac{U(tx)}{U(t)} = x^\rho, \qquad \text{for some } \rho \in \mathbb{R}.$$

The class of all regularly varying functions on $\mathbb{R}_+$ is denoted RV; that is $RV = \cup_{\rho \in \mathbb{R}} RV_\rho$. Set $RV_+ = \cup_{\rho > 0} RV_\rho$.
- For random elements X, Y on the same probability space, $X \perp\!\!\!\perp Y$ means X and Y are independent.
- *Mouthful notation*: Write (page 34),

$$\mu \in \mathrm{MRV}(\alpha, b(t), \eta(\cdot), \mathbb{S} \setminus F),$$

to mean $\mu(\cdot)$ is a multivariate regularly varying measure with index α, scaling function $b(\cdot) \in RV_{1/\alpha}$, limit measure $\eta(\cdot)$, state space $\mathbb{S}$ and forbidden zone F.
- The space $\mathbb{D} = \mathbb{D}[0, 1]$ of real valued right continuous functions on $[0, 1)$ with finite left-hand limits on $(0, 1]$.

    - Λ is the set of homeomorphisms from $[0, 1] \mapsto [0, 1]$ with typical element $\lambda(\cdot)$. The identity in Λ is $e(s) = s$.
    - For $x \in \mathbb{D}$, $\|x\|$ is the sup-norm:

$$\|x\| := \sup_{0 \leq s \leq 1} |x(s)|.$$

- $d_{\text{sk}}(\boldsymbol{x}, \boldsymbol{y})$, Skorohod metric on $\mathbb{D}$ (page 122) defined by

$$d_{\text{sk}}(\boldsymbol{x}, \boldsymbol{y}) = \inf_{\lambda \in \Lambda} \|\lambda - e\| \vee \|\boldsymbol{x} \circ \lambda - \boldsymbol{y}\|.$$

- Let $T_m : \mathbb{R}_+^{\infty \downarrow} \times [0, 1]^\infty \mapsto \mathbb{D}$ be the map $T_m(\boldsymbol{x}, \boldsymbol{u})(t) = \sum_{i=1}^{m} x_i 1_{[u_i, 1]}(t)$, $0 \leq t \leq 1$, (page 127).
- $A_m := \{(\boldsymbol{x}, \boldsymbol{u}) \in \mathbb{R}_+^{\infty \downarrow} \times [0, 1]^\infty : u_i \in (0, 1) \text{ for } 1 \leq i \leq m \text{ and } u_i \neq u_j$ for $i \neq j, 1 \leq i, j \leq m\}$ (page 127).
- Subsets of $\mathbb{D}$:
  · $\mathbb{D}^{\text{posJ}} := \mathbb{D} \cap \{x \in \mathbb{D} : x(0) = 0, x(s) - x(s-) \geq 0, 0 \leq s \leq 1\}$;
  · $\mathbb{D}^\uparrow := \{x \in \mathbb{D} : x(s_1) \leq x(s_2), \forall 0 \leq s_1 \leq s_2 \leq 1\}$.
  · $\mathbb{D}_{\leq m}[0, 1]$: The closed subset of $\mathbb{D}$ consisting of nondecreasing step functions on $[0, 1]$ with at most m jumps, $m \geq 0$, (page 124).
  · $\mathbb{D}_{\leq 0} = \mathbb{D}_0$: non-negative constant functions.
  · $\mathbb{D}_{=m}$: the subset of non-decreasing step functions with exactly m jumps.

- Measures with names:
  - $\mathcal{L}$: Lebesgue measure on $[0, 1]^\infty$ (page 126).
  - $\nu(\cdot)$: Often a measure on $\mathbb{R}_+ \setminus \{0\}$ satisfying $Q(x) := \nu(x, \infty) < \infty$, for $x > 0$. Often $Q(x) \in RV_{-\alpha}$ and $b(t)$ is defined to satisfy $tQ(b(t)x) \to x^{-\alpha}$, $x > 0$. Sometimes ν is the Lévy measure of a Lévy process.
  - $\nu_\alpha(\cdot)$: The Pareto measure on $\mathbb{R}_+ \setminus \{0\}$ satisfying $\nu_\alpha(x, \infty) = x^{-\alpha}$, $x > 0$. So $\nu_\alpha \in \mathbb{M}(\mathbb{R}_+ \setminus \{0\})$.
  - ν_α^j is product measure generated by ν_α with j factors (page 125).
  - $\eta^{(j)}$: the measure concentrated on $\mathbb{H}_{=j}$ given by (109),

$$\eta^{(j)}(dx_1, dx_2, \ldots) = \prod_{i=1}^{j} \nu_\alpha(dx_i) 1_{[x_1 \geq x_2 \geq \cdots \geq x_j > 0]} \prod_{i=j+1}^{\infty} \epsilon_0(dx_i).$$

  - $\mu^{(j)}$: the measure concentrated on $\mathbb{M}(\mathbb{R}_+^\infty \setminus F_{=j})$ given by (page 101),

$$\mu^{(j)}\bigl((dx_1, dx_2, \ldots)\bigr) := \sum_{i_1 < i_2 < \cdots < i_{j+1}} \Bigl(\prod_{j \notin \{i_1, \ldots, i_{j+1}\}} \epsilon_0(dx_j) \Bigr)$$
$$\nu_\alpha(dx_{i_1}) \nu_\alpha(dx_{i_2}) \ldots \nu_\alpha(dx_{i_{j+1}}).$$

  - $P_t^{(j)}(\cdot) := t P[X/b(t^{1/j}) \in \cdot]$.
  - $P_\infty^{(j)}(\cdot) := (\eta^{(j)} \times \mathcal{L}) \circ T_j^{-1}(\cdot)$ (page 125).

- Order statistics.

B Notation Summary

- Decreasing order: If $R_1, \ldots, R_n$ is a sample, the order statistics in decreasing order are

$$R_{(1)} \geq \cdots \geq R_{(n)}.$$

In particular, the kth largest is $R_{(k)}$.

References

1. D. Applebaum. *Lévy Processes and Stochastic Calculus*, volume 116 of *Cambridge Studies in Advanced Mathematics*. Cambridge University Press, Cambridge, second edition, 2009.
2. A.A. Balkema and N. Nolde. Asymptotic independence for unimodal densities. *Adv. in Appl. Probab.*, 42(2):411–432, 2010.
3. B. Basrak, R. Davis, and T. Mikosch. Regular variation of GARCH processes. *Stochastic Process. Appl.*, 99(1):95–115, 2002.
4. B. Basrak, R.A. Davis, and T. Mikosch. A characterization of multivariate regular variation. *Ann. Appl. Probab.*, 12(3):908–920, 2002.
5. B. Basrak and H. Planinić. A note on vague convergence of measures. *Statist. Probab. Lett.*, 153:180–186, 2019.
6. B. Basrak and J. Segers. Regularly varying multivariate time series. *Stochastic Processes and their Applications*, 119(4):1055–1080, 2009. https://doi.org/10.1016/j.spa.2008.05.004
7. J. Beirlant, Y. Goegebeur, J. Teugels, and J. Segers. *Statistics of Extremes*. Wiley Series in Probability and Statistics. John Wiley & Sons Ltd., Chichester, 2004. Theory and applications, With contributions from Daniel De Waal and Chris Ferro.
8. H. Bernhard and B. Das. Heavy-tailed random walks, buffered queues and hidden large deviations. *Bernoulli*, 26(1):61–92, 2020.
9. J. Bertoin. *Lévy Processes*, volume 121 of *Cambridge Tracts in Mathematics*. Cambridge University Press, Cambridge, 1996.
10. A. Bhattacharya, B. Chen, R. van der Hofstad, and B. Zwart. Consistency of the PLFit estimator for power-law data, 2020. ArXiv eprint: 2002.06870.
11. P. Billingsley. *Convergence of Probability Measures*. John Wiley & Sons Inc., New York, 1968.
12. P. Billingsley. *Weak Convergence of Measures: Applications in Probability*. Society for Industrial and Applied Mathematics, Philadelphia, Pa., 1971. Conference Board of the Mathematical Sciences Regional Conference Series in Applied Mathematics, No. 5.
13. P. Billingsley. *Convergence of Probability Measures*. John Wiley & Sons Inc., New York, second edition, 1999. A Wiley-Interscience Publication.
14. P. Billingsley. *Probability and Measure*. Wiley Series in Probability and Statistics. John Wiley & Sons, Inc., Hoboken, NJ, 2012. Anniversary edition [of MR1324786], With a foreword by Steve Lalley and a brief biography of Billingsley by Steve Koppes.
15. N. H. Bingham, C. M. Goldie, and J. L. Teugels. *Regular Variation*, volume 27 of *Encyclopedia of Mathematics and its Applications*. Cambridge University Press, Cambridge, 1989.

16. M. Bladt, E. Hashorva, and G. Shevchenko. Tail measures and regular variation. *Electron. J. Probab.*, 27, 2022.
17. L. Breiman. On some limit theorems similar to the arc-sin law. *Theory Probab. Appl.*, 10:323–331, 1965.
18. L. Breiman. *Probability*, volume 7 of *Classics in Applied Mathematics*. Society for Industrial and Applied Mathematics (SIAM), Philadelphia, PA, 1992. Corrected reprint of the 1968 original.
19. T. Britton. Directed preferential attachment models: Limiting degree distributions and their tails. *Journal of Applied Probability*, 57(1):122–136, 2020.
20. D. Buraczewski, E. Damek, and T. Mikosch. *Stochastic models with power-law tails*. Springer Series in Operations Research and Financial Engineering. Springer, [Cham], 2016. The equation $X = AX + B$.
21. S. Chakraborty and R.S. Hazra. Boolean convolutions and regular variation. *ALEA Lat. Am. J. Probab. Math. Stat.*, 15(2):961–991, 2018.
22. Y. Chen, D. Chen, and W. Gao. Extensions of Breiman's theorem of product of dependent random variables with applications to ruin theory. *Commun. Math. Stat.*, 7(1):1–23, 2019.
23. D. Cirkovic, T. Wang, and S.I. Resnick. Preferential attachment with reciprocity: Properties and estimation. *Journal of Complex Networks*, 11(5), 2023.
24. A. Clauset, C.R. Shalizi, and M.E.J. Newman. Power-law distributions in empirical data. *SIAM Rev.*, 51(4):661–703, 2009.
25. D.B.H. Cline. *Estimation and linear prediction for regression, autoregression and ARMA with infinite variance data*. ProQuest LLC, Ann Arbor, MI, 1983. Thesis (Ph.D.)–Colorado State University.
26. D.B.H. Cline. Convolutions tails, product tails and domain of attraction. *Probab. Theory Rel. Fields*, 72:529–557, 1986.
27. S.G. Coles. *An Introduction to Statistical Modeling of Extreme Values*. Springer Series in Statistics. London: Springer. xiv, 210 p., 2001.
28. D. Cooley and E. Thibaud. Decompositions of dependence for high-dimensional extremes. *Biometrika*, 106(3):587–604, 2019.
29. G. Csardi and T. Nepusz. The igraph software package for complex network research. *InterJournal, Complex Systems*, 1695(5):1–9, 2006.
30. S. Csörgő, P. Deheuvels, and D. Mason. Kernel estimates for the tail index of a distribution. *Ann. Statist.*, 13:1050–1077, 1985.
31. S. Csörgő and D. Mason. Central limit theorems for sums of extreme values. *Math. Proc. Camb. Phil. Soc.*, 98:547–558, 1985.
32. S. Csörgö, E. Haeusler, and D.M. Mason. The quantile-transform–empirical-process approach to limit theorems for sums of order statistics. In *Sums, Trimmed Sums and Extremes*, volume 23 of *Progr. Probab.*, pages 215–267. Birkhäuser Boston, Boston, MA, 1991a.
33. S. Csörgö, E. Haeusler, and D.M. Mason. The asymptotic distribution of extreme sums. *Ann. Probab.*, 19(2):783–811, 1991b.
34. D. J. Daley and D. Vere-Jones. *An Introduction to the Theory of Point Processes*. Springer Series in Statistics. Springer-Verlag, New York, 1988.
35. D. J. Daley and D. Vere-Jones. *An Introduction to the Theory of Point Processes. Vol. I.* Probability and its Applications (New York). Springer-Verlag, New York, second edition, 2003. Elementary theory and methods.
36. J. Danielsson, L. de Haan, L. Peng, and C. G. de Vries. Using a bootstrap method to choose the sample fraction in tail index estimation. *J. Multivariate Anal.*, 76(2):226–248, 2001.
37. B. Das, V. Fasen-Hartmann, and C. Klüppelberg. Tail probabilities of random linear functions of regularly varying random vectors. *Extremes*, 25(4):721–758, 2022.
38. B. Das, A. Mitra, and S.I. Resnick. Living on the multi-dimensional edge: Seeking hidden risks using regular variation. *Advances in Applied Probability*, 45(1):139–163, 2013.
39. B. Das and S.I. Resnick. Conditioning on an extreme component: Model consistency with regular variation on cones. *Bernoulli*, 17(1):226–252, 2011.

40. B. Das and S.I. Resnick. Detecting a conditional extreme value model. *Extremes*, 14(1):29–61, 2011.
41. B. Das and S.I. Resnick. Models with hidden regular variation: Generation and detection. *Stochastic Systems*, 5(2):195–238, 2015.
42. B. Das and S.I. Resnick. Hidden regular variation under full and strong asymptotic dependence. *Extremes*, 20(4):873–904, 2017.
43. B. D'Auria and S.I. Resnick. The influence of dependence on data network models. *Advances in Applied Probability*, 40(1):60–94, 2008.
44. R. Davis and T. Mikosch. The extremogram: A correlogram for extreme events. *Bernoulli*, 15(4):977–1009, 2009.
45. R.A. Davis and T. Mikosch. Extreme value theory for space-time processes with heavy-tailed distributions. *Stochastic Process. Appl.*, 118(4):560–584, 2008.
46. R.A. Davis, T. Mikosch, and I. Cribben. Towards estimating extremal serial dependence via the bootstrapped extremogram. *Journal of Econometrics*, 170(1):142–152, 2012.
47. R.A. Davis and S.I. Resnick. More limit theory for the sample correlation function of moving averages. *Stochastic Process. Appl.*, 20(2):257–279, 1985.
48. R.A. Davis and S.I. Resnick. Limit theory for the sample covariance and correlation functions of moving averages. *Ann. Statist.*, 14(2):533–558, 1986.
49. L. de Haan. *On Regular Variation and Its Application to the Weak Convergence of Sample Extremes*. Mathematisch Centrum Amsterdam, 1970.
50. L. de Haan. A characterization of multidimensional extreme-value distributions. *Sankhyā Ser. A*, 40(1):85–88, 1978.
51. L. de Haan and A. Ferreira. *Extreme Value Theory: An Introduction*. Springer-Verlag, New York, 2006.
52. L. de Haan and E. Omey. Integrals and derivatives of regularly varying functions in $\mathbb{R}^d$ and domains of attraction of stable distributions II. *Stochastic Process. Appl.*, 16(2):157–170, 1984.
53. L. de Haan, E. Omey, and S.I. Resnick. Domains of attraction and regular variation in $\mathbb{R}^d$. *J. Multivariate Anal.*, 14(1):17–33, 1984.
54. L. de Haan and S.I. Resnick. Derivatives of regularly varying functions in $\mathbb{R}^d$ and domains of attraction of stable distributions. *Stochastic Process. Appl.*, 8(3):349–355, 1979.
55. A.L.M. Dekkers and L. de Haan. Optimal choice of sample fraction in extreme-value estimation. *J. Multivariate Anal.*, 47(2):173–195, 1993.
56. D. Denisov and B. Zwart. On a theorem of Breiman and a class of random difference equations. *J. Appl. Probab.*, 44(4):1031–1046, 2007.
57. C. Dombry, E. Hashorva, and P. Soulier. Tail measure and spectral tail process of regularly varying time series. *Ann. Appl. Probab.*, 28(6):3884–3921, 2018.
58. C. Dombry, C. Tillier, and O. Wintenberger. Hidden regular variation for point processes and the single/multiple large point heuristic. *The Annals of Applied Probability*, 32(1):191 – 234, 2022.
59. G. Draisma, L. de Haan, L. Peng, and T. T. Pereira. A bootstrap-based method to achieve optimality in estimating the extreme-value index. *Extremes*, 2(4):367–404 (2000), 1999.
60. H. Drees. Weighted approximations of tail processes for β-mixing random variables. *Ann. Appl. Probab.*, 10(4):1274–1301, 2000.
61. H. Drees, L. de Haan, and S.I. Resnick. How to make a Hill plot. *Ann. Statist.*, 28(1):254–274, 2000.
62. H. Drees and A. Janßen. Conditional extreme value models: fallacies and pitfalls. *Extremes*, 20(4):777–805, Dec 2017.
63. H. Drees, A. Janßen, S.I. Resnick, and T. Wang. On a minimum distance procedure for threshold selection in tail analysis. *Siam J. Math. Data Sci.*, pages 75–102, 2020.
64. H. Drees and E. Kaufmann. Selecting the optimal sample fraction in univariate extreme value statistics. *Stochastic Processes and Their Applications*, 75:149–172, 1988.
65. H. Drees and H. Rootzén. Limit theorems for empirical processes of cluster functionals. *Ann. Statist.*, 38(4):2145–2186, 2010.

66. R. M. Dudley. *Real Analysis and Probability*, volume 74 of *Cambridge Studies in Advanced Mathematics*. Cambridge University Press, Cambridge, 2002. Revised reprint of the 1989 original.
67. M. Dwass. Extremal processes. *Ann. Math. Statist*, 35:1718–1725, 1964.
68. M. Dwass. Extremal processes. II. *Illinois J. Math.*, 10:381–391, 1966.
69. M. Dwass. Extremal processes. III. *Bull. Inst. Math. Acad. Sinica*, 2:255–265, 1974. Collection of articles in celebration of the sixtieth birthday of Ky Fan.
70. J.H.J. Einmahl and D.M. Mason. Generalized quantile processes. *Ann. Statist.*, 20(2):1062–1078, 1992.
71. P. Embrechts and C.M. Goldie. On closure and factorization properties of subexponential and related distributions. *J. Austral. Math. Soc. Ser. A*, 29(2):243–256, 1980.
72. P. Embrechts and C.M. Goldie. Comparing the tail of an infinitely divisible distribution with integrals of its Lévy measure. *Ann. Probab.*, 9(3):468–481, 1981.
73. P. Embrechts, C. Klüppelberg, and T. Mikosch. *Modelling Extremal Events for Insurance and Finance*. Springer-Verlag, Berlin, 2003. 4th corrected printing.
74. F. Eyi-Minko and C. Dombry. Extremes of independent stochastic processes: a point process approach. *Extremes*, 19(2):197–218, 2016.
75. P.D. Feigin, M.F. Kratz, and S.I. Resnick. Parameter estimation for moving averages with positive innovations. *Ann. Appl. Probab.*, 6(4):1157–1190, 1996.
76. W. Feller. *An Introduction to Probability Theory and Its Applications*, volume 2. Wiley, New York, 2nd edition, 1971.
77. A. Ferreira, L. de Haan, and L. Peng. On optimising the estimation of high quantiles of a probability distribution. *Statistics*, 37(5):401–434, 2003.
78. A.L. Fougeres and C. Mercadier. Risk measures and multivariate extensions of breiman's theorem. *Journal of Applied Probability*, 49(2):364–384, 2012.
79. J. Galambos. *The Asymptotic Theory of Extreme Order Statistics*. John Wiley & Sons, New York-Chichester-Brisbane, 1978. Wiley Series in Probability and Mathematical Statistics.
80. D. Garlaschelli and M.I. Loffredo. Patterns of link reciprocity in directed networks. *Phys. Rev. Lett.*, 93:268701, 2004.
81. J. L. Geluk and L. de Haan. *Regular Variation, Extensions and Tauberian Theorems*, volume 40 of *CWI Tract*. Stichting Mathematisch Centrum, Centrum voor Wiskunde en Informatica, Amsterdam, 1987.
82. J. L. Geluk and L. Peng. An adaptive optimal estimate of the tail index for ma(1) time series. *Statist. Probab. Lett.*, 46(3):217–227, 2000.
83. C.S. Gillespie. Fitting heavy tailed distributions: The poweRlaw package. *Journal of Statistical Software*, 64(2):1–16, 2015. http://www.jstatsoft.org/v64/i02/.
84. A.V. Gnedin. On a best-choice problem with dependent criteria. *J. Appl. Probab.*, 31(1):221–234, 1994.
85. N. Goix, A. Sabourin, and S. Clémençon. Sparse representation of multivariate extremes with applications to anomaly detection. *J. Multivariate Anal.*, 161:12–31, 2017.
86. P. Hall. On some simple estimates of an exponent of regular variation. *J. Roy. Statist. Soc. Ser. B*, 44(1):37–42, 1982.
87. J.E. Heffernan and S.I. Resnick. Hidden regular variation and the rank transform. *Adv. Appl. Prob.*, 37(2):393–414, 2005.
88. J.E. Heffernan and S.I. Resnick. Limit laws for random vectors with an extreme component. *Ann. Appl. Probab.*, 17(2):537–571, 2007.
89. J.E. Heffernan and J.A. Tawn. A conditional approach for multivariate extreme values (with discussion). *JRSS B*, 66(3):497–546, 2004.
90. J. Hemelrijk. The statistical work of David van Dantzig (1900-1959). *The Annals of Mathematical Statistics*, 31(2):269–275, 1960.
91. F. Hernández-Campos, J.S. Marron, C. Park, S.I. Resnick, and K. Jaffay. Extremal dependence: Internet traffic applications. *Stochastic Models*, 21(1):1–35, 2005.
92. B.M. Hill. A simple general approach to inference about the tail of a distribution. *Ann. Statist.*, 3:1163–1174, 1975.

93. T. Hsing. On tail estimation using dependent data. *Ann. Statist.*, 19:1547–1569, 1991.
94. X. Huang. *Statistics of Bivariate Extreme Values*. Ph.D. thesis, Tinbergen Institute Research Series 22, Erasmus University Rotterdam, Postbus 1735, 3000DR, Rotterdam, The Netherlands, 1992.
95. H. Hult and F. Lindskog. Extremal behavior of regularly varying stochastic processes. *Stochastic Process. Appl.*, 115(2):249–274, 2005.
96. H. Hult and F. Lindskog. Regular variation for measures on metric spaces. *Publ. Inst. Math. (Beograd) (N.S.)*, 80(94):121–140, 2006.
97. H. Hult and F. Lindskog. Extremal behavior of stochastic integrals driven by regularly varying Lévy processes. *Ann. Probab.*, 35(1):309–339, 2007.
98. H. Hult, F. Lindskog, T. Mikosch, and G. Samorodnitsky. Functional large deviations for multivariate regularly varying random walks. *Ann. Appl. Probab.*, 15(4):2651–2680, 2005.
99. A. Janßen. Spectral tail processes and max-stable approximations of multivariate regularly varying time series. *Stochastic Process. Appl.*, 129(6):1993–2009, 2019.
100. A. Janßen and H. Drees. A stochastic volatility model with flexible extremal dependence structure. *Bernoulli*, 22(3):1448–1490, 2016.
101. A. Janßen and J. Segers. Markov tail chains. *J. Appl. Probab.*, 51(4):1133–1153, 2014.
102. A.H. Jessen and T. Mikosch. Regularly varying functions. *Publ. Inst. Math. (Beograd) (N.S.)*, 80(94):171–192, 2006.
103. B. Jiang, Z.L. Zhang, and D. Towsley. Reciprocity in social networks with capacity constraints. In *Proceedings of the 21th ACM SIGKDD International Conference on Knowledge Discovery and Data Mining*, KDD '15, pages 457–466, New York, NY, USA, 2015. Association for Computing Machinery.
104. H. Joe. *Multivariate models and dependence concepts*, volume 73 of *Monographs on Statistics and Applied Probability*. Chapman & Hall, London, 1997.
105. H. Joe. *Dependence modeling with copulas*, volume 134 of *Monographs on Statistics and Applied Probability*. CRC Press, Boca Raton, FL, 2015.
106. H. Joe and H. Li. Tail risk of multivariate regular variation. *Methodology and Computing in Applied Probability*, pages 1–23, 2010. 10.1007/s11009-010-9183-x.
107. O. Kallenberg. *Random Measures, Theory and Applications*, volume 77 of *Probability Theory and Stochastic Modelling*. Springer, Cham, 2017.
108. J. Karamata. Sur un mode de croissance régulière des fonctions. *Mathematica (Cluj)*, 4:38–53, 1930.
109. J. Karamata. Sur un mode de croissance régulière. Théorèmes fondamentaux. *Bull. Soc. Math. France*, 61:55–62, 1933.
110. R.W. Katz. Weather and climate disasters. In *Extreme value modeling and risk analysis*, pages 439–460. CRC Press, Boca Raton, FL, 2016.
111. L. Kleinrock. *Queueing systems, volume 2: Computer applications*, volume 66. wiley New York, 1976.
112. E.D. Kolaczyk and G. Csárdi. *Statistical Analysis of Network Data with R*. Use R! Springer, NY, 2014.
113. M. Kratz and S.I. Resnick. The qq–estimator and heavy tails. *Stochastic Models*, 12:699–724, 1996.
114. R. Kulik and P. Soulier. *Heavy-Tailed Time Series*. Springer Series in Operations Research and Financial Engineering. Springer, New York, NY, 2020.
115. J. Kunegis. Konect: the Koblenz network collection. In *Proceedings of the 22nd International Conference on World Wide Web*, pages 1343–1350. ACM, 2013.
116. J. Kunegis. Handbook of network analysis; the konect project. Github, May 2021. https://github.com/kunegis/konect-handbook/raw/master/konect-handbook.pdf.
117. A.E. Kyprianou. *Fluctuations of Lévy Processes with Applications*. Universitext. Springer, Heidelberg, second edition, 2014. Introductory lectures.
118. Jure L. and Andrej K. SNAP Datasets: Stanford large network dataset collection. http://snap.stanford.edu/data, June 2014.
119. J. Lamperti. On extreme order statistics. *Ann. Math. Statist*, 35:1726–1737, 1964.

120. E. Landau. *Darstellung und Begrundung Einiger Neueren Ergebnisse der Funktionentheorie*. Springer-Verlag, Berlin, 1916.
121. M. Larsson and S.I. Resnick. Extremal dependence measure and extremogram: the regularly varying case. *Extremes*, 15(2):231–256, 2012.
122. J. Lehtomaa and S.I. Resnick. Asymptotic independence and support detection techniques for heavy-tailed multivariate data. *Insurance: Mathematics and Economics*, 93:262 – 277, 2020.
123. J. Leskovec and R. Sosič. Snap: A general-purpose network analysis and graph-mining library. *ACM Transactions on Intelligent Systems and Technology (TIST)*, 8(1):1, 2016.
124. F. Lindskog. *Multivariate Extremes and Regular Variation for Stochastic Processes*. Ph.d. thesis, Department of Mathematics, Swiss Federal Institute ofTechnology, 2004. https://doi.org/10.3929/ethz-a-004669275.
125. F. Lindskog, S.I. Resnick, and J. Roy. Regularly varying measures on metric spaces: Hidden regular variation and hidden jumps. *Probab. Surv.*, 11:270–314, 2014.
126. D. Mason. Laws of large numbers for sums of extreme values. *Ann. Probab.*, 10:754–764, 1982.
127. D. Mason. A strong invariance theorem for the tail empirical process. *Ann. Inst. Henri Poincaré*, 24:491–506, 1988.
128. D. Mason and T. Turova. Weak convergence of the Hill estimator process. In J. Galambos, J. Lechner, and E. Simiu, editors, *Extreme Value Theory and Applications*, pages 419–432. Kluwer Academic Publishers, Dordrecht, Holland, 1994.
129. G. Matheron. *Random Sets and Integral Geometry*. John Wiley & Sons, New York-London-Sydney, 1975. With a foreword by G.S. Watson, Wiley Series in Probability and Mathematical Statistics.
130. K. Maulik and S.I. Resnick. Characterizations and examples of hidden regular variation. *Extremes*, 7(1):31–67, 2005.
131. K. Maulik, S.I. Resnick, and H. Rootzén. Asymptotic independence and a network traffic model. *J. Appl. Probab.*, 39(4):671–699, 2002.
132. A.J. McNeil, R. Frey, and P. Embrechts. *Quantitative risk management*. Princeton Series in Finance. Princeton University Press, Princeton, NJ, revised edition, 2015. Concepts, techniques and tools.
133. N. Mhatre and D. Cooley. Transformed-linear models for time series extremes. *arXiv*, 2021.
134. T. Mikosch and O. Wintenberger. *Extremes for Time Series*. Springer Series in Operations Research and Financial Engineering. Springer-Verlag, New York, 2024. To appear.
135. A. Mitra and S. I. Resnick. Hidden Regular Variation: Detection and Estimation. *ArXiv e-prints 1001.5058*, 2010.
136. A. Mitra and S.I. Resnick. Hidden regular variation and detection of hidden risks. *Stochastic Models*, 27(4):591–614, 2011.
137. A. Mitra and S.I. Resnick. Modeling multiple risks: hidden domain of attraction. *Extremes*, 16(4):507–538, 2013.
138. I. Molchanov. *Theory of Random Sets*. Probability and its Applications (New York). Springer-Verlag London Ltd., London, 2005.
139. T. Mori and H. Oodaira. A functional law of the iterated logarithm for sample sequences. *Yokohama Math. J.*, 24(1-2):35–49, 1976.
140. P.J. Northrop and C.L. Coleman. Improved threshold diagnostic plots for extreme value analyses. *Extremes*, 17(2):289–303, 2014.
141. L. Peng. *Second Order Condition and Extreme Value Theory*. PhD thesis, Tinbergen Institute, Erasmus University, Rotterdam, 1998.
142. H. Planinić and P. Soulier. The tail process revisited. *Extremes*, 21(4):551–579, 2018.
143. Yu. V. Prohorov. Convergence of random processes and limit theorems in probability theory. *Teor. Veroyatnost. i Primenen.*, 1:177–238, 1956.
144. R Core Team. *R: A Language and Environment for Statistical Computing*. R Foundation for Statistical Computing, Vienna, Austria, 2019.
145. Lennart Råde. On the use of generating functions and laplace transforms in applied probability theory. *International Journal of Mathematical Educational in Science and Technology*, 3(1):25–33, 1972.

146. A. Rényi. On the theory of order statistics. *Acta Math. Acad. Sci. Hungar.*, 4:191–231, 1953.
147. S.I. Resnick. Inverses of extremal processes. *Adv. in Appl. Probab.*, 6:392–406, 1974.
148. S.I. Resnick. Weak convergence to extremal processes. *Ann. Probab.*, 3(6):951–960, 1975.
149. S.I. Resnick. Point processes, regular variation and weak convergence. *Adv. Applied Probability*, 18:66–138, 1986.
150. S.I. Resnick. *Extreme Values, Regular Variation and Point Processes*. Springer-Verlag, New York, 1987.
151. S.I. Resnick. Point processes and Tauberian theory. *Math. Sci.*, 16(2):83–106, 1991.
152. S.I. Resnick. *Adventures in Stochastic Processes*. Birkhäuser, Boston, 1992.
153. S.I. Resnick. *A Probability Path*. Birkhäuser, Boston, 1999.
154. S.I. Resnick. Hidden regular variation, second order regular variation and asymptotic independence. *Extremes*, 5(4):303–336 (2003), 2002.
155. S.I. Resnick. The extremal dependence measure and asymptotic independence. *Stochastic Models*, 20(2):205–227, 2004.
156. S.I. Resnick. *Heavy Tail Phenomena: Probabilistic and Statistical Modeling*. Springer Series in Operations Research and Financial Engineering. Springer-Verlag, New York, 2007. ISBN: 0-387-24272-4.
157. S.I. Resnick. *Extreme Values, Regular Variation and Point Processes*. Springer, New York, 2008. Reprint of the 1987 original.
158. S.I. Resnick. Multivariate regular variation on cones: application to extreme values, hidden regular variation and conditioned limit laws. *Stochastics: An International Journal of Probability and Stochastic Processes*, 80(2):269–298, 2008. http://www.informaworld.com/10.1080/17442500701830423.
159. S.I. Resnick and M. Rubinovitch. The structure of extremal processes. *Adv. in Appl. Probability*, 5:287–307, 1973.
160. S.I. Resnick and G. Samorodnitsky. Tauberian theory for multivariate regularly varying distributions with application to preferential attachment networks. *Extremes*, 18(3):349–367, 2015.
161. S.I. Resnick and C. Stărică. Consistency of Hill's estimator for dependent data. *J. Appl. Probab.*, 32(1):139–167, 1995.
162. S.I. Resnick and C. Stărică. Smoothing the moment estimator of the extreme value parameter. *Extremes*, 1(3), 1998.
163. S.I. Resnick and C. Stărică. Tail index estimation for dependent data. *Ann. Appl. Probab.*, 8(4):1156–1183, 1998.
164. S.I. Resnick and C. Stărică. Smoothing the Hill estimator. *Adv. Applied Probab.*, 29:271–293, 1997.
165. S.I. Resnick and D. Zeber. Asymptotics of Markov kernels and the tail chain. *Adv. in Appl. Probab.*, 45(1):186–213, 2013.
166. S.I. Resnick and D. Zeber. Transition kernels and the conditional extreme value model. *Extremes*, 17(2):263–287, 2014.
167. C.H. Rhee, J. Blanchet, and B. Zwart. Sample path large deviations for Lévy processes and random walks with regularly varying increments. *Ann. Probab.*, 47(6):3551–3605, 2019.
168. J. Th. Runnenburg. On the use of the method of collective marks in queueing theory. In *Proc. Sympos. Congestion Theory (Chapel Hill, N.C., 1964)*, pages 399–438. Univ. North Carolina Press, Chapel Hill, N.C., 1965.
169. J.A. Ryan and J.M. Ulrich. *quantmod: Quantitative Financial Modelling Framework*, 2019. R package version 0.4-15.
170. G. Samorodnitsky, S.I. Resnick, D. Towsley, R. Davis, A. Willis, and P. Wan. Nonstandard regular variation of in-degree and out-degree in the preferential attachment model. *J. Appl. Probab.*, 53(1):146–161, 2016.
171. C. Scarrott and A. MacDonald. A review of extreme value threshold estimation and uncertainty quantification. *REVSTAT*, 10(1):33–60, 2012.
172. E. Seneta. *Regularly Varying Functions*. Springer-Verlag, New York, 1976. Lecture Notes in Mathematics, 508.

173. R. Serfozo. Functional limit theorems for extreme values of arrays of independent random variables. *Ann. Probab.*, 10(1):172–177, 1982.
174. A.K. Sinha. *Estimating Failure Probability when Failure is Rare: Multidimensional Case*. PhD thesis, Tinbergen Institute, Erasmus University Rotterdam, 1997. PhD Thesis # 165, Promotor: Prof. Dr. L.F.M. de Haan, published by Thela Thesis, Amsterdam.
175. A.V. Skorohod. Limit theorems for stochastic processes. *Theory of Probability and Its Aplications*, 1(3):261–290, 1956.
176. U. Stadtmüller and R. Trautner. Tauberian theorems for Laplace transforms in dimension $D > 1$. *J. Reine Angew. Math.*, 323:127–138, 1981.
177. A. Stam. Regular variation in $\mathbb{R}_+^d$ and the Abel-Tauber theorem. Technical Report, unpublished, Mathematisch Instituut, Rijksuniversiteit Groningen, August 1977.
178. W. Vervaat and H. Holwerda, editors. *Probability and Lattices*, volume 110 of *CWI Tract*. Stichting Mathematisch Centrum, Centrum voor Wiskunde en Informatica, Amsterdam, 1997.
179. Y. Virkar and A. Clauset. Power-law distributions in binned empirical data. *Ann. Appl. Stat.*, 8(1):89–119, 2014.
180. B. Viswanath, A. Mislove, M. Cha, and K.P. Gummadi. On the evolution of user interaction in facebook. In *Proceedings of the 2nd ACM SIGCOMM Workshop on Social Networks (WOSN'09)*, August 2009.
181. R. von Mises. *Selected papers of Richard von Mises. Vol. Two: Probability and statistics, general*. American Mathematical Society, Providence, R.I., 1964.
182. J. L. Wadsworth and J. A. Tawn. Likelihood-based procedures for threshold diagnostics and uncertainty in extreme value modelling. *J. R. Stat. Soc. Ser. B. Stat. Methodol.*, 74(3):543–567, 2012.
183. T. Wang and S.I. Resnick. Multivariate regular variation of discrete mass functions with applications to preferential attachment networks. *Methodology and Computing in Applied Probability*, 20:1029–1042, 2018.
184. T. Wang and S.I. Resnick. Consistency of Hill estimators in a linear preferential attachment model. *Extremes*, 22(1):1–28, 2019.
185. T. Wang and S.I. Resnick. Degree growth rates and index estimation in a directed preferential attachment model. *Stochastic Process. Appl.*, 130(2):878–906, 2020.
186. T. Wang and S.I. Resnick. Asymptotic dependence of in- and out-degrees in a preferential attachment model with reciprocity. *Extremes*, 25:417–450, 2022.
187. T. Wang and S.I. Resnick. Measuring reciprocity in a directed preferential attachment network. *Adv. in Appl. Probab.*, 54(3):718–742, 2022.
188. T. Wang and S.I. Resnick. Common growth patterns for regional social networks: A point process approach. *Journal of Data Science*, 21(3):446–469, 2023.
189. T. Wang and S.I. Resnick. Random networks with heterogeneous reciprocity. *Extremes*, 27:123–161, 2024.
190. T. Wang and S.I.Resnick. 2RV+HRV and testing for strong vs full dependence, 2023. https://arxiv.org/abs/2312.16332; submitted.
191. W. Whitt. Some useful functions for functional limit theorems. *Math. Oper. Res.*, 5(1):67–85, 1980.
192. W. Whitt. *Stochastic Processs Limits: An Introduction to Stochastic-Process Limits And their Application to Queues*. Springer-Verlag, New York, 2002.
193. WOSN. Facebook friendships network dataset – KONECT, October 2017.
194. A. L. Yakimiv. Tauberian theorems and the asymptotics of infinitely divisible distributions in a cone. *Teor. Veroyatnost. i Primenen.*, 48(3):487–502, 2003.
195. A.L. Yakimiv. *Probabilistic Applications of Tauberian Theorems*. Modern Probability and Statistics. VSP, Leiden, The Netherlands, 2005.
196. S. Yun. The extremal index of a higher-order stationary Markov chain. *Ann. Appl. Probab.*, 8(2):408–437, 1998.

Index

Symbols
$\mathbb{BA}(\mathbb{S}_F)$, 6
$\mathbb{C}(\mathbb{S}_F)$, 7
$\mathbb{D}[0, 1]$, 4, 121
$\mathbb{D}_\sqcap$, 5
    products, 55
$\mathbb{M}(\mathbb{S}_F)$, 7
CUMSUM, 18
GPOLAR, 40
PROJ$_p$, 20

A
AltHillalpha, 156, 183, 186
Angular measure, 36
Aretha, 15, 127
Asymptotic full dependence, 119, 223
    examples, 212
Asymptotic independence, 50, 82, 151, 187
    equivalences, 187, 190
    examples, 114–116, 211
    and mappings, 191
    pairwise criterion, 188
    partial converse, 192
Asymptotic strong dependence, 163

B
BA sets, 7
Binding, 47, 49, 83
Bounded away, 6
Breiman's theorem, 57
    converse, 90, 91
    iid factors, 90
    matrix extension, 89
    multivariate extension, 88
    no independence, 89
    process extension, 88
    Tauberian theorem, 59

C
Càdlàg functions, 4
Cauchy, 229
Clauset, 152
Clunk function, 198
    asymptotic independence, 204
    how to unclunk, 204, 205
        full dependence, 207
        strong dependence, 205
Collective marks, 62
Comfort, 1
Conditional extreme value, 85
Cone, 31, 178
    choice, 182
    parameterization, 179
    ray, 179
    wedge, 179

D
Découpage de Lévy, 166, 200
Diagnostics, 95, 96, 151, 171
    asymptotic independence, 151
    diamond plot, 161, 176, 220
        mischief, 176
    EDM, 194
    Hill plot, 96, 97, 151

Diagnostics (*cont.*)
  Hillish analysis, 171
    definition, 172
    examples, 176, 178, 182
    modify data, 175
    wedge, 182
  HRV, 95
  igraph, 183
  MDSP, 154
  QQ plot, 96, 97, 151
  sliders, 170
    diamond plot, 170
Diamond plot, 161, 176, 220
  definition, 161
  examples, 161
  mischief, 163
Domain of attraction
  regular variation, 231

E
Extremal dependence measure (EDM), 193
  asymptotic normality
    examples, 210, 212
    tightness, 203
  clunk function
    asymptotic independence, 204
  definition, 194
  estimator, 195
    asymptotic normality, 197
    clunk function, 198
    ineffective for degree data, 221
  product moments, 223

F
Finite dimensional convergence, 65
Forbidden zone, 2, 31
  empty set, 26
  respect, 15, 127

G
Generalized polar coordinates, 39, 178
  definition, 39
Great oaks from little acorns, 46

H
Hausdorff metric, 164
  support estimation, 164
Heavy tail analysis
  Foundation, 1
Hidden regular variation, 93

diagnostics, 95, 96
examples, 94, 98, 99, 113, 115, 117, 119
and risk, 95
steroidal, 99
Hillish plot, 171
  definition, 172
  examples, 176, 178, 182
  modified data, 175
Hill plot, 96, 97, 151, 153

I
Igraph, 152, 183
Inverse, 226
Inversion, 241

K
Karamata representation, 237
Karamata's theorem, 60, 233
  multivariate Tauberian, 62
KONECT, 152

L
Lévy measure, 122
Lévy process, 121
  compound Poisson, 123
    steroidal regular variation, 131
  cruise time, 138
    bonus, 143
    converse, 143
    extremal process, 141
    largest jump, 141
    smoothing, 145
    supremum, 138
  Itô representation, 123
  and Poisson random measure, 123
  steroidal regular variation, 124, 125, 136
  where did the jumps go?, 125
Lebron *vs* Peewee, 53
Limit measure, 34, 36, 43, 48, 49, 55, 77, 78,
    83, 84, 94, 96, 104, 108, 116–118,
    121
  and angular measure, 36
  and polar coordinates, 39
  and scaling, 34
  and support estimation, 164

M
M-convergence, 2, 7
  weak convergence, 23
Mapping theorem, 15, 17, 18, 127

compactness, 18
respect, 16
restriction, 22
uniform continuity, 17
  examples, 18
variant, 21
Map to happiness, 53
Metric space, 2
Minimum distance method, 152
Minimum distance selection procedure (MDSP)
definition, 153
examples, 155
mischief, 154
Mount Sinai tablets, 158
not asymptotically normal, 155
optimal k, 153
  mathematical kvetching, 154
Mouthful notation, 34
Multivariate reduction, 96, 151
in $\mathbb{R}_+^p$, 96
in $\mathbb{R}_+^p$, 96

O
One-jump principle, 53
variants, 54, 87
Oops, 174

P
Pareto measure, 36, 124, 151
Parfit, 156, 183
Polar coordinate transformation, 36
generalized, 39
Portmanteau theorem, 9
Potter bounds, 241
PoweRlaw, 152, 156, 183
Power laws, 1
Preferential attachment, 214
and heavy tails, 217
model definition, 215
tail indices, 220
with reciprocity, 215
Products, 55
Breiman's theorem, 57
Projection map, 20, 65

Q
QQ plot, 96, 97, 151
Quantile function, 230

R
Rast operator, 222
Reciprocity, 214
Regular variation
composition, 241
and data, 151
examples, 229
integration, 233
inversion, 241
Karamata representation, 237
Karamata theorem, 233
measures, 31
  apply CUMSUM, 71
  asymptotic independence, 50, 82
  construction, 37, 41
  definition, 33
  distribution tails, 74
  in $\mathbb{D}_\sqcap$, 78
  examples, 35, 42
  how to prove, 46, 47, 49, 53, 65, 74, 78, 80
  in $\mathbb{R}_+^2 \setminus$ [axes], 80
  in $\mathbb{R}_+^\infty$, 70, 71
  scaling function, 34
  strong dependence, 45
Potter bounds, 241
properties, 241
sequential form, 230
and sets, 74
smooth version, 242
uniform convergence, 232

S
Scaling function, 31
examples, 33
postulates, 32
Skorohod metric, 4, 122
Sliders, 170
Steroidal regular variation, 99, 120, 131
Dr. Spock, 102
extremal process, 143
iid, 100
Lévy process, 124
lost in notation, 103
Poisson points, 103
  big reveal, 108
in $\mathbb{R}_+^\infty$, 100, 103
uncountably many, 111
Strong dependence, 163
support estimation, 163
Support estimation, 163

T
TABOF space, 2
  examples, 3, 5
  generation, 4
  products, 4
Tail process, 72
Tail regions, 2
Tauberian theorem, 59
Test drive, 13, 155
Test functions, 8
  uniformly continuous, 9
Threshold selection, 152

U
Uncouth, 35
  what they say, 35
Unit sphere, 25
  generalized, 40
    wedge, 180

Y
Yoda, 209

Z
Zombie, 150

SPRINGER NATURE

GPSR Compliance

The European Union's (EU) General Product Safety Regulation (GPSR) is a set of rules that requires consumer products to be safe and our obligations to ensure this.

If you have any concerns about our products, you can contact us on ProductSafety@springernature.com

In case Publisher is established outside the EU, the EU authorized representative is:

Springer Nature Customer Service Center GmbH
Europaplatz 3
69115 Heidelberg, Germany

The manufacturer's authorised representative in the EU is Springer Nature Customer Service Centre GmbH, Europaplatz 3, 69115 Heidelberg, Germany. If you have any concerns regarding our products, please contact ProductSafety@springernature.com

Printed and bound by CPI Group (UK) Ltd, Croydon, CR0 4YY

25/03/2026

02078171-0009